N-Nitroso Compounds

N-Nitroso Compounds

Richard A. Scanlan, EDITOR
Oregon State University

Steven R. Tannenbaum, EDITOR
Massachusetts Institute of Technology

Based on a symposium
cosponsored by the Divisions of
Agricultural and Food Chemistry
and Pesticide Chemistry
at the 181st Meeting of the
American Chemical Society,
Atlanta, Georgia,
March 31–April 1, 1981.

ACS SYMPOSIUM SERIES 174

AMERICAN CHEMICAL SOCIETY
WASHINGTON, D. C.   1981

Library of Congress CIP Data

N-nitroso compounds.

(ACS symposium series, ISSN 0097–6156; 174)

"Based on a symposium cosponsored by the Divisions of Agricultural and Food Chemistry and Pesticide Chemistry at the 181st meeting of the American Chemical Society, Atlanta, Georgia, March 31–April 1, 1981."

Includes bibliographies and index.

1. Nitroso-compounds—Congresses. 2. Nitroso-amines—Congresses. 3. Carcinogens—Congresses. 4. Nitroso-compounds—Metabolism—Congresses.
I. Scanlan, Richard A., 1937– . II. Tannenbaum, Steven R., 1937– . III. American Chemical Society. Division of Agricultural and Food Chemistry. IV. American Chemical Society. Division of Pesticide Chemistry. V. Series.

RC268.7.N58N18	616.99'4071	81–19047
ISBN 0–8412–0667–8		AACR2
	ACSMC8 174 1–400 1981	

ACS Symposium Series

M. Joan Comstock, *Series Editor*

FOREWORD

The ACS SYMPOSIUM SERIES was founded in 1974 to provide a medium for publishing symposia quickly in book form. The format of the Series parallels that of the continuing ADVANCES IN CHEMISTRY SERIES except that in order to save time the papers are not typeset but are reproduced as they are submitted by the authors in camera-ready form. Papers are reviewed under the supervision of the Editors with the assistance of the Series Advisory Board and are selected to maintain the integrity of the symposia; however, verbatim reproductions of previously published papers are not accepted. Both reviews and reports of research are acceptable since symposia may embrace both types of presentation.

CONTENTS

PREFACE

It is approximately twenty-five years since John Barnes and Peter Magee discovered that dimethylnitrosamine was a carcinogen for the rat. Since that time N-nitroso compounds have come to be among the most important experimental mutagens and carcinogens for laboratory investigation. Equally, if not more importantly, they have come to be viewed as one of the most important classes of environmental carcinogens, and the potential for endogenous formation from ubiquitous precursors has been recognized. Their presence and the presence of their precursors in foods, water, air, and industrial and agricultural products has led to frantic calls for legislative and regulatory action, and a never-ending search for more sensitive and specific methods of analysis.

The symposium upon which this volume is based was organized at a turning point in nitrosamine research. Almost all types of commercial products have been tested for volatile nitrosamines, and there have been a number of outstanding accomplishments of combined university–government–private industry actions to lower or eliminate volatile nitrosamines in those products found to be contaminated. However, there is still a major gap of knowledge with regard to compounds that are not amenable to analysis by gas chromatography, and this is clearly a frontier of current research. There are also many important questions regarding chemistry, mechanism of action, and relation to human disease whose answers lie in the future of research in this field.

It is the purpose of this volume to summarize the current state of knowledge of nitrosamine research with a chemical orientation, and to help lead the way into the future.

RICHARD A. SCANLAN
Oregon State University
Corvallis, OR 97331

STEVEN R. TANNENBAUM
Massachusetts Institute of Technology
Cambridge, MA 02139

July 31, 1981

CHEMISTRY AND METABOLISM

Activation of Nitrosamines to Biological Alkylating Agents

C. J. MICHEJDA, M. B. KROEGER-KOEPKE, S. R. KOEPKE, and D. H. SIEH

Chemical Carcinogenesis Program, Frederick Cancer Research Center, Frederick, MD 21701

N-Nitrosamines require metabolic activation to generate the reactive species which result in tumor induction. The most commonly accepted hypothesis for this activation is the α-hydroxylation hypothesis. Nitrosamines which have hydrogens on the alpha carbons are hydroxylated in that position by a mixed function oxygenase. The resultant α-hydroxyalkyl nitrosamine breaks down to an alkyldiazonium ion and the corresponding carbonyl compound. The diazonium ion alkylates a variety of nucleophiles and releases molecular nitrogen. The determination of the amount of molecular nitrogen provides a measure of the extent of α-hydroxylation. This concept was applied to the in vitro metabolism of doubly [^{15}N]-labeled dimethylnitrosamine (DMN) and N-nitrosomethylaniline (NMA) by uninduced rat liver S-9 (the postmitochondrial fraction). The amount of total metabolism was determined by the measurement of nitrosamine loss. It was found that measurement of formaldehyde formation gave an artificially low value of nitrosamine metabolism. The in vitro results indicated that about 34% of the DMN metabolism proceeded by a α-hydroxylation, while only 19% of the theoretical nitrogen was released by NMA. A similar experiment in vivo, using [^{15}N]-labeled DMN showed that about 66% of the metabolism proceeded by α-hydroxylation.

Alternative pathways of activation of nitrosamines, including β-hydroxylation followed by sulfate conjugation and the formation of alkoxydiazenium ions are discussed. The formation of alkyldiazonium ions from trialkyltriazenes is presented to show that the formation of the putative ultimate carcinogens from nitrosamines can be studied in a system not requiring metabolic activation.

In 1956 Magee and Barnes (1) reported that dimethylnitrosamine (DMN) was a potent carcinogen in rats. This discovery
initiated a mighty outburst of research activity. There are
now thousands of papers dealing with the chemistry, metabolism,
mutagenicity, teratogenicity, and carcinogenicity of nitrosamines and other N-nitroso compounds. There are several reasons for this continuing interest. Practically all nitrosamines are carcinogenic, making them the largest single chemical group which has that property. In fact, non-carcinogens
among them may provide important clues to what makes nitrosamines carcinogenic. Nitrosamines are remarkably site specific; essentially all the organs are effected by one or another
nitrosamine (2). This makes nitrosamines excellent tools for
the study of mechanisms of chemical carcinogenesis. Perhaps
most importantly, nitrosamines can be and are formed in the
human environment. The reason for this is that the two precursors to nitrosamines, nitrite and amines, are ubiquitous components of the environmental mix. Moreover, nitrosamines can
be formed in the stomach by ingesting the precursor amines and
nitrite. Several chapters in this volume concern themselves
with environmental and dietary aspects of nitrosamines.

The first ten years of nitrosamine research, particularly
with respect to carcinogenesis, were summarized in an admirable
review by Magee and Barnes (3). Since that time an enormous
amount of work has been carried out on those substances. Some
of this work up to 1974 was reviewed by Magee, Montesano and
Preussmann (4). Recently, Lai and Arcos (5) provided a useful
synopsis of contemporary work on the bioactivation of some
selected dialkylnitrosamines.

Nitrosamines, in common with many other organic carcinogens, have to be metabolized before their carcinogenic potential
can be expressed. Thus, Magee (6) in 1956 showed that dimethylnitrosamine (DMN) was cleared rapidly from the bodies of rats,
mice, and rabbits, with very little being excreted in the urine
or the feces. Dutton and Heath (7) reported that [^{14}C]-DMN
was metabolized in rats and mice, with a major portion of ^{14}C
appearing in expired carbon dioxide, with the rest of the radioactivity being evenly distributed in the tissues. Only about
7% of the radioactivity was found in the urine. These workers
also postulated that the metabolism proceeded by a demethylation
reaction and concluded that the lesions produced by DMN were
the result of the interaction of metabolites of DMN with cellular components, rather than with the intact DMN. Since the
liver was the principal target of DMN, Magee and Vandekar (8)
used liver slices and subcellular fractions to study the in
vitro metabolism of DMN. They found that the metabolizing activity was localized in the microsomes and in the cytosol and that
it required molecular oxygen and the presence of NADPH. The
formation of formaldehyde from DMN was demonstrated by Brouwers
and Emmelot (9). The principal enzyme system responsible for

DMN oxidative demethylation has been shown to be a liver microsome cytochrome P-450 monooxygenase (10). Lotlikar et al. (11) found that a reconstituted enzyme system, consisting of cytochrome P-450, NADPH-cytochrome P-450 reductase and phosphatidyl choline was effective in catalyzing the demethylation of DMN. The most commonly accepted mechanism for the oxidative demethylation of DMN and, by extension, of other dialkylnitrosamines is shown in Scheme 1.

Scheme 1

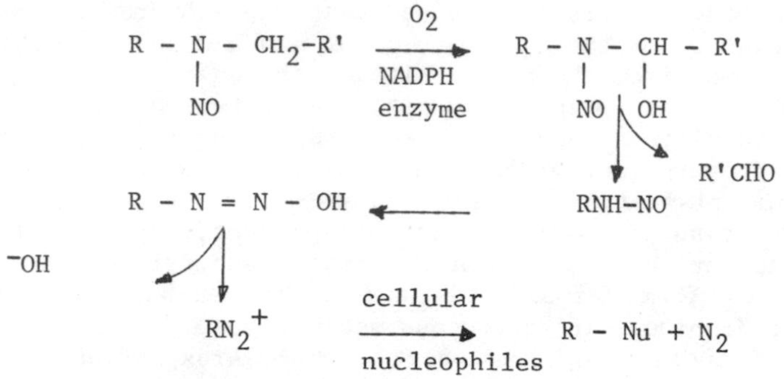

The α-Hydroxylation Hypothesis

The above scheme satisfies much of the metabolic data; however, some of it is speculative, and it is certainly incomplete. The evidence for the formation of the α-hydroxylated intermediate is circumstantial. The acetate ester of α-hydroxylated dimethylnitrosamine has been prepared (12,13) and has been found to be a potent, directly acting carcinogen (14). Other esters of a variety of α-hydroxylated nitrosamines have also been prepared (15). While it has been shown that DMN acetate is hydrolyzed to hydroxymethylmethylnitrosamine by an esterase enzyme, it has been pointed out that these derivatives of the α-hydroxylated nitrosamines also dissociate to N-nitrosoimmonium ions (15, 16).

Wiessler (15) suggested that the nitrosoimmonium ion could act as the electrophile which alkylates cellular nucleophiles.

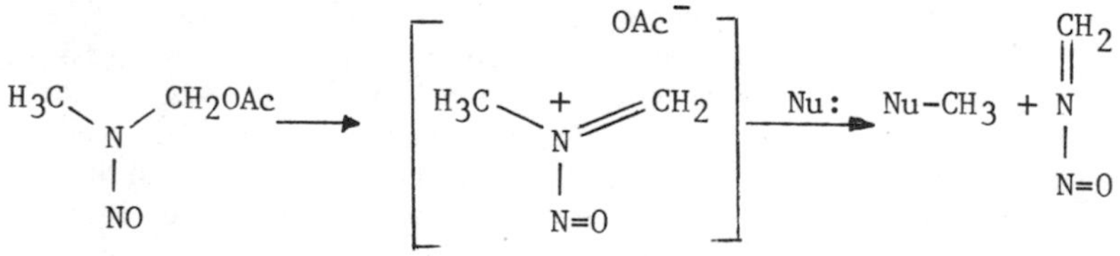

Gold and Linder (17) studied the esterase catalyzed hydrolysis of ℓ-(-)-acetoxymethyl-(1-phenylethyl)nitrosamine. They found that the stereochemistry of 1-phenylethanol produced in the reaction was the same as that observed in the base catalyzed hydrolysis of the nitrosamine and also of N-(1-phenylethyl)nitrosocarbamate. These results indicated that the same diazotate was produced in all three reactions. The fact that no irreversible inhibition of the enzymatic hydrolysis of the nitrosamine was observed, while extensive irreversible inhibition was obtained with the nitrosocarbamate, led these workers to conclude that the α-hydroxynitrosamine produced by the hydrolysis had sufficient stability to diffuse away from the active site of the enzyme.

Recently, the preparation of the first authentic α-hydroxylated nitrosamines has been reported (18,19). N-Butyl-N-hydroperoxymethylnitrosamine (20,21) was reduced with with sodium bisulfite or deoxygenated with triphenylphosphine. The resulting, rather unstable product, was converted to the α-acetoxy derivative with acetic anhydride and it was shown to form formaldehyde and 1- and 2-butanol when allowed to decompose in an aqueous medium.

All of these results support the α-hydroxylation hypothesis, but they do not prove it. No α-hydroxy nitrosamine or a derivative of one has ever been isolated as a metabolite of any nitrosamine. Nevertheless, α-hydroxylation is an attractive working hypothesis and it seems to account for many observed facts, at least in the case of the simplest nitrosamines. One of the salient features of Scheme 1 is that the α-hydroxylated nitrosamines decompose non-enzymatically to the unstable monoalkylnitrosamine, which then undergoes a prototropic shift to the corresponding diazotic acid. This substance (or its dissociation product, the alkyldiazonium ion) is the electrophile that alkylates cellular nucleophiles. Many authors write the electrophile as the corresponding carbenium ion. In most cases, this is incorrect; carbenium ions such as methyl or ethyl are high energy molecules and it is highly unlikely that they are formed in an aqueous medium. Exceptions may exist for such stabilized ions as benzyl or t-butyl, but not for primary, unstabilized carbenium ions. In fact, it was shown that the alkylation of hepatic DNA in rats by di-n-propylnitrosamine gave exclusively 7-n-propylguanine, rather than 7-isopropylguanine, which would have been predicted if free carbenium ions were involved (22). The RNA from these livers, however, did show that about 5% of the 7-propylguanine was the isopropyl isomer.

Whatever the true details of the metabolic pathway shown in Scheme 1 might be, there are certain facts which are very secure. Among these are that the nitrosamines are oxidatively dealkylated, that electrophilic intermediates which alkylate proteins and nucleic acids are formed, and that one of the

products of the dealkylation is a carbonyl compound, formaldehyde in the case of DMN. One consequence of Scheme 1 is
that α-hydroxylation obligatorily results in the formation
of molecular nitrogen. The measurement of the released
nitrogen, in principle, allows an accurate determination of
the extent of α-hydroxylation. Magee (23) was the first
to indicate the possibility of use of [^{15}N]-labeled nitrosamines in the measurement of the extent of α-hydroxylation.
In this work, Holsman, Haliday and Magee measured the formation in rats of ^{15}N-^{15}N gas. The animals were dosed with
doubly [^{15}N]-labeled dimethylnitrosamine, while being maintained in a respiration chamber whose atmosphere was 50%
oxygen and 50% sulfur hexafluoride. After 5 hrs, the rats
were killed and the expired gas was put through a set of
scrubbers and cold traps to remove O_2, CO_2, SF_6, and other
condensibles. The residual gas, consisting largely of nitrogen, was analyzed by mass spectrometry. They found that a
major portion of the DMN was metabolized by the α-hydroxyla-
tion pathway, since up to 90% of the total applied dose was
found in the expired nitrogen. These results, however, were
considered to be preliminary.

Cottrell et al (24) studied the in vitro metabolism of
[^{15}N]-labeled dimethylnitrosamine. They found that the
10,000g supernatant fraction (usually called the S-10 fraction) of rat liver homogenates produced labeled nitrogen at
the rate of 5% of the total metabolism of the nitrosamine as
measured by the production of formaldehyde and methanol.
These workers concluded that α-hydroxylation was relatively
unimportant during in vitro metabolism. In stark contrast
to these results, Milstein and Guttenplan (25), using unlabeled dimethylnitrosamine, found that essentially 100% of the
microsome-catalyzed metabolism of DMN proceeded by α-hydroxy-
lation. Against the backdrop of these seemingly contra-
dictory results, we have undertaken a systematic examination
of both the in vivo and in vitro metabolism of [^{15}N]-labeled
nitrosamines. These experiments are being carried out in
collaboration with Dr. P.N. Magee of the Fels Institute. In
the initial experiments we chose to concentrate on doubly
[^{15}N]-labeled dimethylnitrosamine and N-nitroso-N-methylani-
line (NMA).

The in vitro experiments, using the S-9 fraction from
livers of uninduced Fisher 344 rats, was complicated by the
fact that it became apparent that formaldehyde production was
a poor measure of the extent of metabolism. The reason for
that was that the S-9 fraction apparently catalyzed the oxidation of formaldehyde to formate. Consequently, determination of formaldehyde in an S-9 catalyzed reaction consistently gave low values of nitrosamine metabolism. Many workers
use semicarbazide to suppress formaldehyde loss. We found,
however, that semicarbazide is not a neutral bystander,

because it itself acts as a substrate for the S-9 enzymes, giving molecular nitrogen as a product. The difficulties with semicarbazide as a formaldehyde protector in S-9 catalyzed reactions were also independently noted by Savenije-Chapel and Noordhoek (26). Since formaldehyde formation was inadequate as an indicator of metabolism, we chose to follow the metabolism by determining substrate loss. This is an inherently imprecise measurement because the amount of total metabolism is small. Nevertheless, we found that with appropriate sample pre-treatment, reasonably reproducible data could be obtained. The S-9 catalyzed reactions were carried out in calibrated flasks. The reaction atmosphere was oxygen containing a precisely known amount of neon. After one hour reaction, the gas was transferred by means of a Toepler pump into a bulb, the contents of which were analyzed on a mass spectrometer. The instrument was calibrated against precisely known mixtures of $^{15}N-^{15}N$ and neon in oxygen. By knowing the total amount of gas in the reaction mixture, the absolute amount of $^{15}N-^{15}N$ evolved during the reaction could be calculated. The reaction mixture itself was assayed at the start of the reaction and after one hour by high pressure liquid chromatography (hplc). In order to get reproducible results in the hplc experiment it was necessary to filter the reaction mixture through a membrane ultrafilter, which excluded the high molecular weight materials. The results of these experiments on doubly [^{15}N]-labeled DMN and NMA are presented in Table I.

Table I.

Analysis of Metabolism of [^{15}N]-Labeled
DMN and NMA by the S-9 Fraction[a]

Substrate	Substrate Lost (μ moles/g liver/hr)	HCOH Detected (μ moles/g liver/hr)	$^{15}N-^{15}N$-Produced (μ moles/g liver/hr)
DMN	0.732 ± .184(5)	0.169 ± .050(6)	0.240 ± .031(4)
NMA	1.570 ± .173(5)	0.039 ± .010(4)	0.296 ± .074(3)

[a]Results expressed as the average ± standard deviation (number of observations)

These results indicate that in the in vitro reaction, the α-hydroxylation pathway accounts for about 34% of the metabolism of DMN and about 19% of the metabolism of NMA, when uninduced Fisher 344 rat liver S-9 is used. Thus, our data for DMN fall roughly in-between the two previously published

values (24,25). The data also clearly indicate that the
formaldehyde assay gives erroneous results for the total
amount of metabolism, when the S-9 fraction is used as the
oxidation catalyst.

The present data give us no indication of what the
other pathways of DMN metabolism might be. Other pathways,
including denitrosation and reduction of the NO group to the
unsymmetrical hydrazine have been reported (27).

The in vivo metabolism of [^{15}N]-labeled DMN was carried
out by a method similar to the one developed by Holsman,
Haliday and Magee (23). A rat was injected IP with 276 μ
mole/kg of doubly [^{15}N]-labeled DMN and was placed in a
respiration chamber equipped with a collapsible top. The
chamber was charged with an atmosphere of SF_6, oxygen and a
known amount of neon as an internal standard. The entire
system was calibrated with precisely known mixtures of ^{15}N–
^{15}N, neon, and oxygen. After 6 hrs, the rat was killed by
an injection of chloroform into the chamber and the gas in
the chamber was pumped through a series of traps to remove
the condensible gases. The residual gas was analyzed by
mass spectrometry. This in vivo experiment was carried out
at the Fels Institute by Ms. Cecilia Chu and Dr. Magee. The
gas analysis was carried out at Frederick. The results are
presented in Table II.

Table II.

Analysis of the in vivo Metabolism
of [^{15}N]-Labeled DMN

moles[a] of ^{15}N-DMN injected per rat x 10^5	moles of ^{15}N–^{15}N evolved[b] x 10^5	% of metabolism resulting in ^{15}N–^{15}N
3.04	1.81	60
4.03	3.18	79
4.03	2.08	52
4.14	3.08	74
	Average ± SD =	66 ± 12

[a]Each animal received an equimolar dose of 276 μmole/kg. This
dose was determined by Magee to be metabolized completely
within 6 hrs.
[b]The determination was made using neon as an internal standard.

It is clear from the results in Table II that α-hydroxylation
accounts for a major part of the metabolism of DMN in vivo.
The percentage is a little lower than the preliminary values
obtained by Halsman, Haliday and Magee (23), but it is

interesting to note that this pathway is more important in
vivo that in vitro by about a factor of two.

The α-hydroxylation reaction of cyclic nitrosamines has
been shown to be a significant metabolic pathway in several
studies (28-33). Its likely importance in determining carcinogenicity was demonstrated by the finding that deuteration
of DMN lowered its carcinogenicity (34) and that 3,3,5,5-
tetradeuteronitrosomorpholine was less carcinogenic than its
protio-analog (35).

The data in Table I indicate that only 19% of nitrogen
was released during the in vitro metabolism of N-nitroso-N-
methylaniline (NMA). Control experiments showed that phenyl-
ring hydroxylation was negligible under those conditions.
It is possible that some of this nitrosamine was degraded by
denitrosation and some may have been reduced to 1-methyl-1-
phenylhydrazine. If, however, demethylation of NMA occured,
the product of this reaction would have been the phenyldiazonium ion. The aromatic diazonium ions are substantially
more stable than their alkyl analogs, hence, it is not auto-
matically certain that the phenyldiazonium ion would release
molecular nitrogen. In fact, it is known that aromatic diazonium ions enter into azo coupling reactions with nucleic
acid bases (36). We found that addition of phenol after the
S-9 catalyzed oxidation of NMA was quenched, resulted in the
formation of a detectable amount of 4-hydroxyazobenzene.
Thus, it appears that the extent of α-hydroxylation of NMA
may be higher than the value of 19% $^{15}N-^{15}N$ would indicate.

$$PhN_2^+ \quad + \langle\bigcirc\rangle-OH \longrightarrow Ph - N = N -\langle\bigcirc\rangle-OH$$

As was stated above, the putative alkylating intermediate
which results from the α-hydroxylation of dialkylnitrosamines
is the alkyldiazonium ion. These ions can be generated
chemically by the reaction of primary amines with nitrosating
agents, but that reaction is difficult to carry out cleanly
in the presence of biological materials. However, N-nitroso-
N-alkylureas, N-alkyl-nitrosamides and N-nitroso-N-alkylcar-
bamates, have been shown to decompose non-enzymatically to
alkyldiazonium ions (37).

$$R - \underset{\underset{NO}{|}}{\overset{\overset{O}{\|}}{N}} - C - Z + X^- \longrightarrow R - \underset{\underset{N}{\|}}{N} - \underset{\underset{X}{|}}{C} - Z \xrightarrow{\left[\overset{\overset{O}{\|}}{X - C - Z}\right]} R - N = N - O^-$$

$$R - N_2^+ + OH^-$$

Z = NH$_2$, NHR', NR'$_2$, OR', R'
X$^-$ = nucleophile

These nitrosated amides, in contrast to the nitrosamines, do not require metabolic activation in order to alkylate cellular nucleophilic targets. Thus, many of these substances are directly acting mutagens and also carcinogens (38).

We have recently developed a general method of preparation of the heretofore relatively unknown trialkyltriazenes (39). These substances are readily prepared by reaction of a Grignard reagent or an alkyllithium with an alkyl azide, followed by the reaction of the intermediate dialkyltriazene with base, and alkylation of the resulting triazenyl anion with an alkyl halide.

$$RN_3 + R'Li(or\ R'MgBr) \longrightarrow R-N=N-NHR' \longrightarrow R-N=N-N-R'$$

$$R''X$$

$$\begin{array}{c} R'' \\ \diagdown \\ N-N=N-R' \\ \diagup \\ R \end{array} + R-N=N-N\begin{array}{c} R'' \\ \diagdown \\ \diagup \\ R' \end{array}$$

These trialkyltriazenes were found to be very acid sensitive, decomposing with evolution of nitrogen. Representative triazenes were tested in the Ames assay for mutagenicity. They were found to be directly acting mutagens (40). The mutagenic activity was correlatable with the expected reactivities of the alkyldiazonium ions produced by the decomposition of the triazenes. Thus, for example, 3-benzyl-1,3-dimethyltriazene A was about three orders of magnitude more mutagenic than its isomer, 1-benzyl-3,3-dimethyltriazene B. Triazene A, on decomposition, would yield the methyldiazonium ion,

$$CH_3 - N = N - N\begin{array}{c} CH_2Ph \\ \diagdown \\ CH_3 \end{array}$$

$$PhCH_2 - N = N - N\begin{array}{c} CH_3 \\ \diagdown \\ CH_3 \end{array}$$

A

B

$$CH_3N_2^+$$

$$PhCH_2N_2^+(PhCH_2^+)$$

while triazene B would form the benzyldiazonium ion (or the benzyl cation, since the diazonium ion would be unstable relative to the stabilized carbenium ion). Interestingly, triazenes which gave rise to the n-butyldiazonium ion were less mutagenic than triazene A. The reason for this is that the n-butyldiazonium ion has several reaction paths open to it (rearrangement to the more stable 2-butyl cation followed by elimination to form butenes, etc.) and hence, less is available for reaction with the genetic material of the bacterium. It is significant in this regard that Mochizuki

et al. (19) found that the trapping of the alkyldiazonium ion
with thiophenolate anion was much more efficient from N-me-
thyl-N-(hydroxymethyl)nitrosamine (which gives the methyldiazonium ion) than from N-n-butyl-N-(hydroxymethyl)nitrosamine
(which gives the n-butyldiazonium ion).

The trialkyltriazenes are essentially protected diazonium ions. They decompose cleanly and quantitatively to the
diazonium ions and the corresponding amines over a wide pH
range (41). Good kinetic data were obtained over the range of
pH 6.9 - 8.3. In more acid solutions, the reactions are too
rapid to measure by conventional kinetics. The decomposition
reaction is subject to general acid catalysis. Thus, the
trialkyltriazenes will be a useful tool for the study of the
reactive intermediates produced by the metabolism of dialkylnitrosamines.

Alternative Pathways of Nitrosamine Activation

Although the α-hydroxylation hypothesis has been given
considerable support, it is clear that nitrosamines with
alkyl side chains longer than methyl present additional
possibilities for biological activation. It was stated earlier that even the α-hydroxylation hypothesis is based on
considerable circumstantial evidence. This means that the
other possible pathways are even more speculative, and hence,
much more research needs to be carried out before some of the
ideas advanced below will gain acceptance, or be put to rest.

It is known that nitrosamines with side-chains longer
than methyl are metabolically hydroxylated on virtually every
carbon. For example, the urinary metabolites in the rat of
di-n-butylnitrosamine include the glucuronides of the following hydroxylated nitrosamines (42).

$$
\begin{array}{ccc}
\quad C_4H_9 & \quad C_4H_9 & \quad C_4H_9 \\
\diagup & \diagup & \diagup \\
ON-N & ON-N & ON-N \\
\diagdown & \diagdown & \diagdown \\
C-C-C-C & C-C-C-C & C-C-C-C-OH \\
\ \ | & \ \ | & \\
OH & OH &
\end{array}
$$

In this particular instance, ω and ω-1 hydroxylation, which
is then followed by further oxidation and chain degradation,
are the principal reactions (43). In the case of di-n-propylnitrosamine both α- and β-oxidation occur, with the latter
being about 15% of the former (44). The α-hydroxylation leads
to the formation of the n-propyldiazonium ion (22), while
the β-hydroxylation results, at least in part, in oxidation
to N-propyl-N-(2-oxopropyl)nitrosamine. Krüger and Bertrum
(45) suggested that this product can be cleaved to methyl-

propylnitrosamine, which allows the formation of a methylating agent by subsequent α-hydroxylation of the propyl moiety. Recently, Singer et al. (46) studied the rat urinary metabolites of a series of N-methyl-N-alkylnitrosamines with alkylchain lengths from C_4 to C_{14}. The principal metabolite of nitrosamines with odd-numbered chains was N-methyl-N-(2-carboxyethyl)nitrosamine. The even-numbered chain nitrosamines also gave some nitrosamino acids (major components were N-nitrososarcosine and N-methyl-N-(3-carboxypropyl)nitrosamine. The striking feature, however, was that the even-numbered chain nitrosamines gave considerable amounts of of N-methyl-N-(2-oxopropyl)nitrosamine, while the odd gave relatively small amounts. Since the even-numbered chain nitrosamines are bladder carcinogens in the rat (47), Singer et al. suggested that N-methyl-N-(2-oxopropyl)nitrosamine might be the proximate carcinogen in the bladder. That hypothesis is being tested.

The foregoing data suggest that metabolic oxidation of nitrosamines at carbons other than the α-carbon may be important in the activation of longer chain nitrosamines. The problem of β-hydroxylated nitrosamines is discussed in a separate chapter (by R. Loeppky). However, it is worth pointing out that substitution of the β-carbon by a group which is capable of acting as a leaving group is an activation process. We have pointed out (48,49) that tosylates of β-hydroxylated nitrosamines solvolyze extremely rapidly, owing to the neighboring group effect of the N-nitroso group. The reaction proceeds through the intermediate formation of an oxadiazolium ion which can be isolated when the tosylate is warmed in a nonnucleophilic solvent.

The β-hydroxylated nitrosamines do not appear to be very mutagenic in the Ames assay, with or without S-9 activation (50). We have found, on the other hand, that their tosylate derivatives are potent directly acting mutagens (51), and that the oxadiazolium ion in the above equation has essentially the same dose-response behavior in the mutagenicity assay as the parent tosylate. It is interesting to note that the next higher homolog, N-methyl-N-(3-tosyloxypropyl)nitrosamine is not a directly acting mutagen (51).

Tosylate derivatives are not usually found in nature, but sulfate derivatives of alcohols are common (52). They are formed by the reaction of an alcohol with sulfate, catalyzed by a sulfotransferase enzyme.

$$\underset{\underset{NO}{|}}{CH_3\diagdown N\diagup\diagdown\diagup OH} + SO_4^{=} + ATP \xrightarrow[\text{transferase}]{\text{sulfo-}} \left[\underset{\underset{NO}{|}}{CH_3\diagdown N\diagup\diagdown\diagup OSO_3^-} \right]$$

$$CH_3\diagdown\underset{N}{\overset{+}{N}}\diagdown\diagup \quad SO_4^{=}$$

We have found that the chemical sulfation of N-methyl-N-(2-hydroxyethyl)nitrosamine resulted in the direct formation of the oxadiazolium ion.

The reactions of the oxadiazolium ions with nucleophiles such as thiophenols, guanine, and guanosine resulted in methylation of the nucleophiles (53). Thus, the formation of the oxadiazolium ion creates an electrophilic species not

$$CH_3\diagdown\underset{\underset{N}{||}}{\overset{+}{N}}\diagdown\diagup \quad X^- \quad + \quad Nu: \longrightarrow \quad CH_3 \overset{+}{-} Nu \quad + \quad \left[\underset{\underset{N}{||}}{N}\diagdown\diagup \right]$$

unlike the nitrosoimmonium ions suggested by Wiessler (15).
The acyclic analogs of oxadiazolium ions have been known for many years (54). They are prepared by the reaction of trialkyloxonium tetrafluoroborates (the Meerwein reagent), or of alkyl halides catalyzed by silver perchlorate, with di-alkylnitrosamines. We have found that "Magic Methyl or Ethyl" (methyl or ethyl fluorosulfonates) are particularly useful for alkylation of the nitrosamine oxygen.

$$\underset{\underset{N}{\overset{|}{\diagdown}}O}{CH_3\diagdown\underset{N}{\diagup}CH_3} + EtOSO_2F \xrightarrow{CH_2Cl_2} \underset{\underset{N}{\overset{||}{\diagdown}}OEt}{CH_3\diagdown\overset{+}{N}\diagup CH_3} \quad FSO_3^-$$

The resulting alkoxydiazenium fluorosulfonates are stable, highly hygroscopic, crystalline solids, which are freely soluble in organic solvents, such as dichloromethane (in contrast, the fluoroborate salts are relatively insoluble).
We have studied the reactions of the alkoxydiazenium salts with a variety of nucleophiles in various media. The

reaction in aprotic solvents is relatively straightforward, but the result was completely unexpected. The following examples illustrate this dramatically.

It would appear that all the alkyl groups on both alkoxydiaze- nium ions become randomized. The product distribution is statistical, that is, since both diazenium ions have two ethyl groups and one methyl group, the ratio (ethyl/methyl) of the thioethers formed is precisely 2:1. The by-products of these reactions are the corresponding nitrosamines, diethyl and methylethyl, formed in a 1:2 ratio, respectively. The overall yield is high (virtually quantitative). Thus, this reaction offers the possibility of interconversion of nitros- amines (eg. diethyl to methylethyl) but also raises the possibility of activation of nitrosamines to become alkylating agents, by a totally non-oxidative route. Thus, one can imagine that a dialkylnitrosamine would react with S-adenosyl- methionine, catalyzed by an appropriate methyltransferase, to give an alkoxydiazenium ion, which could alkylate cellular nucleophiles and regenerate the nitrosamine. The cycle could then be repeated. Unfortunately, at the present time, there is no evidence that nitrosamines are O-alkylated in vivo and consequently, this particular mechanism of nitrosamine activ- ation must be considered totally speculative. Druckrey et al. (55) found that administration of N-ethoxyiminomorpho- linium tetrafluoroborate, by SC injection to rats, produced only local sarcomas at the site of injection. Evidently the parent N-nitrosomorpholine was not released in sufficiently high doses to produce the hepatic tumors characteristic of that nitrosamine.

We have recently reported the carcinogenicity in rats of the isomers, N-nitroso-2,3-dehydropiperidine (NDP-$^2\Delta$) and N- nitroso-3,4-dehydropiperidine (NDP-$^3\Delta$)(56). We found that the $^3\Delta$-isomer rearranged to the $^2\Delta$-isomer in vivo, a reaction which occurs readily by base catalysis (57). This study, however, did not identify the proximate carcinogen responsible for the development of the liver and esophageal tumors observed. In an attempt to find some likely candidates for the proximate carcinogen, we tried to prepare the epoxide of NDP-$^2\Delta$ by reaction with m-chloroperbenzoic acid and also with iodosobenzene. This reaction took an entirely unexpected course, because the principal product turned out to be 3-nitro-2,3-dehydropiperidine (58). This product was also formed by a light-catalyzed photooxygenation of NDP-$^2\Delta$. This substance, however, is not formed during the rat-liver micro-

DNP-$^2\Delta$

some catalyzed oxidation of NDP-$^2\Delta$ and, hence, it is an un-likely candidate for the proximate carcinogen. The microsomal oxidation, however, did result in the formation of a metabo-lite, which was identified to be 4-hydroxy-2,3-dehydropiperi-dine.

The interesting aspect of this oxidation is that it is a vinylogous α-hydroxylation, and it offers the possibility that the following rearrangement could occur.

This reaction has not yet been demonstrated in this system, but examples of similar allylic rearrangements are well known.

The purpose of the discussion of alternative pathways of nitrosamine activation (alternative to the α-hydroxylation hypothesis) is to demonstrate how fundamental organic chemis-try may help classical metabolism studies to provide clues to the behavior of carcinogens in biological systems. Frequently, the carcinogenically significant metabolites are too unstable or too reactive to be detected as urinary metabolites. Thus, it is important to be able to assess the chemical transformations of carcinogens within the context of metabolism. Living systems are remarkable in the richness of their chemistry. This makes it difficult, in fact, to dismiss the metabolic importance of any chemical reaction a priori. In this brief chapter, we have ignored a large number of reactions of nitrosamines which may have metabolic significance. Some of these are discussed in other chapters. Many of them, however, are waiting to be discovered.

Acknowledgement

This work was supported by Contract No. N01-CO-75380 with the National Cancer Institute, NIH, Bethesda, Maryland 20205. The authors are particularly grateful to Dr. P.N. Magee and Ms. Cecilia Chu for their stimulating collaboration, to Drs. R. Smith and R. Kupper, our former colleagues, who made substantial contributions to this research, and to Dr. G. McClusky for carrying out the difficult mass spectrometric measurements.

Literature Cited

1. Magee, P.N. and Barnes, J.M., Brit. J. Cancer, 1956, 10, 114-122.
2. Drückrey, H., Preussman, R., Ivankovic, S., and Schmähl, D., Z. Krebsforsch., 1967, 69, 103-201.
3. Magee, P.N. and Barnes, J.M., Adv. Cancer Research, 1967, 10, 163-246.
4. Magee, P.N., Montesano, R. and Preussmann, R., in Searle, C.E. (ed), "Chemical Carcinogens", American Chemical Society, Washington, D.C., 1976, pp. 491-625.
5. Lai, D.Y. and Arcos, J.C., Life Sciences, 1980, 27, 2149-2165.
6. Magee, P.N., Biochem. J., 1956, 64, 676-682.
7. Dutton, A.H. and Heath, D.F., Nature, 1956, 178, 644.
8. Magee, P.N. and Vandekar, M., Biochem. J., 1958, 70, 600-605.
9. Brouwers, J.A. and Emmelot, P., Exptl. Cell. Res., 1960, 19, 467-474.
10. Czygan, P., Greim, H., Garro, A.J., Hutterer, F., Schaffner, F., Popper, H., Rosenthal, O. and Cooper, D.Y., Cancer Res., 1973, 33, 2983-2986.
11. Lotlikar, P.D., Baldy, W.J., Jr. and Dweyer, E.N., Biochem. J., 1975, 152, 705-708.

12. Wiessler, M., Angew. Chemie., 1974, 86, 817.
13. Roller, P.P., Shimp, D.R. and Keefer, L.K., Tetrahedron Lett., 1975, 2065.
14. Joshi, S.R., Rice, J.M., Wenk, M.L., Roller, P.P. and Keefer, L.K., J. Natl. Cancer Inst., 1977, 58, 1531.
15. Weissler, M. in Anselme, J.-P. (ed) "N-Nitrosamines"; ACS Symposium Series 101, American Chemical Society, Washington D.C., 1979, pp. 57-89.
16. Baldwin, J.E., Scott, A., Branz, S.E., Tannenbaum, S.R. and Green, L., J. Org. Chem., 1978, 43, 2427-2431.
17. Gold, B. and Linder, W.B., J. Am. Chem. Soc., 1979, 101, 6772-6773.
18. Okada, M., Mochizuki, M., Anjo, T., Sone, T., Wakabayashi, Y. and Suzuki, E. in Walker, E.A., Griciute, L., Castegnaro, M. and Börzsönyi, M. (eds) "N-Nitroso Compounds: Analysis, Formation and Occurence"; IARC Scientific Publications No. 31, International Agency for Research on Cancer, Lyon, 1980 pp. 71-82.
19. Mochizuki, M., Anjo, T. and Okada, M., Tetrahedron Lett., 1980, 3693-3696.
20. Mochizuki, M., Sone, T., Anjo, T. and Okada, M., Tetrahedron Lett., 1980, 1765-1766.
21. Mochizuki, M., Anjo, T., Wakabayashi, Y., Sone, T. and Okada, M., Tetrahedron Lett., 1980, 1761-1764.
22. Park, K.K., Archer, M.C. and Wishnok, J.S., Chem.-Biol. Interactions, 1980, 29, 139-144.
23. Magee, P.N. in de Serres, F.J., Fouts, J.R., Bend, J.R. and Philpot, R.M. (eds), "In Vitro Metabolic Activation in Mutagenesis Testing", Elsevier/North Holland Biomedical Press, Amsterdam, 1976 pp. 213-216.
24. Cottrell, R.C., Lake, B.G., Phillips, J.C. and Gangolli, S.D., Biochem. Pharmacol., 1977, 26, 809-813.
25. Milstein, S. and Guttenplan, J.B., Biochem. Biophys. Res. Commun., 1979, 87, 337-342.
26. Savenije-Chapel, E.M. and Noordhoek, J., Biochem. Pharmacol, 1980, 29, 2023-2029.
27. Grilli, S. and Prodi, G., Gann, 1975, 66, 473-480.
28. Hecht, S.S., Chen, C.B. and Hoffmann, D., Cancer Res., 1978, 38, 215-218.
29. Hecker, L.I., Farrelly, J.G., Smith, J.H., Saavedra, J.E. and Lyon, P.A., Cancer Res., 1979, 39, 2679-2688.
30. Hecht, S.S., Chen, C.B., McCoy, G.D. and Hoffman, D., Cancer Letters, 1979, 8, 35-41.
31. Chen, C.B., McCoy, G.D., Hecht, S.S., Hoffman, D. and Wynder, E.L., Cancer Res., 1978, 38, 3812-3816.
32. Krüger, F.W., Bertram, B., and Eisenbrand, G., Z. Krebsforsch., 1976, 85, 125-134.
33. Leung, K.H., Park, K.K. and Archer, M.C., Res. Comm. Chem. Pathol. Pharmacol., 1978, 19, 201-211.

34. Keefer, L.K., Lijinsky, W. and Garcia, H., J. Natl. Cancer Inst., 1973, 51, 299-302.
35. Lijinsky, W., Taylor, H.W. and Keefer, L.K., J. Natl. Cancer Inst., 1976, 57, 1311-1313.
36. Kochetkov, N.K. and Budovskii, E.I. (eds)., "Organic Chemistry of Nucleic Acids, Part B", Plenum Press, New York, NY, 1972, p 280.
37. Digenis, G.A. and Issidorides, C.H., Bioorg, Chem., 1979, 8, 97-137.
38. Lijinsky, W. and Andrews, A.W.,, Mutation Res., 1979, 68, 1-8.
39. Sieh, D.H., Wilbur, D.J. and Michejda, C.J., J. Am. Chem. Soc., 1980, 102, 3883-3887.
40. Sieh, D.H., Andrews, A.W. and Michejda, C.J., Mutation Res. 1980, 73, 227-235.
41. Sieh, D.H. and Michejda, C.J., J. Am. Chem. Soc., 1981, 103 442-445.
42. Okada, M., in Magee, P.N. et al. (eds) "Fundamentals in Cancer Prevention", University of Tokyo Press, Tokyo, 1976, 1976, pp. 251-266.
43. Blattmann, L. and Preussmann, Z. Krebsforsch., 1973, 79, 3-5.
44. Park, K.K. and Archer, M.C., Chem.-Biol. Interactions, 1978, 22, 83-90.
45. Krüger, F.W. and Bertram, B., Z. Krebsforsch., 1973, 80, 189.
46. Singer, G.M., Lijinsky, W., Buettner, L. and McClusky, G.A., submitted for publication (1981).
47. Lijinsky, W., Saavedra, J.E. and Reuber, M.D., Cancer Res., 1981, in press.
48. Michejda, C.J. and Koepke, S.R., J. Am. Chem. Soc., 1978, 100, 1959-1960.
49. Koepke, S.R., Kupper, R. and Michejda, C.J., J. Org. Chem., 1979, 44, 2718-2722.
50. Hsieh, S., Kraft, P.L., Archer, M.C. and Tannenbaum, S.R., Mutation Res., 1976, 35, 23-28.
51. Michejda, C.J., Andrews, A.W. and Koepke, S.R., Mutation Res., 1979, 67, 301-308.
52. DeMeio, R.M., in Greenberg, D.M. (ed) "Metabolism of Sulfur Compounds", Academic Press, New York, NY, 1975, pp. 287-358.
53. Michejda, C.J., Kroeger-Koepke, M.B., Koepke, S.R. and Kupper, R.J. in Anselme, J.-P. (ed), "N-Nitrosamines" American Chemical Society, Washington, DC, 1979, pp 77-89.
54. Hünig, S., Helv. Chim. Acta., 1971, 54, 1721 and references cited therein.
55. Druckrey, H., Stekar, J. and Hünig, S., Z. Krebsforsch., 1973, 80, 17-26.

56.   Kupper, R., Reuber, M.D., Blackwell, B.N., Lijinsky, W.,
      Koepke, S.R. and Michejda, C.J., Carcinogenesis, 1980, 1,
      753–757.
57.   Kupper, R. and Michejda, C.J., J. Org. Chem., 1979, 44,
      2326–2328.
58.   Smith, R.H., Kroeger-Koepke, M.B. and Michejda, C.J.,
      submitted for publication, 1981.

RECEIVED July 31, 1981.

Chemical and Biochemical Transformations of β-Oxidized Nitrosamines

RICHARD N. LOEPPKY, JERRY R. OUTRAM, WITOLD TOMASIK, and WAYNE McKINLEY

Department of Chemistry, University of Missouri-Columbia, Columbia, MO 65211

The chemical and biochemical properties of β-oxidized nitrosamines (alkylnitrosamines bearing oxygen functionality at the carbon β to the NNO group)
is reviewed and new findings are presented with respect to the role that these compounds play both
in the area of environmental carcinogenesis and the
biochemical mechanisms underlying the carcinogenicity of nitrosamines. A review of the effect of
structural features on the relative rates of the
base induced retroaldol like fragmentation of β-hydroxynitrosamines is presented. This reaction produces a smaller nitrosamine and an aldehyde or ketone. A similar transformation of β-ketonitrosamines has been found to be a general, and more rapid reaction which is complicated by condensation
reactions of reactants and products. The various
hypotheses relating to biochemical alkyl chain
shortening of nitrosamines are reviewed and the possible role of β-oxidized nitrosamines is discussed.
Preliminary evidence pointing to the existence of
a new retroaldol like fragmentation of β-hydroxynitrosamines is presented.

During the past several years much of our attention has been
directed at elucidating the chemical and biochemical properties of
β-oxidized nitrosamines in relation to their possible role in environmental carcinogenesis.[1-6] We shall define a β-oxidized nitrosamine as one which bears oxygen functionality at the carbon
beta to the nitrosamino nitrogen. This paper is principally
directed at the chemical and biochemical properties of β-hydroxy-
and β-ketonitrosamines. β-Oxidized nitrosamines are found in the
environment[1] and as metabolites of alkyl nitrosamines containing
no oxygen functionality.[7-18] N-Nitrosodiethanolamine (NDElA) has
been found to occur in large concentrations in fluids used in the

machining of metals as well as in cosmetics and other substances.[3,19-22] 3-Hydroxy-N-nitrosopyrrolidine has been found in
cooked bacon.[23] Recently the Chinese have reported that 3-(iso-
pentylnitrosamino)-2-butanone is found in the mold and pickling
brine of certain Chinese foods and is believed to be a potent
esophogeal carcinogen responsible for an unusually high cancer
incidence in certain parts of China.[24] Other examples of environmentally prevalent β-oxidized nitrosamines have been given in
our other publications.[1-3]

There have been a number of reports demonstrating that β-
hydroxy and β-ketonitrosamines are common metabolites from dialkyl
and heterocyclic nitrosamines. Examples include the β-hydroxy
and β-keto derivatives of dipropylnitrosamine (all potent pancreatic carcinogens in Syrian golden hampsters),[7,15,16] and the
β-hydroxy derivatives of diethylnitrosamine,[9] dibutylnitrosamine,[14] dipentylnitrosamine,[18] N-nitrosopyrrolidine,[13] N-nitrosomorpholine[20] and other unsymmetrical compounds. N-Nitroso-3-
hydroxypyrrolidine has even been found as a metabolite of N,N'-
dinitrosopiperazine.[25]

It is obvious that an understanding of the chemical and biochemical properties of these compounds will not only aid in the
elucidation of their role in nitrosamine carcinogenesis but is
also necessary for devising artifact free methods for their determination in environmental samples and the development of possible schemes for preventing their formation and possible chemical
transformation to other hazardous materials. This paper reviews
briefly the organochemical properties of β-oxidized nitrosamines.
The principal subject of this work, however, is the role that β-
oxidized nitrosamines play in biochemical chain shortening and it
is shown that there is a biochemical retroaldol like cleavage
reaction of β-hydroxynitrosamines.

Organochemical Properties of β-Oxidized Nitrosamines.

D. Seebach and his coworkers have employed nitrosamines
creatively in a number of synthetic transformations.[26-31] Among
these is a synthesis for β-hydroxynitrosamines by condensation of
α-lithionitrosamine with an appropriate aldehyde or ketone. This
transformation is effectively the reverse of the reaction shown
in equation 1. In 1974 we discovered that β-hydroxynitrosamines

undergo the fragmentation reaction shown in equation 1.[1] While our work was ongoing Seebach published several papers containing evidence that his condensation reaction of nitrosamines with aldehydes and ketones was reversible.[28-30] We have conducted kinetic studies of the cleavage of a number of β-hydroxynitros-amines in order to correlate the rate of the transformation with the structural properties of these compounds.[2-6] There are two principal features which control the rates of this retroaldol like cleavage reaction of β-hydroxynitrosamines. The reaction becomes faster as the stability of the incipient carbonyl product increases. Thus, tertiary nitrosaminoalcohols cleave faster than their secondary counterparts and the primary nitrosaminoalcohols are very slow to react.[4] A compound substituted with aromatic groups at the β-carbon cleaves more rapidly than its alphatic counterpart. This is because the tertiary nitrosaminoalcohols give rise to the thermodynamically more stable ketones whereas the other nitrosaminoalcohols give rise to the less stable aldehydes. The second feature of the transformation which controls the rate is the stability of the incipient α-nitrosaminocarbanion. For example, substitution of hydrogen at the α-carbon by an alkyl group reduces the rate of the cleavage reaction as one would anticipate from the decreased stability of the carbanion formed in the reaction. The most striking feature resulting from carbanion stability, however, is the control which the orientation of the N-NO group exerts on the rate of the transformation. When the NO bond is oriented syn to the alcohol function the compound cleaves at a rate 300-500 times greater than its corresponding anti isomer. It has been postulated that the syn nitrosaminocarbanion is much more stable than its anti counterpart.[32] We have demonstrated that the retroaldol like cleavage reaction of β-hydroxynitros-amines exhibit a rate range of 10^5.[4-5]

The chemistry of the ubiquitous NDElA in basic solution has been studied extensively because the metalworking fluids in which this material is frequently found are alkaline. In the presence of alkoxide bases in alcohol solvents NDElA undergoes the retro-aldol like cleavage reaction exceedingly slowly.[6] Another puzzling feature of the retroaldol cleavage of both NDElA and methyl-ethanolnitrosamine is that the reaction appears to only go to about 10% completion and then side transformations occur.[6] In the case of NDElA the prevalent side transformations involve the conversion of NDElA into 2-hydroxyethylvinylnitrosamine and N-nitrosomorpholine. 2-Hydroxyethylmethylnitrosamine is converted into methylvinylnitrosamine after its cleavage reaction stops. Extensive kinetic studies have shown that the retroaldol cleavage reaction of these primary nitrosaminoalcohols is reversible under the conditions of the study and the equilibrium lies far to the side of the hydroxynitrosamines. The most interesting feature to arise from this study however, was the finding that the formaldehyde produced in the cleavage reaction catalyzes the dehydration of the nitrosaminoalcohol (see scheme 1).[6] We have

Scheme 1. *A mechanism for the formaldehyde catalyzed dehydration of β-hydroxynitrosamines.*

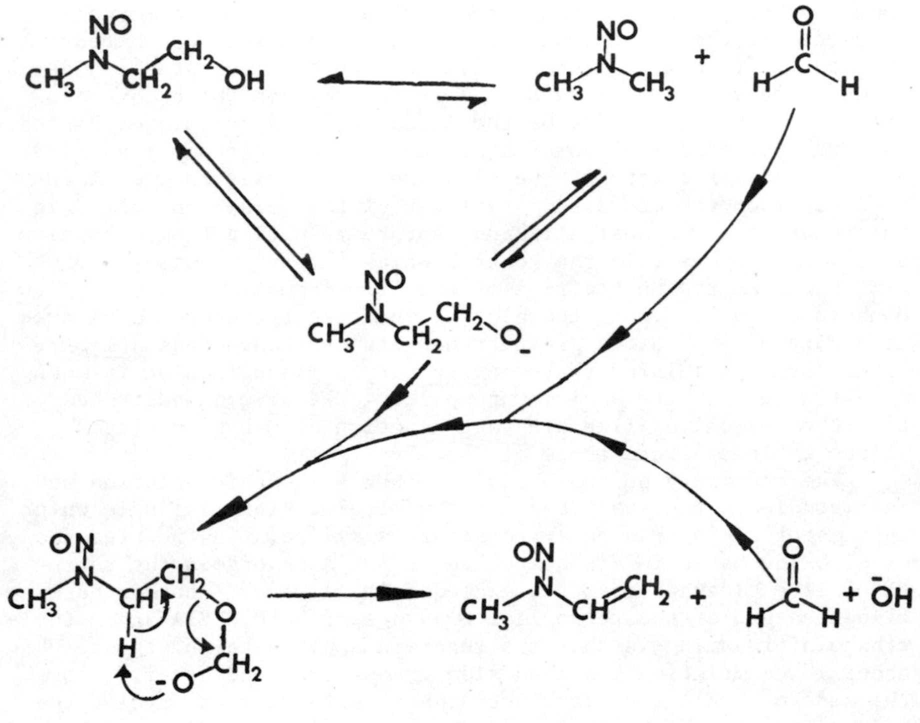

also sought to establish whether these chemical transformations
might occur during the inadvertent employment of NDE1A in metal-
working fluids. A side benefit of these studies has been the development of an efficient method for the analysis of subnanogram
levels of nitrosamine in metalworking fluids and other similar
matrices through the utilization of an extremely simple cleanup
procedure involving an ionic exchange resin.[33]

Michejda and his collaborators have demonstrated a property
of β-hydroxynitrosamines which is important with respect to their
possible biochemical role.[34] These workers demonstrated that
tosylate derivatives of several β-hydroxynitrosamines were extremely reactive compounds and underwent sovolysis reactions by
virtue of neighboring group participation. The nitrosamino oxygen
affects the displacement of the tosyloxy group. The resulting
positively charged heterocycle is a very efficient alkylating agent.

β-Ketonitrosamines are very reactive substances. Hydrogens
attached to the carbon situated between the nitrogen and the ketone function have an acidity at least comparable with those found
in the similar position in β-dicarbonyl compounds. Seebach and
collaborators have sought to prepare these compounds by the condensation reaction of esters for other acyl derivatives with α-metalo-
nitrosamines.[26-28] In several cases Seebach reports the isolation
of double condensation products and often the yields of the β-ketonitrosamines are low. Krüger showed that treatment of 2-oxo-
propylpropylnitrosamine with KOH in alcohol resulted in the production of methylpropylnitrosamine.[46] This retro-Claisen-like
transformation and other chemical properties of β-ketonitrosamines
have been under study in our laboratory. 2-Phenyl-2-oxoethyl-
methylnitrosamine is cleaved by potassium hydroxide in ethanol at
20° to potassium benzoate and dimethylnitrosamine. However, some
of the dimethylnitrosamine which is formed condenses with the
unreactive β-ketonitrosamine to give a bis-nitrosaminoalcohol as
shown in equation 2. The nitrosaminoalcohol product is also found

to be formed when Seebach and collaborators condensed lithiodi-
methylnitrosamine with methyl benzoate.[27] Preliminary experiments

have been performed in our laboratory to compare the cleavage
rates of β-ketonitrosamine shown in equation 2 and its derived
alcohol. The β-ketonitrosamine undergoes a much more rapid
cleavage than the alcohol at 50° but it also enters into other
transformations which have not yet been fully elucidated by our
work.

<u>Biochemical Properties and Physiological Roles of β-Oxidized Ni-
trosamines in Relation to their Carcinogenesis.</u>

The hydroxylation of the alkyl chain of the nitrosamine significantly increases the water solubility of the compound. It
enhances its chances for excretion, and also provides a site which
can be conjugated with glucuronic acid, sulfate or other water
solubilizing substances which are commonly used by mammalian organisms to facilitate excretion and detoxification. The smaller
hydroxynitrosamines are completely miscible in water and several
studies have shown that NDE1A applied to the skin of a rat is
largely excreted unchanged in the urine.[35],[36] Outside of this
rather obvious hypothesis there has been no general agreement as
to the role which β-oxidized nitrosamines might play in nitrosamine carcinogenesis. Schoental[37] and Manson[14] have suggested
that these compounds might be effective crosslinking agents for
nucleic acids and induced tumor formation in this way. In general, the β-hydroxy and keto derivatives of alkyl and heterocyclic nitrosamines possess a carcinogenicity that is comparable
with that of their unoxidized precursors and are sometimes more
carcinogenic. Michejda[34] has proposed that the conjugation of
β-hydroxynitrosamines with sulfate by means of the enzyme sul-
fatase could result in very effective alkylating agents via
neighboring group participation as has been shown in his chemical
studies (see scheme 3). Although this hypothesis has great merit,
most of the speculation on the biological function of β-oxidized
nitrosamines has focussed on their possible role in chain shortening reactions.
Between 1971 and 1976 Krüger published his results which are
summarized in scheme 2.[7],[13],[25],[46] Krüger demonstrated that dipropylnitrosamine which was labeled at the alpha carbon incorporated this labeled carbon as a methyl group in rat liver RNA.[7]
This is an unexpected finding since the established mechanism of
nitrosamine carcinogenesis involves enzymatic α-hydroxylation followed by the decomposition of this unstable species to a carbonyl
compound and an alkylating species (a diazonium ion or a carbocation). In the case of dipropylnitrosamine the expected alkylating agent would be the n-propyl group or the rearranged isopropyl group and the other product of the decomposition is expected to be propanal. Kruger showed that 2-hydroxypropylpropyl-
nitrosamine was a metabolite of dipropylnitrosamine and that this
labeled substance[13] or its derived ketone[46] gave similar results;
methylation of the purine bases in rat liver RNA. All of Krüger's

Scheme 2. *A summary of Krüger's in vivo experiments on the incorporation of radiolabeled carbons into rat liver RNA upon the administration of the compound shown. The position of △ indicates the labeled carbon (see text for references).*

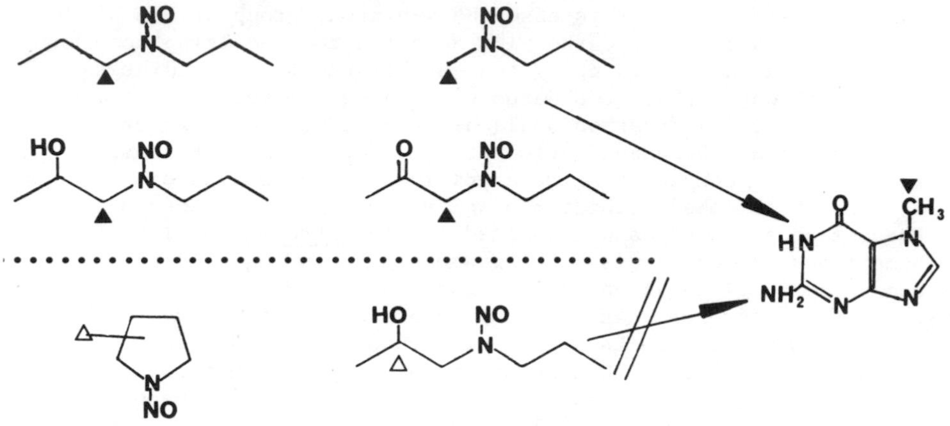

experiments were conducted in vivo, but Preussmann and co-workers[38] showed that the rat liver S-9000 fraction gave rise to extensive methylation of the trapping agent 3,4-dichlorothiophenol
in in vitro experiments when a variety of dialkylnitrosamines
were used. Thus, dipropylnitrosamine produced greater than 90%
methylation and only a small amount of propylation. Methyl-t-
butylnitrosamine, which is not carcinogenic, did not produce any
alkylated 3,4-dichlorothiophenol derivatives. Grandjean and
Althoff[15] reported that 2-hydroxypropylnitrosamine, 2-oxopropyl-
propylnitrosamine and methylpropylnitrosamine were all urinary
metabolites of dipropylnitrosamine in Syrian golden hampsters.

As a hypothesis to explain his observations, Krüger[13] proposed (as is shown in scheme 3 along with other hypothesis) that
the derived β-ketonitrosamine underwent a cleavage reaction to
yield a methylalkylnitrosamine and a carboxylic acid derivative,
in analogy to fatty acid metabolism.

Another chain shortening mechanism has been offered by Okada
and colleagues[14] and Preussmann and Blattmann.[9] This mechanism
stems from these workers studies of the metabolic transformations
of dibutylnitrosamine and the bladder carcinogen, 4-hydroxybutyl-
butylnitrosamine. In this case the 4-hydroxy group of the alkyl
chain is oxidized to a carboxylic acid and the resulting compound
undergoes chain shortening by the established fatty acid degradation pathway. Thus, two carbons would be removed as acetyl-CoA
and the remaining β-carbon would be transformed to a carboxylic
acid function. Decarboxylation of this β-oxidized nitrosamine
would give a methylalkylnitrosamine leading to methylated products
via the established mechanism of nitrosamine carcinogenesis.
Preussmann and Blattmann[9] examined the in vitro metabolism of a
homologous series of dialkylnitrosamines beginning with diethyl
and going to dipentyl and showed that ω-hydroxylation (hydroxylation at the terminal carbon atom) occurred in every case. Taken
all together the results of Okada, Preussmann and Blattmann
strongly suggest that chain shortening of dialkyl nitrosamines
does occur via terminal oxidation and degradation in analogy to or
by the fatty acid pathway. It is important to remember, however,
this is somewhat different from the hypothesis of Krüger.
Blattmann has also suggested an alternative pathway which involves
a terminal oxidation of the chain after β-hydroxylation and has
presented data for the metabolism of 2-hydroxybutylbutylnitros-
amine indicating that this does occur.[44]

Another mechanism for chain shortening has been advanced by
Eisenbrand.[39] Like Kruger's hypothesis Eisenbrand's involves a
β-ketonitrosamine. Eisenbrand proposes that this compound would
undergo hydroxylation at the α' position and subsequent decomposition to yield a diazohydroxide. The resulting β-ketodiazo-
hydroxide would undergo cleavage of C_α-C_β bond with acyl transfer
to the diazohydroxide oxygen. This process would give rise to
either a methyldiazonium ion or diazomethane as shown in scheme 3.
Support for Eisenbrand's hypothesis comes from his studies on
N-nitrosoureas where this type of transformation is observed.[39]

We have proposed that chain shortening of β-hydroxynitrosamines could be accomplished if there were a biochemical retroaldol cleavage analogous to the reaction which occurs when β-
hydroxynitrosamines are treated with a strong base.[2] This hypothesis is outlined in schemes 3 and 4. According to our hypothesis (scheme 4) dipropylnitrosamine first would be β-hydroxylated by an appropriate enzyme system. This substrate would then
be enzymatically converted to methylpropylnitrosamine and propanol
by a biochemical retroaldol cleavage. α-Hydroxylation of methylpropylnitrosamine at the propyl α-carbon would give rise to a
active methylating species. Blattman and Preussmann[45] have shown
that acetaldehyde is a product of the in vitro metabolism of 2-
hydroxybutylbutylnitrosamine and this experiment supports our
hypothesis.

There is good reason why all of these hypothesis should be
seriously considered as viable alternative mechanisms for the
carcinogenic action of β-oxidized nitrosamines. It is a well
known fact of organic chemistry that there is little evidence for
carbocation centers adjacent to hydroxyl or keto bearing carbons.
With few exceptions rearrangement of the carbon skeleton or, in
the case of the hydroxyl moiety hydride migration, accompanies
attempts to generate the carbocation. This property results from
the resonance stabilization of the resulting ion via the un-
shared pairs on the oxygen atom. NDE1A and bis-2-hydroxypropyl-
nitrosamine are examples of carcinogenic nitrosamines which would
not be expected to be a carcinogenic on the basis of the proposed
mechanism of nitrosamine carcinogenesis, because as is shown in
scheme 5 rearrangement should take place to produce carbonyl compounds. It is possible that β-keto nitrosamines could be acyl-
ating agents. This analysis assumes that the diazonium ion produced in the decomposition of the α-hydroxynitrosamines has a
negligible life time. For some nitrosamines this is very likely
a poor assumption, since McGarrity and Smyth have recently shown
the methyldiazonium ion to have a half life of approximately 330
milliseconds in water at pH 7.5 at 25°.[40] Another interesting
point to come from the work of McGarrity and Smyth is the finding
that the methyldiazonium ion has a relatively high acidity with
the pKa ≃ 10. This suggests, particularly for β-ketodiazonium
ions, that diazo compounds could be easily produced by a deproton-
ation of the alpha-carbon. This would greatly add to the lifetime of the proximate carcinogen. Lijinsky and co-workers[41] have
shown that dimethylnitrosamine does not alkylate via a diazo compound, but this hypothesis should be reexamined with respect to
β-ketonitrosamines.

Scheme 3. Hypothetical metabolic transformations of 2-hydroxypropylmethyl-nitrosamine with the name of the originator of each hypothesis. Nameless hypotheses are by the authors.

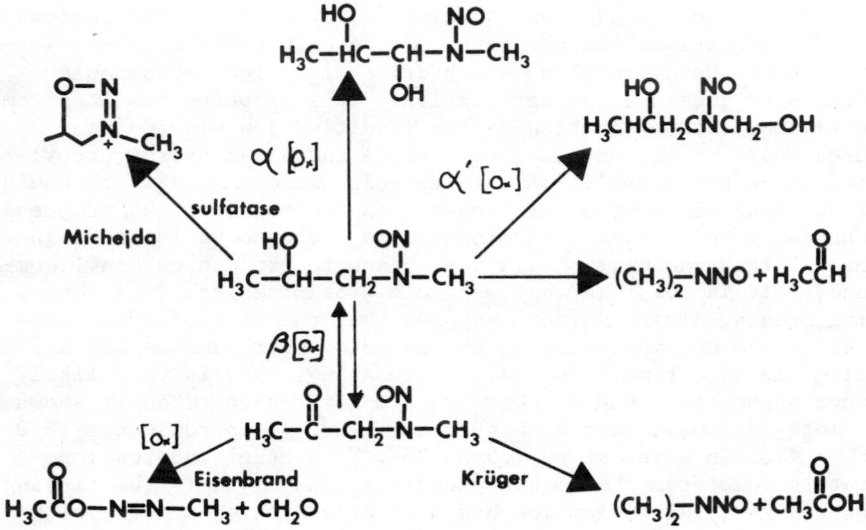

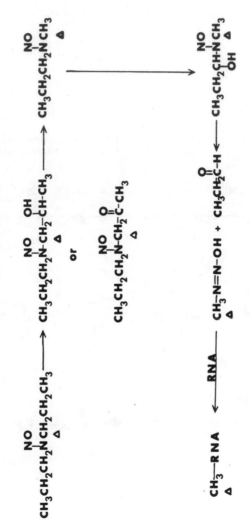

Scheme 4. Authors' explanation for Krüger's experiments.

Scheme 5. *Transformation expected to spontaneously follow enzymatic α′-hydroxylation of a β-hydroxynitrosamine.*

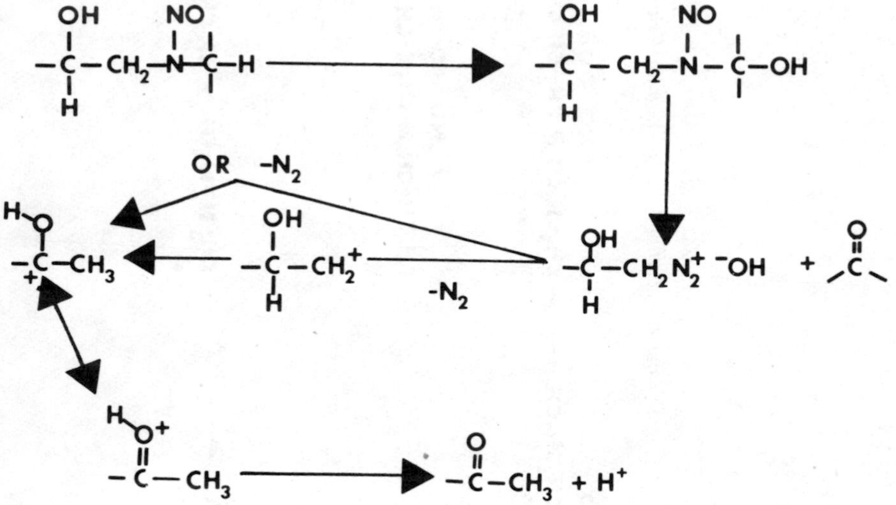

The Discovery of the Biochemical Retroaldol Cleavage of β-Hydroxynitrosamines.

With due consideration to the forgoing we have taken up experiments to demonstrate whether or not our hypothesis relating to a biochemical retroaldol cleavage of β-hydroxynitrosamines has any validity. As a result of our research on the base induced fragmentation of β-hydroxynitrosamines we chose four substrates to investigate this possibility. All of these compounds have been shown to undergo the base induced cleavage with the facility and we reasoned that a biochemical cleavage of these compounds would likely have a similar transition state to the organochemical transformation. Each of the four nitrosamines shown in scheme 6 were incubated with the S-9000 fraction obtained from male Wistar rat livers at 37°. With each of these compounds we were able to detect the formation of a dimethylnitrosamine and the corresponding aldehyde or ketone. No transformation was observed when boiled or acid denatured S-9000 fraction was used in the incubation. The principal means by which these transformations were detected utilized HPLC chromatography of the aqueous mixture derived from the incubation experiments after the precipitation of the protein and centrifigation. These incubation mixtures were also extracted with methylene chloride and the dimethylnitrosamine product was detected by GLC as well. The carbonyl fragments were detected both by HPLC and by the formation of their 2,4-dinitrophenylhydrazones and HPLC analysis according to the procedures of Farrelly[42] or modifications thereof. The 2,4-dinitrophenylhydrazone of formaldehyde was also detected in these experiments. The rats were preinduced with phenobarbitol but controlled experiments demonstrated that this preinduction had no effect on the biochemical cleavage reaction which we are observing.

While we have anticipated that a number of the hypothetical transformations discussed above would occur our experiments are not far enough along to deliniate these at this time. We did make one very interesting observation which is anticipated from the work which has appeared in the literature.[16] There is also oxidation of the secondary nitrosaminoalcohols to their corresponding ketones. This oxidation appears to compete with the cleavage reaction and the competition depends upon the experimental conditions and the substrate structure. Thus, 2-hydroxypropylmethylnitrosamine is converted both into dimethylnitrosamine and the β-oxo derivative. The β-oxo derivative in this case is formed to a greater extent than dimethylnitrosamine at its maximum concentration. On the other hand 2-hydroxy-2-phenyethylmethylnitrosamine undergoes much more extensive cleavage than β-oxidation. Considering that these are in vitro experiments the conversion of the nitrosamines to simpler nitrosamines by the retroaldol type cleavage appears to be a reasonably efficient process. In the case of the 2-hydroxy-2-methylpropylmethylnitrosamine a 9% yield of dimethylnitrosamine was detected at its maxi-

Scheme 6. A summary of some data on the biochemical retroaldol-like fragmentation of four β-hydroxynitrosamines.

OBSERVATIONS

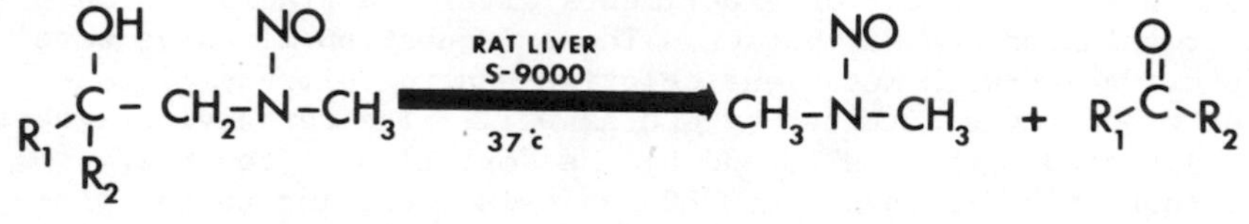

$$R_1 - \underset{\underset{R_2}{|}}{\overset{\overset{OH}{|}}{C}} - CH_2 - \underset{\overset{|}{NO}}{N} - CH_3 \quad \xrightarrow[37°c]{\underset{S\text{-}9000}{RAT\ LIVER}} \quad CH_3 - \underset{\overset{|}{NO}}{N} - CH_3 \quad + \quad R_1 - \overset{\overset{O}{\|}}{C} - R_2$$

R_1	R_2	Maximum % Conversion
H_3C-	$H-$	1.59%
H_3C-	H_3C-	9.50%
H_5C_6-	$H-$	2.83%
H_5C_6-	H_5C_6-	≈1.00%* · (solubility limit)

mum concentration. The phenyl compounds are not appreciably soluble in the aqueous incubation mixture and dimethylsulfoxide had to be added in order to incorporate them into the S-9000 mixture. There is some evidence which suggests that dimethylsulfoxide inhibits the nitrosamine metabolizing enzymes and the lower conversions in these cases may reflect this problem.[43] The yield of acetaldehyde from 2-hydroxypropylmethylnitrosamine was approximately 70% that of dimethylnitrosamine at the maximum yield of the nitrosamine. If the NADPH generating system was omitted from the S-9000 mixture cleavage was observed although at somewhat reduced yields and no oxidation to the β-keto compound was observed. This suggests that dimethylnitrosamine may be arising by oxidative routes as well as the retroaldol cleavage reaction which is not an oxidative transformation.

The course of these transformations has been studied as a function of several variables including time, protein concentration, and substrate concentration. In general the following observations have been made. The dimethylnitrosamine concentration maximizes between 60 and 120 minutes incubation time. The maximum differs somewhat depending upon the substrate. The keto compound yield also goes through a maximum at around a 120 minutes for the 2-hydroxypropylmethylnitrosamine. These data as well as similar results on yield versus protein concentration demonstrate that these nitrosamines are being further metabolized by the S-9000 mixture. The secondary and tertiary nitrosaminoalcohols behave differently as one increases the protein concentration. In the case of 2-hydroxypropylmethylnitrosamine the yield of dimethylnitrosamine reached a maximum at approximately 16 mg/mL protein then decreased. There was very little production of dimethylnitrosamine between 0 and 4 mg/mL protein. On the other hand the yield of the β-oxocompound increased steadily as the protein concentration was increased. With the tertiary nitrosaminoalcohol the yield of dimethylnitrosamine was quite small between 0 and 16 mg/mL of protein and then increased very rapidly between 16–32 mg/mL protein. The product, acetone, was shown to also be increasing in parallel with the nitrosamine. (Even though acetone is a component of our reaction mixture via contamination of the NADP its concentration is clearly seen to increase by our HPLC studies). A general feature of the cleavage reaction is that relatively little cleavage takes place at protein concentrations below 8 mg/mL. The concentration of substrate also influences the yield of cleavage products. In all cases we observed a maximum in the yield showing that additional substrate apparently inhibits the enzyme systems giving rise to the cleavage reaction.

These results support our hypothesis that there is a retroaldol like cleavage of β-hydroxynitrosamines which occurs in biological systems. Our results agree with the data of Kruger, Preussmann, and Blattmann. The fact that the tertiary nitrosaminoalcohols undergo the cleavage as well as their secondary counterparts demonstrates that oxidation to a ketone is not a

requirement for the cleavage. Our data to date do not disprove
any of the other hypothesis with respect to the chain shortening
or roles of β-oxidized nitrosamines in nitrosamine carcinogenesis.
It is obvious that further experiments will be required to put
the existence of this transformation on a firmer foundation.
Labeling studies will be particularly useful in this regard.

We were very surprised at the facility of this transformation
and we had anticipated that a demonstration of its existence would
be more difficult than it has been. This reaction appears to be
a new type of biochemical transformation.

Acknowledgement

The investigators wish to express their gratitude to the PHS
for support of this work under grant numbers CA22289 and contract
number CP75946 awarded by the National Cancer Institute, DHHS.

Literature Cited

1. Loeppky, R.N.; Christiansen, R. in "Environmental Aspects of
    N-Nitroso Compounds"; Walker, E.A.; Castegnaro, M.; Griciute,
    L.; Eds. I.A.R.C. Sci. Publ. No. 19, Lyon, 1978; p. 117.
2. Loeppky, R.N.; Gnewuch, C.T.; Hazlitt, L.; McKinley, W.A. in
    "N-Nitrosamines"; Anselme, J.P.; Ed. Amer. Chemical Society:
    Washington, DC, 1979; p. 109.
3. Loeppky, R.N.; McKinley, W.A.; Hazlitt, L.; Beedle, E.C.;
    DeArman, S.K.; Gnewuch, C.T. in "N-nitroso Compounds:
    Analysis, Formation and Occurence"; Walker, E.A.; Castegnaro,
    M.; Griciute, L.; Börzönyi, M.; Eds., I.A.R.C. Sci. Pub.
    No. 31., Lyon, 1980; p. 15.
4. Loeppky, R.N.; McKinley, W.A.; Gnewuch, C.T.; Outram, J.R.
    manuscript in preparation.
5. Loeppky, R.N.; Hazlitt, L.; McCallister, J.; Outram, J.R.
    manuscript in preparation.
6. Loeppky, R.N.; Outram, J.R. manuscript in preparation.
7. Krüger, F.W. Z. Krebsforch. 1971, 76, 147.
8. Okada, M.; Suzuki, E. Gann. 1972, 63, 391.
9. Blattmann, L.; Preussmann, R. Z. Krebsforsch. 1973, 79, 3.
10. Blattmann, L.; Joswig, N.; Preussmann, R. Z. Krebsforsch.
    1974, 81, 71.
11. Blattmann, L.; Preussmann, R. Z. Krebsforsch. 1974, 81, 25.
12. Blattmann, L.; Preussmann, R. Z. Krebsforsch. 1975, 83, 125.
13. Krüger, F.W. Z. Krebsforsch. 1973, 79, 90.
14. Manson, D.; Cox, P.J.; Jarman, M. Chem. Biol. Interactions
    1978, 20, 341.
15. Althoff, J.; Grandjean, C.; Pour, P.; Bertram, B.
    Z. Krebsforsch. 1977, 90, 141.
16. Park, K.K.; Archer, M.C. Chem. Biol. Interactions 1978,
    22, 83.
17. Hecker, L.I.; Saavedra, J.E. Carcinogenesis 1980, 1, 1017.

18. Hecht, S.S.; Chen, C.B.; Hoffman, D. J. Med. Chem. 1980, 23, 1175.
19. Fan, T.Y.; Morrison, J.; Ross, R.; Fine, D.G.; Miles, W.; Sen, N.P. Science 1977, 196, 70.
20. Zingmark, P.A.; Rappe, C. Anbio 1977, 6, 237.
21. Fan, T.Y.; Goff, U.; Song, L.; Fine, D.H.; Arsenault, G.P.; Biemann, K. Food Cosmet. Toxicol. 1977, 15, 423.
22. Schmeltz, I.; Abidi, S.; Hoffmann, D. Cancer Lett. 1977, 2, 125.
23. Lee, J.S.; Bills, D.D.; Scanlan, R.A.; Libbey, L.M. J. Agric. Food Chem. 1977,25, 422.
24. Li, M.G.; Lu, S.X.; Ji, C.; Wang, M.Y.; Cheng, S.J.; Lin, C.L. Scientia Sinica. 1979, 22, 471.
25. Krüger, F.W.; Bertram, B.; Eisenbrand, G. Z. Krebsforsch. 1976, 85, 125.
26. Seebach, D.; Enders, D. Angew. Chem. Internat. Ed. 1975, 14, 15.
27. Seebach, D.; Enders, D. Chem. Ber. 1975, 108, 1293.
28. Renger, B.; Kalenowski, H.O.; Seebach, D. Chem. Ber. 1977, 110, 1866.
29. Renger, B.; Hugel, H.; Wykpiel, W.; Seebach, D. Chem. Ber. 1978, 111, 2630.
30. Seebach, D.; Wykpiel, W. Synthesis 1979, 1979, 423.
31. Wykpiel, W.; Seebach, D. Tetrahedron Lett. 1980, 21, 1927.
32. Houk, K.N.; Strozier, R.W.; Rondan, N.G.; Frazer, R.R.; Chuaqui-Offermanns, N. J. Amer. Chem. Soc. 1980, 102, 1426.
33. Loeppky, R.N.; Faulconer, J.; McKinley, W.A.; Beedle, E.C.; Outram, J.R. manuscript in preparation.
34. Koepke, S.R.; Kupper, R.; Michejda, C.J. J. Org. Chem. 1979, 44, 2718.
35. Edwards, G.S.; Ping, M.; Fine, D.H.; Spiegelhalder, B.; Kann, J. Tox. Lett. 1979, 4, 217.
36. Sansone, E.B.; Losikoff, A.M.; Lijinsky, W. "N-Nitroso Compounds: Analysis, Formation and Occurence", Walker, E.A.; Griciute, L.; Castegnaro, M.; Börzsönyi, Eds. I.A.R.C. Sci. Pub. No. 31, Lyon, 1980 pp. 705-713.
37. Schoental, R. Brit. J. Cancer 1975, 28, 436.
38. Preussmann, R.; Arjungi, K.N.; Ebers, G. Cancer Research 1976, 36, 2459.
39. Eisenbrand, G. presentation Orlando, Fl. 1976.
40. McGarrity, J.F.; Smyth, T. J. Amer. Chem. Soc. 1980, 102, 7303.
41. Lijinsky, W.; Loo, J.; Ross, A.E. Nature 1968, 218, 1174.
42. Farralley, J.G. Cancer Research 1980, 40, 3241.
43. Anderson, D.; McGregor, D.B. Carcinogenesis 1980, 1, 363.
44. Blattmann, L. Z. Krebsforsch. 1977, 88, 315.
45. Blattmann, L.; Preussmann, R. Z. Krebsforsch. 1977, 88, 311.
46. Krüger, F.W.; Bertram, B. Z. Krebsforsch. 1973, 80, 189.

RECEIVED August 11, 1981.

Mechanisms of Alkylation of DNA by
N-Nitrosodialkylamines

MICHAEL C. ARCHER and KWANG-HANG LEUNG

Department of Medical Biophysics, University of Toronto,
Ontario Cancer Institute, Toronto, Canada, M4X 1K9

Isolated rat liver fractions metabolize N-nitrosodipropylamine to yield both 1-propanol and 2-propanol,
providing evidence for a reaction sequence, initiated by formation of the α-hydroxynitrosamine, that
leads to the formation of carbonium ions. Administration of N-nitrosodipropylamine to rats, however,
leads to formation of 7-propylguanine but not 7-
isopropylguanine in hepatic DNA. Thus in the intact
cell, the reaction sequence is intercepted by
nucleophilic sites in DNA before a carbocation is
formed. Rat liver microsomes also metabolize N-
nitrosodipropylamine by two consecutive β-oxidation
reactions to yield N-nitroso-2-oxopropylpropylamine.
The latter agent leads to production of 7-methyl-
guanine and O^6-methylguanine in hepatic DNA following its administration to rats. Rats administered
N-nitrosodipropylamine, N-nitroso-2-hydroxypropyl-
propylamine, or N-nitroso-2-oxopropylpropylamine
excrete the β-glucuronide of N-nitroso-2-hydroxy-
propylpropylamine as the major urinary product.
Neither N-nitrosomethylpropylamine nor N-nitrosodimethylamine is detected in any of the *in vitro* incubation mixtures or in the urines. The mechanism of
methylation of rat liver DNA by nitrosodipropylamine therefore remains unclear.

Nitrosamines require metabolic activation in order to produce
a chemical species that will alkylate nucleophilic sites on a bio-
molecule such as DNA (1, 2). The crucial initial step in the
formation of this alkylating agent is microsomal α-hydroxylation
of the nitrosamine as shown in Figure 1. The α-hydroxynitrosamine
(II) so formed, spontaneously cleaves to yield an aldehyde fragment and an alkyldiazohydroxide (III). In this hypothetical
scheme, the diazohydroxide loses a mole of water to produce either

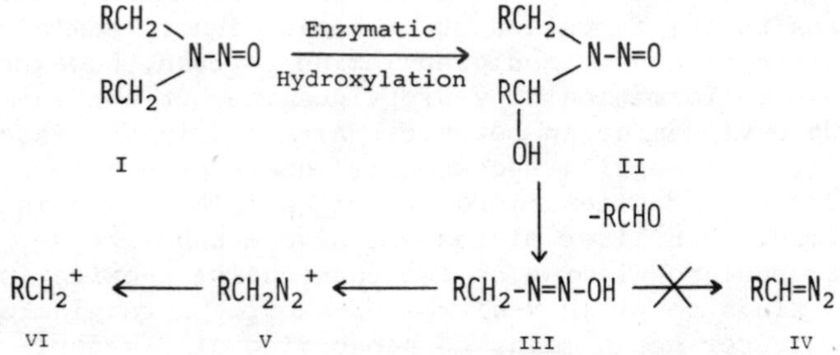

Plenum Publishing Corporation

Figure 1. Reaction scheme for production of alkylating agent following microsomal metabolism of N-nitrosodialkylamine (3).

a diazoalkane (IV), or an alkyl diazonium ion (V) and ultimately
an alkyl carbonium ion (VI). The diazoalkane or the cationic
species may then react with water to form alcohols, or with cellular nucleophiles to form alkylated products.

Experiments by Lijinsky and coworkers (4,5) with nitrosamines
labeled with deuterium atoms have shown that alkylation of nucleic
acids does not proceed via formation of the diazoalkane. There
is, however, only indirect evidence, based on an analysis of
minor alkylation products, for participation of carbonium ion
intermediates in the alkylation of DNA by nitrosamines (5). We
have endeavoured to provide more direct evidence for the production of carbonium ions by the oxidative metabolism of nitrosamines. We have made use of the well characterized property of
primary alkyl cations to rearrange by either a hydride or an
alkyl shift to form a more stable secondary or tertiary ion (6,7).
If a carbonium ion is the ultimate alkylating agent produced by
metabolism of N-nitrosodipropylamine (NDPA), products containing an isopropyl group will be formed in addition to those
containing an n-propyl group. Thus, in the presence of isolated
rat liver preparations *in vitro*, with water as the nucleophile,
NDPA should lead to production of 2-propanol, while direct administration of NDPA to rats, should lead to formation of bases
containing isopropyl groups in hepatic nucleic acids.

Analysis of reaction mixtures for 1-propanol and 2-propanol
following incubation of NDPA with various rat liver fractions in
the presence of an NADPH-generating system is shown in Table I
(8). Presence of microsomes leads to production of both alcohols,
but there was no propanol formed with either the soluble enzyme
fraction or with microsomes incubated with SKF-525A (an inhibitor
of cytochrome P450-dependent oxidations). The combined yield of
propanols from 280 µmoles of NDPA was 6.1 µmoles and 28.5 µmoles
for the microsomal pellet and the 9000 g supernatant respectively.
The difference in the ratio of 1- to 2-propanol in the two rat
liver fractions may be due to differences in the chemical composition of the reaction mixtures (9). Subsequent experiments
have shown that these ratios are quite reproducible. For comparison, Table I also shows formation of propanols following
base catalyzed decomposition of N-propyl-N-nitrosourea. As
expected (10,11), both propanol isomers were formed, the total
yield in this case being almost quantitative.

Formation of both 2-propanol and 1-propanol following metabolism of NDPA by rat liver fractions *in vitro* with water as the
nucleophile, indicates that the propyl cation is indeed formed as
shown in Figure 1.

Next, four male Sprague-Dawley rats were administered NDPA-
[2,3-^{3}H] by intraperitoneal injection (12). After 12 hours, the
animals were sacrificed, and RNA and DNA were isolated from the
combined livers by standard procedures. Following addition of
unlabeled, authentic 7-propylguanine and 7-isopropylguanine as
markers, the nucleic acids were hydrolyzed in perchloric acid at

Table I

Analysis of 1-propanol and 2-propanol obtained by
incubation of N-nitrosodi-n-propylamine (NDPA) with
rat liver fractions and incubation of N-propyl-N-
nitrosourea (PNU) in 0.1 M tris buffer, pH 7.4 at
37°C. (3).

	Proportion of 1- and 2- forms in propanol mixture (%)	
System[a]	1-propanol	2-propanol
NDPA + 105,000 g pellet	61	39
NDPA + 9,000 g supernatant	82	18
NDPA + 105,000 g supernatant	ND	ND
NDPA + 105,000 g pellet + SKF 525A	ND	ND
PNU + buffer	61	39

ND = none detected
[a]Reaction mixtures contained the following components in a final
volume of 20 ml: 15 ml of rat liver fraction prepared from a 25%
(wv) liver homogenate;   0.1 M Tris buffer, pH 7.4;   61.5 mM $MgCl_2$;
2.1 mM NADP;   21.8 mM glucose-6-phosphate;   and 50 units of
glucose-6-phosphate dehydrogenase. 0.28 mmole of N-nitrosodipropylamine were added to start the reaction. After 90 min., the
reaction was terminated by adding 20 ml of 20% $ZnSO_4$ followed by
20 ml of sat. $Ba(OH)_2$. Following removal of the precipitate by
centrifugation, the supernatant was distilled at atmospheric pressure through a short path micro-distillation head. The distillate
which was collected on ice was analyzed for propanal, 2-propanol
and propanol by gas chromatography with 2m Porapak-N column at
135°. Product identities were confirmed by gas chromatography-
mass spectrometry.

$100°$. The chromatographic profile of the DNA hydrolysate using a reverse phase column is shown in Figure 2. The 7-propylguanine isomers were separated from each other and from the major bases and 7-methylguanine which eluted close to the solvent front. When radioactivity in chromatographic fractions was measured, it was clear that the DNA hydrolysate contained 7-propylguanine but no 7-isopropylguanine (Figure 2). The RNA hydrolysate did contain a small quantity of 7-isopropylguanine, but the amount represented less than 5% of the 7-propylguanine that was present.

Since the results of our experiments with isolated rat liver fractions supported a reaction sequence initiated by microsomal oxidation of the nitrosamine leading to formation of a carbonium ion, the results of the animal experiment suggested that in the intact hepatocyte, one of the earlier electrophilic intermediates (II, III or V, Figure 1) is intercepted by nucleophilic sites in DNA (exemplified here by the N7 position of guanine) before a carbocation is formed.

In 1971, Krüger showed that, in addition to formation of 7-propylguanine, administration of NDPA to rats leads to production of 7-methylguanine in hepatic nucleic acids (13). Krüger subsequently showed that 7-methylguanine is actually the major alkylation product in hepatic nucleic acids when NDPA or its β-hydroxy or β-oxo derivatives are administered to rats (14,15). As a possible mechanism to explain the methylating activity of NDPA therefore, Krüger suggested that the nitrosamine is metabolized by two consecutive β-oxidation reactions to yield first N-nitroso-2-hydroxypropylpropylamine (NHPPA), and then N-nitroso-2-oxo-propylpropylamine (NOPPA). Cleavage of the acyl fragment in NOPPA would yield N-nitrosomethylpropylamine (NMPA), which then could act as a methylating agent.

We have confirmed and extended Krüger's observations on the methylating properties of NOPPA (16). We first synthesized NOPPA-[1,3-^{3}H] by tritium exchange as described by Krüger and Bertram (15). We were particularly careful to show that the tritiated nitrosamine contained no N-nitrosodimethylamine (NDMA) and no NMPA as contaminants. Three male albino Sprague-Dawley rats, average weight 145 g, were administered NOPPA-[1,3-^{3}H] (310 mg/23 mCi/kg body weight) intraperitoneally. After 12 hours, DNA was prepared from the pooled livers by standard methods. Following hydrolysis of the DNA in 0.1 N HCl at $70°$ for 30 minutes, the hydrolysate was chromatographed on a Sephadex G-10 column eluted with 0.05 M ammonium formate, pH 6.8, containing 0.02% sodium azide. When fractions were counted for radioactivity, we were able to show the presence of both 7-methylguanine and O^6-methylguanine in the DNA hydrolysate. The ratio of O^6-methylguanine: 7-methylguanine was 0.07, which is similar to the ratio of 0.12 for these bases found by Nicoll et al. (17) at the same time following administration of NDMA at a comparable dose   The ratio of O^6-methylguanine: N7-methylguanine has been shown by Lawley (5) to depend on the reactivity of the methylating agent.

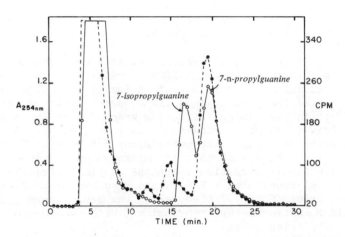

Chemico-Biological Interactions

Figure 2. Chromatographic profile of DNA hydrolysate from rat liver 12 h after application of 133 mg/18.5 mCi/kg N-[2,3-³H]-nitrosodipropylamine. Key: O, A; ●, CPM. Conditions included: column, 30-cm μ-Bondapak-C18; eluant, 3% methanol/0.05M ammonium formate (pH 3.5); flow rate, 1 mL/min; and sample, 10 mg hydrolyzed DNA containing unlabeled 7-propylguanine (41 μg/mg DNA) and 7-isopropylguanine (34 μg/mg DNA), which was injected five times in five equal portions (12).

Our result, therefore, suggests that the methylating agent formed from NOPPA is similar in reactivity to that formed from NDMA, and would support the hypothesis of Krüger (13) that NMPA may be the methylating agent formed from NOPPA.

Further experiments on the metabolism of NDPA by rat liver fractions have also provided support for the β-oxidation mechanism of Krüger. In addition to the products of α-oxidation, we have isolated and characterized NHPPA as a major product of the microsomal oxidation of NDPA (18). We have also shown that NHPPA is further oxidized to NOPPA by microsomal preparations from rat liver (18). Finally, with NOPPA as substrate, we have shown that metabolism takes place principally by reduction with the microsomal or soluble fraction of rat liver to yield NHPPA, although microsomal α-oxidation also takes place to some extent (19).

In recent experiments on the metabolism of NDPA, NHPPA and NOPPA *in vivo* in rats (20) we have shown that the major urinary excretion product in each case is the β-glucuronide of NHPPA (Table II).

Table II

Analysis of nitrosamines in urine of rats collected for 24 hours following administration of 0.59 μmoles/kg NDPA, NHPPA or NOPPA. Results are presented as the mean ± S.E.M. for 3 separate experiments, each using the combined urine from two rats. (20).

Nitrosamine Administered	Urinary Products (% Administered Dose)			
	NDPA	NHPPA	NOPPA	NHPPA[a]
NDPA	0.08 ± 0.02	0.10 ± 0.03	trace[b]	5.1 ± 0.3
NHPPA	–	1.73 ± 1.33	trace	79.3 ± 4.6
NOPPA	–	0.23 ± 0.13	0.17 ± 0.03	49.7 ± 7.1

[a]After exhaustive hydrolysis with bovine liver β-glucuronidase.

[b]Trace = < 0.01% administered dose.

These results and also the presence of small amounts of unconjugated NHPPA following administration of NDPA or NOPPA, confirm our experimental results using rat liver fractions *in vitro*.

According to Krüger's original hypothesis to explain methylation of rat liver DNA following administration of NDPA, NMPA should be formed from NOPPA. NOPPA is converted into NMPA in a base-catalyzed, non-enzymatic reaction, but this takes place only at high pH (16). We observed no detectable reaction at physio-

logical pH at 37° even after 16 hours (16). We have been unable
to detect NMPA in rat urine (20) or rat liver (Leung and Archer,
unpublished results) following intraperitoneal or subcutaneous
administration of NDPA, NHPPA or NOPPA. In an extensive search
we have also found no evidence of NMPA formation following incubation of NDPA, NHPPA or NOPPA with rat liver slices or any
subcellular fraction of rat liver (16, Leung and Archer, unpublished results). It is, of course, possible that we have not investigated appropriate conditions that would allow us to detect
such an enzyme. The mechanism whereby NDPA acts to methylate DNA
in the rat therefore remains unclear.

Acknowledgements

These investigations were supported by Grants CA 26651 and
CA 21951 awarded by the National Cancer Institute, DHHS, Research
Career Development Award ES 00033 awarded by the National
Institute of Environmental Health Sciences, DHHS, and the Ontario
Cancer Research and Treatment Foundation. The collaboration of
John S. Wishnok and Kwanghee K. Park are gratefully acknowledged.

Literature Cited

1.   Magee, P. N.; Barnes, J. M. Adv. Cancer Res. 1967, 10, 164.
2.   Druckrey, H.; Preussmann, R.; Ivankovic, S.; Schmäl, D.
     Z. Krebsforsch. 1967, 69, 103.
3.   Archer, M.C., in "Biological Reactive Intermediates 2: Chem-
     ical Mechanisms and Biological Effects"; Snyder, R., Parke,
     D.V., Kocsis, J., Jollow D. J., and Gibson, G. G., Eds.;
     Plenum Publishing Corp.; New York, (in press).
4.   Lijinsky, W.; Loo, J.; Ross, A. E. Nature. 1968, 218, 1174.
5.   Lawley, P. D., in "Chemical Carcinogens"; Searle, C. E.,
     Ed.; American Chemical Society Monograph 173, Washington,
     DC, 1976; p83.
6.   Collins, C. J. Accts. Chem. Res. 1971, 4, 315.
7.   March, J. "Advanced Organic Chemistry: Reaction Mechanisms
     and Structure"; McGraw-Hill, New York, 1977.
8.   Park, K. K.; Wishnok, J. S., Archer, M.C. Chem. -Biol.
     Interact. 1977, 18, 349.
9.   Zollinger, H. "Diazo and Azo Chemistry: Aliphatic and
     Aromatic Compounds"; Interscience, New York, 1961.
10. Moss, R. A. Acc. Chem. Res. 1974, 7, 421.
11. Kirmse, W. Angew. Chem. Int. Ed. Engl. 1976, 15, 251.
12. Park, K.K.; Archer, M. C.; Wishnok, J. S. Chem. -Biol.
     Interact. 1980, 29, 139.
13. Krüger, F. W. Z. Krebsforsch. 1971, 76, 145.
14. Krüger, F. W. Z. Krebsforsch. 1973, 79, 90.

15. Krüger, F.W.; Bertram, B. Z. Krebsforsch. 1973, 80, 189.
16. Leung, K. H.; Park, K. K.; Archer, M. C. Tox. Appl. Pharmacol. 1980, 53, 29.
17. Nicoll, J.W.; Swann, P.F.; Pegg, A.E. Nature. 1975, 254, 201.
18. Park, K.K.; Archer, M.C. Chem. -Biol. Interact. 1978a, 22, 83.
19. Park, K. K.; Archer, M. C. Cancer Biochem. Biophys. 1978b, 3, 37.
20. Leung, K. H.; Archer, M.C. Carcinogenesis, (in press).

RECEIVED August 10, 1981.

The Metabolism of Cyclic Nitrosamines

STEPHEN S. HECHT, G. DAVID McCOY, CHI-HONG B. CHEN, and DIETRICH HOFFMANN

Naylor Dana Institute for Disease Prevention, American Health Foundation, Valhalla, NY 10595

Investigations of the in vitro and in vivo metabolism of the cyclic nitrosamines N-nitrosopyrrolidine, N'-nitrosonornicotine, N-nitrosopiperidine, N-nitrosohexamethyleneimine, N-nitrosomorpholine, and related compounds are reviewed. Each of these compounds undergoes metabolic α-hydroxylation leading to electrophilic diazohydroxide intermediates which may act as ultimate carcinogens. Metabolism by β-hydroxylation has been established for N-nitrosopyrrolidine, N'-nitrosonornicotine, N-nitrosohexamethyleneimine, N-nitrosomorpholine, and 2,6-dimethyl-N-nitrosomorpholine.With the possible exception of 2,6-dimethyl-N-nitrosomorpholine, available evidence suggests that β-hydroxylation is not a major activation pathway . γ -Hydroxylation has been observed in those cases where it is possible, namely with N-nitrosopiperidine and N-nitrosohexamethyleneimine. Its role in carcinogen activation has not been established. The general patterns of cyclic nitrosamine metabolism in laboratory animals have been established, but little data are available on the carcinogen–DNA adducts formed from these compounds or on the mechanisms of their organ specificity.

Cyclic nitrosamines are among the most potent and environmentally significant nitrosamine carcinogens. Like the acyclic nitrosamines, metabolism is necessary for their carcinogenicity. Elucidation of the specific metabolic pathways of cyclic nitrosamine activation and detoxification is a challenging problem, and considerable progress has been achieved in recent years. In this chapter, we will review metabolic studies on N-nitrosopyrrolidine (NPYR), N'-nitrosonornicotine (NNN), N-nitrosopiperidine (NPIP), N-nitrosohexamethyleneimine (NHEX), N-nitrosomorpholine (NMOR),

0097-6156/81/0174-0049$06.75/0
© 1981 American Chemical Society

and related compounds and will consider the relationship of
metabolic pathways to carcinogenicity and mutagenicity in those
cases where data are available.   Studies on cyclic nitrosamine
metabolism in vivo and in vitro are summarized in Tables I and
II.

NPYR   NNN   NPIP

NHEX   NMOR

N-Nitrosopyrrolidine (NPYR)

NPYR occurs in processed meats and, most commonly, in cooked
bacon (1-3).   NPYR is one of the principal volatile nitrosamines
detected in mainstream and sidestream cigarette smoke (4,5).   It
induces hepatocellular carcinomas in rats, even at relatively low
doses, and tumors of the nasal cavity and trachea in Syrian
golden hamsters (6,7,8).

The metabolism of NPYR is summarized in Figure 1.   α-Hydroxylation (2 or 5 position) leads to the unstable intermediates
1 and 4; decomposition of 4 gives 4-hydroxybutyraldehyde [6].
The latter, which exists predominantly as the cyclic hemiacetal
1, has been detected as a hepatic microsomal metabolite in rats,
hamsters, and humans and from lung microsomes in rats (9-13).
The role of 1 and 4 as intermediates in the formation of 6 and 7
is supported by studies of the hydrolysis of 2-acetoxyNPYR and
4-(N-carbethoxy-N-nitrosamino)butanal, which both gave high
yields of 7 (9,14).   In microsomal incubations, 6 can be readily
quantified as its 2,4-dinitrophenylhydrazone derivative (15).
The latter has also been detected in the urine of rats treated
with NPYR (9).

Post-microsomal supernatants convert 6 and 7 to 8, 9, and
10; 9 is generally the major metabolite observed from NPYR under
these conditions (12,13).   Compounds 8 and 9, as well as succinic

Table I.

METABOLISM OF CYCLIC NITROSAMINES IN VIVO.

Compound	Species	Dose (mg/kg)	% of Dose Excreted		Urinary Metabolites (% of dose, hr)	Reference
			CO_2 (hr)	Urine (hr)		
NPYR	Rat	6	20 (6)	7 (16)	3-hydroxyNPYR (<1)	20
NPYR	Rat	16 648	70 (24) 14 (24)	11 (24) N.R.	N.R.[a]	17
NPYR	Rat	582	N.R.	N.R.	4-hydroxybutanal (as 2,4-di-nitrophenylhydrazone) (0.09, 48)	9
NPYR	Rat	6	49 (6)	N.R.	N.R.	18
NPYR	Rat	6-500	24 (24)	N.R.	dimethylamine (1.3-0.5, 24) 3-hydroxyNPYR (0.30-0.78, 24) 2-pyrrolidinone (0.04-.0.78, 24) succinic semialdehyde (0.1, 24) 4-hydroxybutyric acid γ-butyrolactone 2-pyrrolidinone oxime (0.9,24)[b]	16

Table I (continued).

Compound	Species	Dose (mg/kg)	% of Dose Excreted CO$_2$ (hr)	% of Dose Excreted Urine (hr)	Urinary Metabolites (% of dose, hr)	Reference
NNN	Rat	329	<1 (48)	73-85 (48)	4-hydroxy-4-(3-pyridyl)-butyric acid (30-40, 48) 4-oxo-4-(3-pyridyl)butyric acid (1-2, 48) 5-(3-pyridyl)tetrahydrofuran-2-one (1-2, 48) 5-(3-pyridyl)-2-pyrrolidinone (2, 48) myosmine (<1, 48) NNN (4, 48)	34
NNN	Rat	300 3-300	N.R.	N.R.	3'-hydroxyNNN (<0.1, 48) 4'-hydroxyNNN (<0.1, 48) NNN-1-N-oxide (6.7-9.4, 48)	36
NNN	Rat	3-300	N.R.	80-90 (48)	4-hydroxy-4-(3-pyridyl)-butyric acid (37.1-53.3, 48) NNN-1-N-oxide (6.7-10.7, 48) 5-(3-pyridyl)-2-pyrrolidinone (3.2-5.1, 48)	37

Table I (continued).

Compound	Species	Dose (mg/kg)	% of Dose Excreted CO$_2$ (hr)	% of Dose Excreted Urine (hr)	Urinary Metabolites (% of dose, hr)	Reference
NNN (cont'd)					4-oxo-4-(3-pyridyl)butyric acid (31.1.-12.8, 48) NNN (3.3-5.2, 48)	37
NNN	Syrian golden hamster	59 20	<0.5 (48) N.R.	62-78 (48) 80 (48)	4-hydroxy-4-(3-pyridyl)butyric acid (38.5, 48) NNN-1-N-oxide (2.5, 48) 4-oxo-4-(3-pyridyl)butyric acid (14.6, 48) NNN (0.4, 48)	42
NHEX	Rat	12	18 (24)	37 (24)	-caprolactam (11, 24) -aminocaproic acid (5, 24) -aminocaprohydroxamic acid (6, 24)	52
NHEX	Rat	8 576	45 (24) 4 (24)	33 (24) N.R.	N.R.	17

Table I (continued).

Compound	Species	Dose (mg/kg)	% of Dose Excreted		Urinary Metabolites (% of dose, hr)	Reference
			CO_2 (hr)	Urine (hr)		
NHEPT[c]	Rat	8 280	28 (24) 8 (24)	43 (24) N.R.	N.R.	17
NMOR	Rat	400	3.3 (24)	81 (24)	N-nitrosodiethanolamine (15, 24) NMOR (24, 24)	61
NMOR	Rat	≈120	N.R.	N.R.	N-Nitrosodiethanolamine (N.R.) NMOR (N.R.)	60
NMOR	Rat	125	N.R.	N.R.	(2-hydroxyethoxy)acetic acid (16, 24) N-nitrosodiethanolamine (12, 24) N-nitroso-(2-hydroxyethyl)-glycine (33, 24) NMOR (1.5, 24)	59

Table I (continued).

Compound	Species	Dose (mg/kg)	% of Dose Excreted		Urinary Metabolites (% of dose, hr)	Reference
			CO_2 (hr)	Urine (hr)		
NDMMOR	Syrian golden hamster	100	N.R.	N.R.	N-nitroso(2-hydroxypropyl-2-oxopropyl)amine (2, 24) N-nitroso-bis(2-hydroxypropyl)amine (2, 24)	68
D-NPZ[d]	Rat	10	1 (6)	43 (16)	3-hydroxyNPYR (1.7, 16) 1-N-nitrosopiperazine-3-one (4.4, 16) D-NPZ (3.4, 16)	70

[a]N.R. = not reported
[b]after 500 mg/kg dose
[c]N-Nitrosoheptamethyleneimine
[d]Dinitrosopiperazine

Table II.

METABOLISM OF CYCLIC NITROSAMINES __IN VITRO__.

Compound	Species	Tissue Preparation	Comments	Reference
NPYR	Rat	Liver microsomes	identification of 4-hydroxybutyr-aldehyde (as its 2,4-dinitrophenyl-hydrazone) from α-hydroxylation	9
NPYR	Rat	Liver microsomes	HPLC assay for α-hydroxylation	15
NPYR	Syrian golden hamster	Liver microsomes	enhanced α-hydroxylation and muta-genicity in ethanol-consuming hamsters	10
NPYR	Rat	Liver microsomes and postmitochondrial super-natant	identification of 1,4-butanediol, and 4-hydroxybutyric acid	12
NPYR, NNN	Human	Liver microsomes	α-hydroxylation	11
NPYR	Rat	Liver and lung micro-somes and postmitochon-drial supernatant	α-hydroxylation but not β-hydroxy-lation in both tissues	13

Table II (continued)

Compound	Species	Tissue Preparation	Comments	Reference
NPYR	Rat, Hamster	Liver microsomes	effects of inducers on α-hydroxylation and mutagenicity	19
NPYR, NPIP, D-NPZ[a]	Human	Cultured bronchi	metabolism to CO_2 and binding to DNA and protein	26
NPYR, NNN, NPIP D-NPZ[a]	Human	Colon organ culture	metabolism to CO_2 and binding to DNA and protein	28
NPYR	Human	Cultured esophagus	binding to protein	27
NNN	Rat	Liver microsomes	identification of products of α-hydroxylation	34
NNN	Rat	Liver microsomes	assay for α-hydroxylation and the effect of α-deuterium substitution	33
NNN	Rat	Liver microsomes	identification of products of β-hydroxylation and pyridine-N-oxidation	36

Table II (continued)

Compounds	Species	Tissue Preparation	Comments	Reference
NNN	Rat, hamster	Liver microsomes	effects of inducers on rates of α-hydroxylation	35
NNN	Rat	Liver microsomes	metabolism in the presence of DNA or guanosine	39
NPIP	Rat	Liver microsomes	identification of 4-hydroxyNPIP	47
NPIP	Rat	Liver microsomes	identification of 5-hydroxypenta-nal from α-hydroxylation	46
NHEX	Rat	Liver microsomes and post-mitochondrial supernatant	identification of 3-hydroxyNHEX and 4-hydroxyNHEX	53
NMOR	Rat	Liver microsomes	identification of 2-hydroxyNMOR	60
NMOR	Rat	Liver microsomes	formation of N-aminomorpholine	63

Table II (continued)

Compounds	Species	Tissue Preparation	Comments	Reference
NMOR	Rat, mouse	Liver microsomes	denitrosation	62
NMOR	Rat	Liver microsomes	identification of (2-hydroxyethoxy)-acetaldehyde (as its 2,4-dinitro-phenylhydrazone) from α-hydroxy-lation	59

aD-NPZ, dinitrosopiperazine

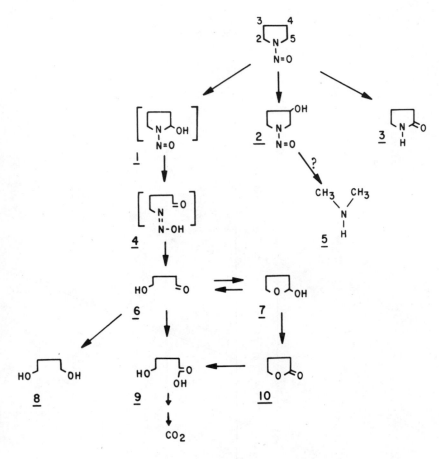

Figure 1.   Metabolism of N-*nitrosopyrrolidine in the rat.*

semialdehyde, have been detected as urinary metabolites of NPYR (16). Metabolism of 9 leads to CO_2, which is the major metabolite of NPYR in the rat (9,16,17,18).

The hepatic microsomal α-hydroxylase activity for NPYR is inducible in rats by pretreatment with Aroclor, and in hamsters by pretreatment with Aroclor, 3-methylcholanthrene, phenobarbital, and ethanol (10,15,19). In contrast, pretreatment of rats with 3-methylcholanthrene or phenobarbital causes no change or a slight decrease in microsomal NPYR α-hydroxylase activity (19).

3-HydroxyNPYR [2] has been identified as a urinary metabolite of NPYR in the rat (up to 1% of the dose), but was not detected when NPYR was incubated with subcellular fractions from rat liver and lung (13, 20). It has been proposed that further metabolism of 2 leads to dimethylamine,[5], another urinary metabolite of NPYR. 2-Pyrrolidinone [3] has also been detected in the urine of rats treated with NPYR. Its origin has not been conclusively established, but it may form from pyrrolidinone-2-oxime (16).

Current evidence favors α-hydroxylation as a major activation pathway for NPYR. The model compounds, 2-acetoxyNPYR and 4-(N-carbethoxy-N-nitrosamino)butanal are both highly mutagenic towards S. typhimurium without enzymatic activation (21, 22). This activity is probably a result of conversion to the electrophilic diazohydroxide 1, or to the corresponding carbonium ion. Inducers which increase the rates of microsomal α-hydroxylation of NPYR also increase its mutagenicity toward S. typhimurium (10, 19) and, in the case of ethanol, its carcinogenicity toward Syrian golden hamsters (8). In addition, 2,5-dimethylNPYR is less carcinogenic than NPYR (23). This observation may support the role of α-hydroxylation in the activation of NPYR but is limited by the fact that different enzymes may catalyze the hydroxylation of secondary and tertiary α-carbon atoms, as described below for NNN. 3-HydroxyNPYR is less carcinogenic in rats than is NPYR, which indicates that this metabolite is a product of detoxification (24).

If α-hydroxylation of NPYR is its mechanism of activation, one would expect the formation of carcinogen-DNA adducts containing a 4-oxobutyl- or related residue. Adducts have been isolated from the liver RNA of NPYR treated rats, but their structures have not been determined (25). Carcinogen DNA adducts have also been isolated from cultured human esophagus, colon, and bronchus (26, 27, 28).

These studies on NPYR are typical of the state of the art in cyclic nitrosamine metabolism and activation. The major metabolic pathways have been rather well characterized, but data on the relationship of these pathways to carcinogenesis are limited. This is especially true of the organospecific effects of NPYR and the other cyclic nitrosamines. For example, the main target organs for NPYR in the Syrian golden hamster are the trachea and nasal cavity rather than the liver. This is in spite of the fact that electrophilic agents are generated in the hamster liver by α-hydroxylation. To understand the reasons for this, more work is necessary on target organ metabolism of NPYR and on the formation and persistence of specific DNA adducts.

N'-Nitrosonornicotine (NNN)

NNN is formally a derivative of NPYR, but the pyridine ring has a marked effect on its metabolism and carcinogenicity. NNN induces lung adenomas in mice, esophageal and nasal cavity tumors in rats, and tracheal and nasal cavity tumors in Syrian golden hamsters (29, 30, 31). Its tumorigenic activity in Syrian golden hamsters is only slightly less than that of NPYR (8). NNN is important because of its relatively high concentrations in mainstream and sidestream tobacco smoke and in unburned tobacco (32). The occurrence and carcinogenicity of NNN and related tobacco specific nitrosamines is reviewed in another chapter of this volume.

The metabolism of NNN in the F-344 rat is summarized in Figure 2. Liver microsomal α-hydroxylation of NNN leads to 2'-hydroxyNNN [2] and 5'-hydroxyNNN [5]. These unstable intermediates open to diazohydroxides 8 and 9 which undergo solvolysis to keto alcohol 10 and lactol 12. In addition, 2 gives rise to myosmine [7]. The chemistry of the intermediates 2, 5, 8, and 9 has been established through studies of the corresponding model compounds, 2'-acetoxyNNN, 5'-acetoxyNNN, and 4-(N-carbethoxy-N-nitrosamino)-1-(3-pyridyl)-1-butanone (14). Keto alcohol 10 and lactol 12 can be assayed in microsomal incubations as their corresponding 2,4-dinitrophenylhydrazone derivatives (33, 34). α-Hydroxylation of NNN has also been demonstrated in human and Syrian golden hamster liver microsomes (11, 35). Species differences in rates of 2'-, and 5'-hydroxylation and in their inducibility have been observed. In the rat, 2'-hydroxylation but not 5'-hydroxylation is induced by pretreatment with phenobarbital or 3-methylcholanthrene whereas in the Syrian golden hamster 5'-hydroxylation is induced by phenobarbital but 2'-hydroxylation is unchanged by pretreatments (35). Evidently, different enzymes catalyze the 2'- and 5'-hydroxylations of NNN and the distribution of these enzymes differs in the livers of rats and hamsters. There may also be differences in distribution between organs of a given species. For example, in preliminary studies, we have observed that in cultured rat esophagus, a target organ, 2'-hydroxylation is the major metabolic pathway, in contrast to results obtained in liver, a non-target organ for NNN.

β-Hydroxylation of NNN by rat liver microsomes to give 3'-hydroxyNNN [3] and 4'-hydroxyNNN [4] has also been observed, but the rates are lower than those of α-hydroxylation (36). Another microsomal metabolite of NNN is NNN-1-N-oxide [1] (36).

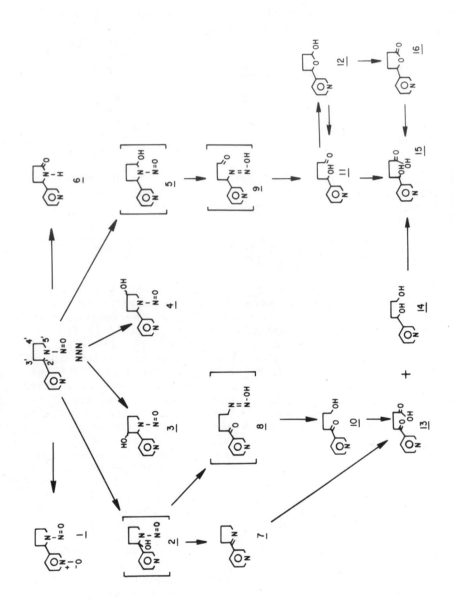

Figure 2. Metabolism of N'-nitrosonornicotine in the rat (37).

The rates of formation of these metabolites in liver microsomes from Aroclor pretreated rats is summarized in Table III.

Table III. Rates of Formation of NNN Metabolites by F-344
Rat Liver Microsomes[a]

product	source	rate, nmol min^{-1} (mg of protein)$^{-1}$
4-hydroxy-1-(3-pyridyl)-1-butanone [10][b]	2'-hydroxylation	0.40+0.02
3'-hydroxy-N'-nitroso-nornicotine [3]	3'-hydroxylation	<0.01
4'-hydroxy-N'-nitroso-nornicotine [4]	4'-hydroxylation	<0.01
5-(3-pyridyl)-2-hydroxy-tetrahydrofuran [12]	5'-hydroxylation	0.76+0.02
N'-nitrosonornicotine 1-N-oxide [1]	N-oxidation	0.57

[a]Liver microsomes from Aroclor 1254 pretreated rats were used.
[b]See Figure 2.

In intact cell systems or in vivo, the primary products of α-hydroxylation, 10 and 12, have not been detected. The principal urinary metabolites of NNN resulting from α-hydroxylation are keto acid 13 from 2'-hydroxylation and hydroxy acid 15 from 5'-hydroxylation. Trace amounts of 7, 14, and 16 have also been detected as urinary metabolites (34). The interrelationships of these metabolites as shown in Figure 2 have been confirmed by administration of each metabolite to F-344 rats (37). The other metabolites which are routinely observed in the urine are NNN-1-N-oxide [1] and 5-(3-pyridyl)-2-pyrrolidinone [norcotinine, 6]. The β-hydroxy derivatives 3 and 4 were also detected in the urine of NNN treated rats, but at less than 0.1% of the dose (36). An HPLC trace of the urinary metabolites of NNN is shown in Figure 3. Urine is the major route of excretion (80-90% of the dose) of NNN and its metabolites in the F-344 rat in contrast to NPYR which appears primarily as CO_2 (70%) after a dose of 16 mg/kg (17). This is because the major urinary metabolite of NNN, hydroxy acid 15, is not metabolized further, in contrast to 4-hydroxybutyric acid [9, Figure 1] which is converted to CO_2. In addition, a significant portion of NNN is excreted as NNN-1-N-oxide [1], a pathway not open to NPYR.

The effects of deuterium substitution on the rates of α-hydroxylation of NNN have been measured. The results obtained in vitro, with rat liver microsomes, showed only a small deuterium isotope effect of 1.2 for 2'-hydroxylation, whereas a significant effect of 2.4-2.7 was observed for 5'-hydroxylation (33). Analogous results were obtained in vivo when the urinary metabolites

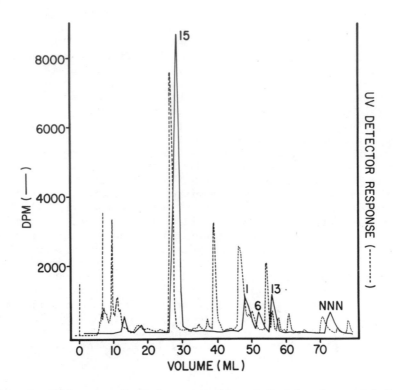

Figure 3. High pressure liquid chromatographic analysis of the urinary metabolites of 2'-¹⁴C-N'-nitrosonornicotine in the rat after a dose of 300 mg/kg (37). Peaks are identified by numbers corresponding to structures in Figure 2. Key: (———), decompositions per minute; (– – –), UV absorbance. Radioactivity was recorded 2 mL after UV absorbance.

resulting from α-hydroxylation of NNN and [2',5',5'-D]NNN were compared. In addition, urinary excretion of NNN-1-N-oxide was greater from [2',5',5'-D]NNN than from NNN. Thus, in comparing the carcinogenicity of NNN and [2',5',5'-D]NNN one will have to take into account both the increased rates of N-oxidation and the decreased rates of 5'-hydroxylation. Bioassays of deuterated NNN derivatives are currently in progress.

The ability of NNN to damage mammalian DNA has been demonstrated by its activity in the hepatocyte primary culture DNA repair assay (38). In contrast, NNN did not bind to exogenous DNA in the presence of rat liver microsomes (39). Binding of NNN-2'-^{14}C and its metabolites to various tissues of mice has been observed by whole body autoradiography (40, 41). After 24 hours, the highest levels were in the nasal cavity, bronchi, esophagus, and salivary glands as well as in the pigmented tissues. In Syrian golden hamsters, the highest levels of binding to the trichloroacetic acid insoluble materials, 1 hour after administration of NNN, were found in liver, lung, kidney, and adrenals (42).

α-Hydroxylation may be the major pathway leading to DNA binding of NNN. Both 4-(N-carbethoxy-N-nitrosamino)-1-(3-pyridyl)-1-butanone, a model compound for diazohydroxide 8, and 5'-acetoxyNNN, a model compound for 5'-hydroxyNNN [5] were highly mutagenic toward S. typhimurium without activation; lower activity was observed for the unstable 2'-acetoxyNNN (14, 34). These results indicate that diazohydroxides such as 8 and 9 can damage bacterial DNA. However, the structures of the DNA adducts formed from NNN have not yet been determined. The low rates of formation of 3'-hydroxyNNN and 4'-hydroxyNNN in vitro and in vivo suggest a minor role for these metabolites in the activation of NNN. NNN-1-N-oxide may also be a detoxification product of NNN, since it is excreted mainly unchanged after administration to F-344 rats (37). Since NNN effects primarily the esophagus and nasal cavity in rats, and the trachea and nasal cavity in hamsters, further studies are necessary to determine the basis for its organospecificity.

N-Nitrosopiperidine (NPIP) and Related Compounds

NPIP induces esophageal and nasal cavity tumors in the rat, forestomach, liver and lung tumors in the mouse, and tracheal tumors in the Syrian golden hamster (43, 44, 45). Its potent carcinogenicity is indicated by the fact that a single dose of only 22 mg/kg was sufficient to induce tumors in 20% of Syrian golden hamsters (45). The environmental occurrence of NPIP appears to be less frequent than that of NPYR, but it has been detected in food (3, 44).

Only limited metabolic studies have been carried out on NPIP. It undergoes α-hydroxylation by rat liver microsomes to give 5-hydroxypentanal, a process analogous to the formation of

4-hydroxybutanal from NPYR (46). 4-HydroxyNPIP has also been identified as a metabolite of NPIP formed by rat liver microsomes (47). In the Udenfriend system, N-nitroso-4-piperidone was observed (47,48). No studies have been reported on the metabolism of NPIP in the rat esophagus, a target organ.

The related unsaturated compounds, N-nitroso-1,2,3,6-tetrahydropyridine and N-nitroso-1,2,3,4-tetrahydropyridine, have been tested for carcinogenicity and both were potent esophageal carcinogens, as is nitrosopiperidine. However, N-nitroso-1,2,3,6-tetrahydropyridine also produced liver tumors, whereas N-nitroso-1,2,3,4-tetrahydropyridine also gave tumors of the forestomach and oropharynx. The difference in tumor spectrum between the two unsaturated isomers may be related to differences in metabolism. N-Nitroso-1,2,3,6-tetrahydropyridine isomerized to N-nitroso-1,2,3,4-tetrahydropyridine in vivo, but the reverse reaction was not observed (49).

N-Nitrosohexamethyleneimine (NHEX)

NHEX is a potent carcinogen which induces tumors of the liver and esophagus in rats, and tumors of the trachea in Syrian golden hamsters (50, 51). It has not been detected in the environment.

Metabolism of NHEX-^{14}C in the rat results in dose dependent formation of $^{14}CO_2$, with 45% exhaled after a dose of 8 mg/kg NHEX but only 4% after 576 mg/kg (17). Similar results were obtained for NPYR and nitrosoheptamethyleneimine. At doses of 8-12 mg/kg NHEX, 33-37% of the radioactivity was excreted in the urine (17, 52). Urinary metabolites of NHEX were ε-caprolactam, ε-aminocaproic acid, and ε-aminocaprohydroxamic acid (52). The formation of ε-caprolactam is analogous to results with NPYR and NNN, in which 2-pyrrolidinone and norcotinine were observed as urinary metabolites. Caprolactam did not originate from hexamethyleneimine, a product of denitrosation.

Analysis of the products formed from NHEX in vitro using rat liver microsomes and postmitochondrial supernatant resulted in the identification of 3-hydroxyNHEX from β-hydroxylation and 4-hydroxyNHEX from γ-hydroxylation. The ratio of 4-hydroxyNHEX to 3-hydroxyNHEX was 3 to 1. Both conformeric forms of each of these metabolites were detected (53). From in vitro data which are available so far for NPYR, NNN, NPIP, and NHEX, it does appear that the rates of α- and γ-hydroxylation (when possible), exceed those of β-hydroxylation.

The liver RNA isolated from rats treated with ^{14}C-NHEX yielded 1,6-hexanediol upon hydrolysis with acid (54). This result indicates that NHEX underwent metabolic α-hydroxylation to give an adduct that may have been formed at O^6 of guanine. The initially formed adduct would have been expected to be a 6-oxohexyl derivative; reduction of the adduct must have occurred in order to produce the observed 1,6-hexanediol.

N-Nitrosomorpholine (NMOR)

NMOR is a potent hepatocarcinogen in the rat and induces
tracheal and nasal cavity tumors in the Syrian golden hamster
(43, 44, 45). It is formed readily from nitrite and morpholine
in vitro and administration of these precursors to rodents causes
tumors indicative of NMOR formation in vivo (44, 55, 56). NMOR
has been detected in crankcase emissions of diesel engines and in
factories engaged in rubber and tire manufacturing (57, 58).
     The metabolism of NMOR in the rat is outlined in Figure 4.
α-Hydroxylation yields the unstable intermediates 1 and 4; the
latter hydrolyzes to (2-hydroxyethoxy)acetaldehyde [7] which has
been identified as a liver microsomal metabolite by isolation of
the corresponding 2,4-dinitrophenylhydrazone (59). (2-Hydroxy-
ethoxy)acetaldehyde, which exists predominantly as the cyclic
hemiacetal 8, was not detected in the urine of rats gavaged with
125 mg/kg NMOR. However, (2-hydroxyethoxy)acetic acid was a major
urinary metabolite (16% of the dose). These transformations are
analogous to those observed with NPYR and NNN.
     β-Hydroxylation of NMOR to give 2-hydroxyNMOR [2] has been
observed in incubations with rat liver microsomes (60). 2-Hy-
droxyNMOR was not detected as a urinary metabolite, but N-nitro-
so-2-(hydroxyethyl)glycine [9] which can be formed by oxidation
of 2 or 5, was detected in the urine of rats treated with NMOR
(59, 60). The other major urinary metabolite of NMOR is N-nitrosodiethanolamine (12 %) (59, 60, 61). As in the case of NNN, the
major in vivo metabolites of NMOR are apparently resistant to
further degradation, which explains the relatively high (81%)
percentage of activity from ^{14}C-NMOR found in urine. Two other
metabolic processes have been observed for NMOR in vitro; formation of nitrite in the presence of rat liver microsomes and an
NADPH generating system and reduction to N-aminomorpholine under
anaerobic conditions (62, 63). The occurrence of these pathways
in vivo has not been reported.
     Deuterium substitution in the α-positions of NMOR signifi-
cantly decreases its carcinogenicity and mutagenicity. Thus,
3,3,5,5-tetradeuteroNMOR induced fewer liver tumors in Sprague-
Dawley rats than did NMOR and was less mutagenic toward S. typhimurium TA1535 than was NMOR (64, 65). To establish a metabolic
basis for these observations, the urinary metabolites of rats
treated with equivalent doses of NMOR and 3,3,5,5-tetradeutero-
NMOR were compared. Figure 5 shows gas-liquid chromatograms of
the silylated 24 hour urinary metabolites of NMOR and its α-tetradeutero derivative. Peak A, resulting from α-hydroxylation,
was diminished significantly in 3,3,5,5-tetradeuteroNMOR. The
ratio of its concentration in the urine of the NMOR treated animals to that in the urine of the animals treated with 3,3,5,5-
tetradeuteroNMOR was 5. However, peaks B and C, from ring cleavage and β-hydroxylation, did not change in the metabolism of the
α-tetradeutero derivative. These results provide strong evi-

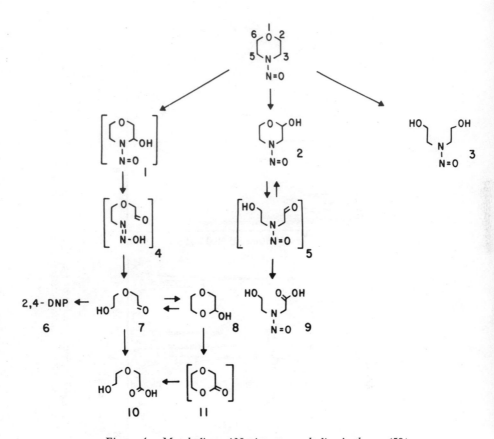

Figure 4. Metabolism of N-nitrosomorpholine in the rat (59).

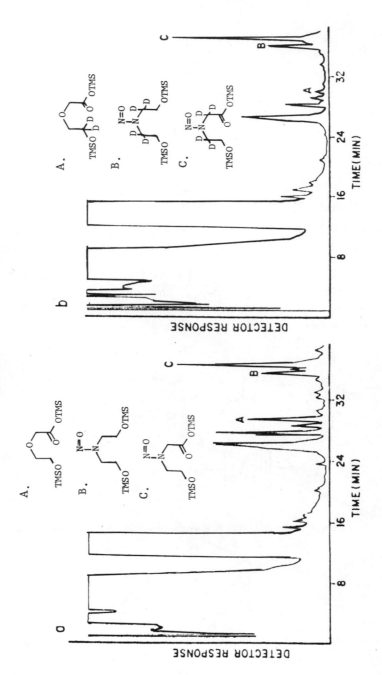

Figure 5.   Gas-liquid chromatograms of the silylated urinary metabolites of N-nitrosomorpholine and 3,3,5,5-tetradeutero-N-nitrosomorpholine in the rat (59).

dence that α-hydroxylation is the primary activation process for NMOR. DNA and RNA adducts are formed in the livers of rats treated with[14]C-NMOR. However, with the possible exception of 7-(2-hydroxyethyl)guanine, the structures of these adducts have not been determined (61).

The carcinogenicity and metabolism of N-nitroso-2,6-dimethylmorpholine (NDMMOR) have been studied. This compound induces respiratory tract and pancreatic tumors in Syrian golden hamsters and tumors of the esophagus in F-344 rats (45, 66, 67). The induction of pancreatic tumors in hamsters is apparently related to its metabolic conversion via β-hydroxylation to N-nitroso(2-hydroxypropyl-2-oxopropyl)amine and N-nitroso-bis-(2-hydroxypropyl)amine, which are pancreatic carcinogens in this species (68). Its carcinogenicity to the rat esophagus may be mediated through α-hydroxylation since 3,3,5,5-tetradeuteroNDMMOR was apparently less carcinogenic than NDMMOR, but 2,6-dideutero-NDMMOR was more active than NDMMOR (69).

Conclusions

As is evident from Tables I and II, most of the metabolic studies on cyclic nitrosamines have been carried out during the past six years. These studies have led to a reasonable understanding of the general pattern of cyclic nitrosamine metabolism. Metabolites resulting from α-hydroxylation, β-hydroxylation, and γ-hydroxylation have been identified in vitro and in vivo and methods have been developed to study the effects of modifiers on these pathways. In certain cases, these metabolic studies have provided leads to the nature of the ultimate carcinogens of cyclic nitrosamines which appear to result primarily from α-hydroxylation. However, further studies on target organ metabolism of cyclic nitrosamines are necessary if a meaningful relationship between metabolism and organospecificity is to be developed.

In contrast, little is known about the DNA adducts formed from cyclic nitrosamines. Early studies indicated methylation of DNA by certain cyclic nitrosamines but these were not confirmed (25, 71, 72). Adducts from NPYR and NMOR have been detected but remain mostly uncharacterized (25, 61). The nature of the carcinogen-DNA adducts from cyclic nitrosamines must be clarified if we are to advance in our understanding of their mechanisms of action. This may require more sensitive methods than have been employed previously. We do know from studies on other nitroso compounds that differences in rates of DNA repair of specific carcinogen-DNA adducts are at least partially responsible for the organospecificity of these compounds (73). The metabolic studies described in this chapter provide the foundation for further work on characterization of DNA adducts.

Another area which needs further investigation is the role of specific forms of cytochrome P-450 in the metabolism of cyclic (and acyclic) nitrosamines. Several studies described in this chapter indicate that different forms of cytochrome P-450 catalyze particular hydroxylation reactions for a given nitrosamine substrate and that the distribution of these forms is species and organ dependent. This may be a significant lead to a better understanding of the organ-specificity of cyclic nitrosamines which may depend both on target tissue metabolism and rates of repair of specific carcinogen-DNA adducts.

Acknowledgements

Our studies on cyclic nitrosamine metabolism are supported by NCI Grants CA-21393, CA23901, and CA12376. This is paper Number 34 in "A Study of Chemical Carcinogenesis."

Literature Cited

1. Fazio, T.; Havery, D.C.; Howard, J.W.; In: Walker, E.A.; Griciute, L.; Castegnaro, M.; Boerzsoenyi, M. (eds.); "N-Nitroso Compounds: Analysis, Formation and Occurrence", International Agency for Research on Cancer, Lyon, France, 1980, 419.

2. Sen, N.P.; Seaman, S.; McPherson, M.; In: Walker, E.A.; Griciute, L.; Castegnaro, M.; Boerzsoenyi, M. (eds.); "N-Nitroso Compounds: Analysis, Formation and Occurrence", International Agency for Research on Cancer, Lyon, France, 1980, 457.

3. Spiegelhalder, B.; Eisenbrand, G.; Preussmann, R.; In: Walker, E.A.; Griciute, L.; Castegnaro, M.; Boerzsoenyi, M. (eds.); "N-Nitroso Compounds: Analysis, Formation and Occurrence", International Agency for Research on Cancer, Lyon, France, 1980, 467.

4. Hoffmann, D.; Adams, J.D.; Piade, J.J.; Hecht, S.S.; In: Walker, E.A.; Griciute, L.; Castegnaro, M.; Boerzsoenyi, M. (eds.); International Agency for Research on Cancer, Lyon, France, 1980, 507.

5. Brunnemann, K.D.; Yu, L.; Hoffmann, D. Cancer Res. 1977, 37, 3218.

6. Preussmann, R.; Schmaehl, D.; Eisenbrand, G. Z. Krebsforsch. 1977, 90, 161.

7. Greenblatt, M.; Lijinsky, W. J. Natl. Cancer Inst. 1972, 48, 1687.

8. McCoy, G.D.; Katayama, S.; Hecht, S.S.; Wynder, E.L. Cancer Res. Submitted.

9. Hecht, S.S.; Chen, C.B.; Hoffmann, D. Cancer Res. 1978, 38, 215.

10. McCoy, G.D.; Chen, C.B.; Hecht, S.S.; McCoy, E.C. Cancer Res. 1979, 39, 793.

11. Hecht, S.S.; Chen, C.B.; McCoy, G.D.; Hoffmann, D.; Domellof, L. Cancer Letters 1979, 8, 35.
12. Hecker, L.I.; Farrelly, J.G.; Smith, J.H.; Saavedra, J.E.; Lyon, P.E. Cancer Res. 1979, 39, 2679.
13. Hecker, L.I. Chem.-Biol. Interactions 1980, 30, 57.
14. Hecht, S.S.; Chen, C.B. J. Org. Chem. 1979, 14, 1563.
15. Chen, C.B.; McCoy, G.D.; Hecht, S.S.; Hoffmann, D.; Wynder, E.L. Cancer Res. 1978, 38, 3812.
16. Cottrell, R.C.; Walters, D.G.; Young, P.J.; Phillips, J.C.; Lake, B.G.; Gangolli, S.D. Toxicol. App. Pharmacol. 1980, 54, 368.
17. Snyder, C.M.; Farrelly, J.G.; Lijinsky, W. Cancer Res. 1977, 37, 3530.
18. Cottrell, R.C.; Young, P.J.; Walters, D.G.; Phillips, J.C.; Lake, B.G.; Gangolli, S.D. Toxicol. App. Pharmacol. 1979, 51, 101.
19. McCoy, G.D.; Hecht, S.S.; McCoy, E. Environ. Mutagenesis Submitted.
20. Kruger, F.W.; Bertram, B. Z. Krebsforsch. 1975, 83, 255.
21. Baldwin, J.E.; Branz, S.E.; Gomez, R.F.; Kraft, P.; Sinskey, A.J.; Tannenbaum, S.R. Tetrahedron Letters 1976, 333.
22. Hecht, S.S.; Chen, C.B.; McCoy, G.D.; Hoffmann, D. In: Anselme, J.P. (ed.) "N-Nitrosamines", American Chemical Society, Washington, D.C., 1979, 125.
23. Lijinsky, W.; Taylor, H.W. Cancer Res. 1976, 36, 1988.
24. Eisenbrand, G.; Habs, M.; Schmaehl, D.; Preussmann, R. In: Walker, E.A.; Griciute, L.; Castegnaro, M.; Boerzsoenyi, M. (eds.) "N-Nitroso Compounds: Analysis, Formation and Occurrence", International Agency for Research on Cancer, Lyon, France, 1980, 657.
25. Krueger, F.W. In: Nakahara, M.W.; Takayama, S.; Sugimura, T.; Odashima, S. (eds.) "Topics in Chemical Carcinogenesis", University of Tokyo Press, Tokyo, 1972, 213.
26. Harris, C.C.; Autrup, H.; Stoner, G.D.; McDowell, E.M.; Trump, B.F., Schafer, P.W. J. Natl. Cancer Inst. 1977, 59, 1401.
27. Harris, C.C.; Autrup, H.; Stoner, G.D.; Trump, B.F.; Hillman, E.; Schafer, P.W.; Jeffrey, A.M. Cancer Res. 1979, 39, 4401.
28. Autrup, H.; Harris, C.C.; Trump, B.F. Proc. Soc. Exptl. Biol. Med. 1978, 159, 111.
29. Hecht, S.S.; Chen, C.B.; Hirota, N.; Ornaf, R.M.; Tso, T.C.; Hoffmann, D. J. Natl. Cancer Inst. 1978, 60, 819.
30. Hoffmann, D.; Raineri, R.; Hecht, S.S.; Maronpot, R.; Wynder, E.L. J. Natl. Cancer Inst. 1975, 55, 977.
31. Hilfrich, J.; Hecht, S.S.; Hoffmann, D. Cancer Letters 1977, 2, 169.
32. Hoffmann, D.; Adams, J.D.; Brunnemann, K.D.; Hecht, S.S. Cancer Res. 1979, 39, 2505.

33. Chen, C.B.; Fung, P.T.; Hecht, S.S. Cancer Res. 1979, 39, 5057.
34. Chen, C.B.; Hecht, S.S.; Hoffmann, D. Cancer Res. 1978, 38, 3639.
35. McCoy, G.D.; Chen, C.B.; Hecht, S.S. Drug Metab. Disp. 1981, in press.
36. Hecht, S.S.; Chen, C.B.; Hoffmann, D. J. Med. Chem. 1980, 23, 1175.
37. Hecht, S.S.; Lin, D.; Chen, C.B. Carcinogenesis, Submitted.
38. Williams, G.M.; Laspia, M.F. Cancer Letters 1979, 6, 199.
39. Lai, D.Y.; Arcos, J.C.; Argus, M.F. Res. Commun. Chem. Pathol. Pharmacol. 1980, 28, 87.
40. Waddell, W.J.; Marlowe, C. Cancer Res. 1980, 40, 3518.
41. Brittebo, E.; Tjaelve, H. J. Cancer Res. Clin. Oncol. 1980, 98, 233.
42. Hoffmann, D.; Castonguay, A.,; Rivenson, A.; Hecht, S.S. Cancer Res. 1981, in press.
43. Druckrey, H.; Preussmann, R.; Ivankovic, S.; Schmaehl, D. Z. Krebsforsch. 1967, 69. 103.
44. International Agency for Research on Cancer, "IARC Monographs on the Evaluation of the Carcinogenic Risk of Chemicals to Humans: Some N-Nitroso Compounds, Vol 17," IARC, Lyon, France, 1978; p 263, 287.
45. Mohr, U. Progr. Exp. Tumor Res. 1979, 24, 235.
46. Leung, K.H.; Park, K.K.; Archer, M.C. Res. Commun. Chem. Pathol. Pharmacol. 1978, 19, 201.
47. Rayman, M.P.; Challis, B.C.; Cox, P.J.; Jarman, M. Biochem. Pharmacol. 1975, 24, 621.
48. Hsieh, S.T.; Kraft, P.L.; Archer, M.C.; Tannenbaum, S.R. Mutat. Res. 1976, 35, 23.
49. Kupper, R.; Reuber, M.D.; Blackwell, B.N.; Lijinsky, W.; Koepke, S.R.; Micheda, C.J. Carcinogenesis 1980, 1, 753.
50. Goodall, C.M.; Lijinsky, W.; Tomatis, L. Cancer Res. 1968, 28, 1217.
51. Althoff, J.; Cardesa, A.; Pour, P. J. Natl. Cancer Inst. 1973, 50, 323.
52. Grandjean, C.J. J. Natl. Cancer Inst. 1976, 57, 181.
53. Hecker, L.I.; Saavedra, J.E. Carcinogenesis, 1980, 1, 1017.
54. Ross, A.E.; Mirvish, S.S. J. Natl. Cancer Inst. 1977, 58, 651.
55. Mirvish, S.S. Toxicol. Appl. Pharmacol. 1975, 31, 325.
56. Sander, J.; Burkle, G. Z Krebsforsch. 1969, 73, 54.
57. Fajen, J.M.; Carson, G.A.; Rounbehler, D.P.; Fan, J.Y.; Vita, R.; Goff, U.E.; Wolf, M.H.; Edwards, G.S.; Fine, D.H.; Reinhold, V.; Biemann, K. Science 1979, 205, 1262.
58. Goff, U.E.; Coombs, J.R.; Fine, D.H.; Baines, J.M. Anal. Chem. 1980, 52, 1833.
59. Hecht, S.S.; Young, R. Cancer Res. 1981, Submitted.
60. Manson, D.; Cox, P.J.; Jarman, M. Chem.-Biol. Interact. 1978, 20, 341.

61. Stewart, B.W.; Swann, P.F.; Holsman, J.W.; Magee, P.N. Z. Krebsforsch. 1974, 82, 1.
62. Appel, K.E.; Schrenk, D.; Schwarz, M., Mahr, B.; Kunz, W. Cancer Letters 1980, 9, 13.
63. Suess, R. Z. Naturforsch. 1965, 20b, 714.
64. Lijinsky, W.; Taylor, H.W.; Keefer, L.K. J. Natl. Cancer Inst. 1976, 57, 1311.
65. Charnley, G.; Archer, M.C. Mutat. Res. 1977, 46, 265.
66. Lijinsky, W.; Taylor, H.W. Cancer Res. 1975, 35, 2123.
67. Reznik, G.; Mohr, U.; Lijinsky, W. J. Natl. Cancer Inst. 1978, 60, 371.
68. Gingell, R.; Wallcave, L.; Nagel, D.; Kupper, R.; Pour, P. Cancer Lett. 1976, 2, 47.
69. Lijinksky, W.; Saavedra, J.E.; Reuber, M.D.; Blackwell, B.N. Cancer Letters 1980, 10, 325.
70. Krueger, F.W.; Bertram, B.; Eisenbrand, G. Z. Krebsforsch. 1976, 85, 125.
71. Lee, K.Y.; Lijinsky, W. J. Natl. Cancer Inst. 1966, 37, 401.
72. Lijinsky, W.; Keefer, L.; Loo, J.; Ross, A.E. Cancer Res. 1973, 33, 1634.
73. Pegg, A.E. Adv. Cancer Res. 1975, 25, 195.

RECEIVED August 10, 1981.

5

Effects of Structure on the Carcinogenic Behavior of Nitrosamines

JOHN S. WISHNOK

Department of Nutrition and Food Science, Massachusetts Institute of Technology, Cambridge, MA 02139

We have developed a quantitative structure–activity
model for the variations in potency among the
nitrosamines and, more recently, a related model
for the variation in target organ for a smaller
set of nitrosamines. We are currently developing
a model for interspecies variation in susceptibility toward carcinogenic nitrosamines. The
model for organ selectivity requires terms for
the parent nitrosamine as well as for the hypothesized metabolites while the model for potency
variations contains terms only for the unmetabolized parent compound.
All of the quantitative models are implicitly
dose–dependent. This is of particular importance
with respect to interspecies comparisons since
it may be possible for the relative susceptibility of species to reverse on going from higher
to lower doses.

Several research groups are currently investigating the
application of quantitative structure–activity relationships to
various problems in nitrosamine carcinogenesis. One approach is
to use pattern recognition techniques to separate large sets of
compounds into carcinogenic or noncarcinogenic clusters in multi-
dimensional space. These methods appear promising as a means of
rapid non-biological pre-screening of new compounds for carcinogenic potential.

Our approach is to examine small, closely-related series of
nitrosamines and to develop structure–activity models based on
molecular descriptors which are explicitly meaningful with respect
to the organic chemistry and biochemistry of the compounds. The
forms of these models can then often be interpreted in terms of
the mechanisms through which these compounds exert their carcinogenic effects.

Introduction

The simple assumption, that a compound containing the NNO
functionality is carcinogenic, may be correct as often as about
3 times out of 4 (1). In addition, with few exceptions, e.g.,
H (2) or HOCH$_2$ (3), virtually any alkyl or aryl group - substituted or unsubstituted, cyclic or acyclic - can be attached to
the two remaining nitrogen valences to give an isolatable compound (4). The nitrosamines consequently constitute one of the
most extensive classes of known chemical carcinogens. Most
nitrosamines are reasonably easy to synthesize and this, along
with their potential importance as environmental carcinogens (5-
7), has lead to the evaluation of nearly 300 of these compounds
for carcinogenicity in a number of different animal species (8-
10).
      The potency of the carcinogenic nitrosamines has been found
to vary by over a thousand-fold (8), and most of them are selective toward specific target organs (8). In an empirical sense,
then, the nitrosamines are useful and versatile laboratory carcinogens. On a more subjective level, however, is the question
of why a given structural variation leads to a variation in
potency or target organ. Equally intriguing is the question of
why different animal species are more or less susceptible to
carcinogenesis by a given nitrosamine or why the target organs
are sometimes different for different species.
      We, (11, 12), along with several other research groups (13-
17), have been attempting to answer some of these questions
through the use of quantitative biological structure-activity
analyses.
      As summarized in a recent report in Chemical and Engineering
News (1), much of the current work in this area is being done
in the area of structure-potency variations among large groups of
N-nitroso compounds using empirical and semi-empirical molecular
descriptors such as most of those shown in Table I. The objective in these investigations is primarily to be able to evaluate
the likelihood that a given compound is carcinogenic or not.
      An alternative viewpoint for structure-activity investiga-
tions is to utilize quantitative models as probes into the mechanism of action of the set of compounds being studied. In this
case it is most useful if the molecular descriptors are explicitly
meaningful in terms of chemical reactivity or physiological
behavior, e.g., distribution of the compound in an organism (see
Table II). In a previous symposium, (18), we described our
application of this approach toward the development of a quantitative structure-potency expression, equation 1,

$$\log (1/D_{50}) = 1.74 - 0.26\pi^2 + 0.92\pi + 0.59\sigma*$$

$$n = 21 \qquad s = 0.31 \qquad R^2 = 0.84$$

(1)

Table I
A Sampling of Molecular Descriptors[a]

Physicochemical descriptors	Geometrical descriptors
Molecular weight	Molecular volume
Density	Molecular shape
Melting point	Molecular surface area
Boiling point	Substructure shape
Logarithm of partition	Taft steric parameter[b]
coefficient	Verloop sterimol constants[b]
Molecular refractivity[b]	

Topological descriptors	Electronic descriptors
Atom and bond fragments	Hammett–Taft sigma constants[b]
Substructures (atom groups)	Electron density
Substructure environment	π-Bond reactivity
Number of carbon atoms	Electron polarizability
Number of rings (in poly-	Dielectric constant
cyclic compounds)	Dipole moments
Molecular connectivity	Ionization potential
(extent of branching)	Electron affinity

[a]Dagani; (1)

[b]These "complex" descriptors could be placed in other categories
as well.

which modeled the variations in carcinogenicity, as the molecular
structure changed, for a series of 29 nitrosamines (11).

We recently reported a structure–activity model for variations in target organs (12) and are currently examining the possible application of the quantitative structure–activity approach
to the problem of species-to-species differences in susceptibility
toward nitrosamine carcinogenesis (19). These two topics will be
discussed in the remainder of this presentation.

Methods

Our basic methods have been detailed in previous reports
(11, 12). In summary, however, our approach is basically the
same as that used by Hansch and co-workers (20–22). A set of
compounds, which can reasonably be expected to elicit their
carcinogenic response via the same general mechanism, is chosen,
and their relative biological activities, along with a set of
molecular descriptors, is entered into a computer. The computer,
using the relative biological response as the dependent variable,
then performs stepwise multiple regression anayses (23) to select

Table II
Molecular Descriptors and Statistical Terms for Multiple
Regression Analyses

D_{50}: Median tumorigenic dose, i.e., the total molar dose of a
compound required to induce tumors in 50% of the test
animals.

σ^*: Taft electronic parameter. The σ^* parameter reflects the
electron-withdrawing or electron-donating ability of a
given substituent. Increasing σ^* reflects increasing
electronegativity.

E_s: Taft steric parameter. E_s reflects the effective volume
of a substituent in the vicinity of a reactive center on
a molecule.

π: Relative partition coefficient. $\pi = \log K - \log K_p$
where K is the water-octanol or water-hexane partition
coefficient for the compound of interest and K_p is the
corresponding partition coefficient for the parent mem-
ber of the series (N-nitrosodimethylamine in this case).

π_s, π_ℓ: Relative π's for the metabolites of the short (s) and
long (ℓ) side chains of a given nitrosamine.

s: Standard error of the estimate. S is an absolute indicator of the reliability of the calculated value of the
dependent variable.

R^2: Multiple correlation coefficient. R^2 indicates the percentage of the variability of the relative biological
response that can be accounted for by the selected independent variables.

which of the molecular descriptors can be correlated with the
relative biological response. The result of this type of analysis
is an equation which quantitatively models the relative biological
response in terms of the molecular descriptors.

We generally try to use as few descriptors as possible
and to relate them wherever possible to the known or expected
physiological and metabolic behavior of the test compounds.
These descriptors include relative partition coefficients (π),
Taft electronic factors (σ^*), and, in some cases, steric parameters (E_s) ([11], [12], [20], [22]).

Partition coefficients are thought to reflect the probability
that a compound can diffuse, from the site of administration to a
site of biological action, within a standard time interval ([21]).
The electronic and steric factors, σ^* and E_s, are assumed to
reflect the reactivity of a crucial reaction site on the molecule,
e.g., in the case of the nitrosamines, the ease of enzymatic
oxidation of an α C-H bond ([8], [24]).

The final form of the quantitative model can then, in principle, be interpreted in terms of the proposed mechanisms of action or to modify various aspects of these mechanisms.

If, for example, the route from parent carcinogen to the actual biologically-active metabolite is considered as a multistep pathway, the terms that appear in the model equation can be thought of as representing the rate-determining steps.

Results and Discussion

A Structure-Selectivity Model. Nitrosamines are usually highly sensitive towards one or a few target organs; some examples are listed in Table III. A casual examination of the structures

Table III

Nitrosamine	Major Target Organ (BD rat)[a]
Dimethyl	Liver
Dipropyl	Liver
Dibutyl	Esophagus & Liver
Methyl Ethyl	Liver
Methyl Vinyl	Esophagus
Methyl Amyl	Esophagus

[a] Druckrey et al.; (8).

associated with a given target organ can lead quickly to some intuitive structure-activity generalizations. Small, symmetrical compounds, for example, tend to attack the liver, while the more unsymmetrical compounds (i.e., greater differences between the smaller and larger side chains) tend to attack the esophagus. More careful examination, however, leads to frustration because of the number of apparent exceptions. The situation becomes more complex still when it is noted that the main target organ for a given compound can vary from species to species (25), or that the tumor distribution for a given compound in a given species can be affected by dose (26).

Nonetheless, while working on a quantitative model to relate variations in potency with variations in structure, we noted that the models were slightly different if only single organs were studied (12). We consequently decided to see if anything could be learned from a structure-activity analysis of target organ selection. We felt that some of the complexities of the problem could be minimized by examining a closely-related set of compounds and only a few target organs. Our test compounds, then, were a series of thirteen dialkylnitrosamines which induced tumors in the liver or in the lung or esophagus (Table IV, columns 1, 2). The latter two categories - mostly esophageal tumors - were combined (12).

Table IV
Observed and Calculated Values for the Fraction of
Liver Tumors (L)[a]

Compound (N-Nitrosamine)	L(observed) (8)	L[b]	L[c]	L[d]
Dimethyl	1.00	0.98	1.00	1.00
Diethyl	0.87	0.36	0.87	0.87
Dipropyl	0.76	0.11	0	0.76
Di-isopropyl	1.00	0.11	1.00	1.00
Dibutyl	0.42	0.29	0.42	0.42
Diamyl	1.00	0.71	1.00	1.00
Methylethyl	1.00	0.81	1.00	1.00
Methylvinyl	0	0.15	0	0
Methylallyl	0	0.40	0	0
Methylamyl	0	0.11	0	0
Methylcyclohexyl	0	0.35	0	0
Methylheptyl	0	0.27	0	0
Methylphenyl	0	0.57	0	0
Methylbenzyl	0	0.14	0	0
Methyl-(2-phenylethyl)	0	0.17	0	0
Ethylvinyl	0	0.10	0	0
Ethylisopropyl	0.39	0.41	1.00	0.39
Ethylbutyl	0.09	0.11	0	0
Butylamyl	1.00	0.62	1.00	1.00

[a]Edelman et al.; (12).

[b]Calculated using Eq. 3.

[c]Calculated using π, σ^*, and E_s for unmetabolized nitrosamine.

[d]Calculated using Eq. 4.

As our index of relative biological response, we chose simply the fraction of liver tumors

$$L = \frac{A}{A + B}$$

where A was the number of animals with liver tumors and B the number of animals with esophageal tumors (plus a few with lung tumors) (12). This variable could then range from 1 to 0 for exclusively liver tumors or exclusively esophagus/lung, respectively. This behavior can be represented in the general sense by a common biological logistical expression

$$\frac{1}{1 + e^{-f}}$$

which varies from 0 to 1 as the function f goes from $-\infty$ to ∞. By assuming that the relative probability that an animal will develop a liver tumor is related to the relative probability that a carcinogen will reach a site of action in the liver, it is possible to show that f can reasonably be a parabolic function of π, e.g.,

$$L = \frac{1}{1 + e^{-f}} \tag{2}$$

where

$$f = k_1 \pi^2 + k_2 \pi + c \tag{3}$$

As might be expected, our assumption, that L is a function solely of the distribution of the parent carcinogen, is incomplete. Regression analysis on equation 2 with f defined by equation 3 did give a crude separation of the test compounds into two groups but did not approach a truly quantitative model. Interestingly, several years ago, Mirvish and coworkers had intuitively noted a rough correlation between partition coefficients and target organ for a different set of N-nitroso compounds (27).

The addition of steric and electronic terms for the parent compound leads to an improved model, and an expression where f contains π's for the metabolites as well as the parent compound in addition to electronic terms for the parent compound (equation 4) constitutes a good model for our set of compounds. The

$$f = 10.64\pi^2 - 34.11\pi - 62.82\sigma^* + 34.35E_s$$
$$- 10.25\pi_s^2 + 42.52\pi_s + 8.04\pi_\ell^2 - 61.65\pi_\ell + 53.98 \tag{4}$$

$$n = 19 \qquad s = 0.02846 \qquad R^2 = 0.9978$$

evolution of this model is detailed in Table IV, columns 3-5.

Interspecies Variations in Susceptibility. An important assumption in the development of both the structure-potency model (eq. 1) and the structure-target-organ model (eq. 2) was that all of the test compounds were administered at equal daily

doses. As noted earlier (12), this was approximately true,
although fortuitously (8). It had been pointed out, however, in
the extensive paper by Druckrey and coworkers (8) that the daily
dose, d, and the mean induction time, t_{50}, were related by the non-
linear expression (5) where n varied from nitrosamine to nitros-

$$d(t_{50})^n = k \tag{5}$$

amine. Since our basic index of carcinogenicity, D_{50}, is given
by dt_{50}, it is apparent that this index is dose-dependent, i.e.,
relative D_{50}'s are meaningful only for a given d for all test
compounds.

We consequently became interested in developing a dose-
dependent quantitative model using t_{50} as the relative biological
response. This model is shown as equation 6.

$$\log(t_{50}) \cong 0.13\pi^2 - 0.5\pi - 0.32\sigma^* - 0.4 \log d - 0.8 \tag{6}$$

$$n = 13 \qquad s = 0.305 \qquad R^2 = 0.895$$

With the available data, it is apparent from R^2 and s that this
relationship offers no particular improvement over equation 1.
It does suggest, however, that our original assumptions concerning d were at least reasonable. In addition, the dose-dependent model can be manipulated to examine - in a general sense -
some aspects of interspecies variations in susceptibility to
nitrosamine carcinogenesis.

With t_{50} as an index of carcinogenicity, an interspecies
relative potency (IRP) can be defined, e.g.,

$$IRP = \log\left(\frac{t_a}{t_b}\right) \tag{7}$$

where t_a is t_{50} for species a and t_b is t_{50} for species b.
Assuming that the general form of equation 6 is the same for each
species, a general expression for interspecies relative potency
can be written based on π, σ^*, and d:

$$IRP = f_{ab} - \left(\frac{1}{n_a} - \frac{1}{n_b}\right) \log d + c \tag{8}$$

where f_{ab} is a linear combination of the π and σ^* components for
species a and b, and n_a and n_b are the exponents from the Druckrey
expression (equation 5). This model suggests that it may be
ultimately possible - in principal - to rationally relate the
effects of a carcinogen in one species to its effects in another
species. On a practical level, however, it indicates that it may
be impossible to do this unless something is known about dose-
response effects in both species. Unless $n_a = n_b$, for example, a
given carcinogen could be observed to be more, less, or equally
carcinogenic in one species as compared to another depending on
the doses used in the evaluations.

We have so far been unable to find an appropriate set of data to adequately test this model. A small set of nitrosamines has been studied by Lijinsky and coworkers in both Fischer and Sprague-Dawley rats (28); although there is considerable scatter, the results are consistent with this model.

A recent series of experiments which may be related to this concept has been reported by Prehn and Lawler (29). They treated 10 strains of mice with two different dose levels (5% and 0.05%) of 3-methylcholanthrene and observed that the rank order of susceptibility, as measured by the average number of tumor-free days, was reversed on going from the higher to the lower dose. They suggested differential stimulation of immune response as an explanation of their results but it is also possible that different dose-responses, as suggested by Druckrey's equation (equation 5), may be important.

It should be noted in closing that the n values in the Druckrey-Preussmann expression probably represent a net contribution of various physiological and metabolic properties of the test animals. Thus, while investigating a series of carcinogens in a given species, these can be assumed to be constant for each chemical. The relative properties of the compounds will then reflect only differences in the molecular structure of the test substances.

The extension of this type of investigation to other animal species, however, will require fairly extensive evaluations of representative chemicals in each species.

Conclusions

There appears now to be ample evidence that the variations in carcinogenicity among the nitrosamines are systematically and rationally related to structure and that several indices of carcinogenic potency can be used as indices of biological response for the generation of quantitative structure-activity models (11-17).

Within a given animal species, in addition, the selection of target organs, at least under certain dosing schedules, may also be an explicit function of molecular structure.

To the extent that these structure-activity models are valid, their forms suggest that the variations in potency depend mainly on the structure of the unmetabolized parent nitrosamine, while the variations in target organ are determined to some extent by the properties of the various possible metabolites, i.e., the DNA-alkylating electrophile arising from one of amine side-chains and the aldehyde and aldehyde-derived metabolites arising from the other side-chain. The rate-determining steps in the metabolic pathway leading from administration of the nitrosamine to a final active metabolite appear - as generally hypothesized - to be the transport of the compound to the site of

metabolism and enzymatic oxidation at the α-carbon. The organ-
selection process might involve the non-alkylating metabolites.
A reasonable role for the metabolites might be as irritants
which stimulate cell proliferation and thus aid in the expression of the lesion created by the alkylating metabolite.
The application of quantitative structure-activity analyses
to the question of interspecies variations of responses to chemical carcinogens, although apparently possible in principle,
appears to require extensive additional physiological and biochemical information about the respective species. These studies,
perhaps in a sense stating the obvious, also suggest that the
extrapolation of relative potency data from one species to another
should only be attempted with the understanding that such evaluations may be strongly dependent on the original experimental
dosing scheme.

Acknowledgements

The structure-activity project at M.I.T. was originated in
collaboration with Dr. Michael C. Archer, and has been carried
out with Mr. Andrew Edelman, Dr. William M. Rand, Dr. Patricia
Kraft, and Ms. Vicki Woolworth. We are grateful to the M.I.T.
Health Sciences Fund, The Marjorie Merriwether Post Foundation,
The M.I.T. Undergraduate Research Opportunities Program, and the
National Institutes of Health (Grants #R01-CA21951, 2-P01-ES00597,
and 1-T32-ES-0702), for support.

Literature Cited

1. Dagani, R.; Chem. & Eng. News, March 9, 1981, p. 26.
2. Kemp, D.S.; and Vellaccio, F., "Organic Chemistry"; Worth:
    New York, 1980; pp. 747, 1257.
3. Baldwin, J.E.; Branz, S.E.; Gomez, R.F.; Kraft, P.L.;
    Sinskey, A.J.; Tannenbaum, S.R.; Tetrahedron Lett. 1976, 333.
4. Wishnok, J.S.; Encyclopedia of Chemical Technology (in press).
5. Wishnok, J.S.; J. Chem. Educ. 1977, 54, 440.
6. Walker, E.A.; Bogovski, P.; and Griciute, L., Eds.;
    "Environmental N-Nitroso Compounds Analysis and Formation";
    International Agency for Research on Cancer: Lyon, France;
    IARC Scientific Publication No. 14, 1976.
7. IARC Working Group on the Evaluation of Carcinogenic Risk of
    Chemicals to Humans, Some N-Nitroso Compounds, IARC Mono-
    graph No. 17: Lyon, France, 1978.
8. Druckrey, H.; Preussmann, R.; Ivankovic, S.; and Schmahl, D.
    Z. Krebsforsch. 1967, 69, 102.
9. Magee, P.N.; and Barnes, J.M. Adv. Cancer Res. 1967, 10,
    163.
10. Magee, P.N.; Montesano, R.; and Preussmann, R. in Searles,
    C., Ed.; "Chemical Carcinogens"; ACS Monograph 173, American
    Chemical Society: Washington, D.C., 1976, pp. 491-625.

11. Wishnok, J.S.; Archer, M.C.; Edelman, A.S.; and Rand, W.M. Chem.-Biol. Interact. 1978, 20, 43.
12. Edelman, A.S.; Kraft, P.L.; Rand, W.M.; and Wishnok, J.S. Chem.-Biol. Interact. 1980, 31, 81.
13. Singer, G.M.; Taylor, H.W.; and Lijinsky, W. Chem.-Biol. Interact. 1977, 19, 133.
14. Wenzel, V.; and Metzner, J. Pharmazie. 1978, 33, 716.
15. Dunn, W.J., III; and Wold, S. Biorg. Chem. (in press).
16. Kier, L.B.; Simons, R.J.; Hall, L.H. J. Pharm. Sci. 1978, 67, 725.
17. Chou, J.T.; and Jurs, P.C. J. Med. Chem. 1979, 22, 792.
18. Wishnok, J.S. in Anselme, J.-P., Ed.; "N-Nitrosamines"; ACS Symposium Series, No. 101, American Chemical Society: Washington, 1979, p. 153.
19. Edelman, A.S.; and Wishnok, J.S. unpublished data.
20. Hansch, C. Accounts Chem. Res. 1969, 2, 232.
21. Hansch, C.; and Fujita, T. J. Amer. Chem. Soc. 1964, 86, 1616.
22. Hansch, C.; Smith, N.; Engle, R.; and Wood, H. Cancer Chemother. Rep. 1972, 56, 443.
23. Dixon, W.J., Ed.; "BMDP Biomedical Computer Programs"; University of California Press: Berkeley, 1975, pp. 491-539.
24. Magee, P.N.; and Barnes, J.B. Adv. Cancer Res. 1967, 10, 163.
25. Magee, P.N.; Montesano, R.; and Preussmann, R.; in Searle, C.E., Ed.; "Chemical Carcinogens"; ACS Monograph 173, American Chemical Society: Washington, 1976, pp. 504, 557.
26. Magee, P.N.; and Barnes, J.M. Brit. J. Cancer. 1956, 10, 114; J. Path. Bact. 1962, 84, 19.
27. Mirvish, S.S.; Issenberg, P.; and Sornson, H.C. J. Natl. Cancer Inst. 1976, 56, 1125.
28. Lijinsky, W.; personal communication.
29. Prehn, L.M.; and Lawler, E.M. Science. 1979, 204, 310.

RECEIVED August 10, 1981.

Structure—Activity Relationships Among N-Nitroso Compounds

WILLIAM LIJINSKY

Chemical Carcinogenesis Program, Frederick Cancer Research Center, Frederick, MD 21701

Early studies of the simple nitrosodialkylamines, especially of nitrosodimethylamine and nitrosodiethylamine, indicated that their carcinogenic action derived from oxidation at the alpha carbon to an aldehyde, with release of the other alkyl group attached to the nitrogen atom as an alkylating moiety, which alkylated some key macromolecules (1). Very reasonably it was assumed that the most important macromolecule was DNA (2). This suggestion was fortified by the observation that the unstable nitrosoalkylamides (such as nitrosoalkylureas and nitrosoalkylnitroguanidines) were locally acting carcinogens, presumably not requiring metabolic activation (3). In support of the concept that DNA was the principal target, the nitrosoalkylamides are directly acting mutagens, while the simple nitrosodialkylamines require metabolic activation, usually with rat-liver microsomal preparations, to mutate bacteria as in the Salmonella mutagenesis test developed by Ames (4). The cyclic nitrosamines, such as nitrosopyrrolidine and nitrosomorpholine pose a more difficult problem, not seeming to conform to any simple mechanism of carcinogenic action. While most of the cyclic nitrosamines studied are mutagenic to bacteria when suitably activated, those that have been examined do not interact measurably with DNA so as to form an identifiable alkylated component, analogous to the 7-alkylguanines and O^6-alkylguanines isolated from the liver DNA of animals treated with nitrosodimethylamine or nitrosodiethylamine (5). A great variety of sites of alkylation in DNA by aliphatic N-nitroso compounds has been reported by Singer (6), but analogous derivatives have not been found in experiments with cyclic nitrosamines (7). Although alkylation might occur at very low, undetectable levels in the latter case, it is hard to sustain the concept that the detection of such interactions can be related quantitatively to the carcinogenic potency of nitrosamines, since many of the cyclic compounds are of comparable potency with the aliphatic nitrosamines which alkylate DNA extensively. It seems, therefore, that other reactions mediate the process by which nitrosamines are transformed into the

0097-6156/81/0174-0089$05.00/0
© 1981 American Chemical Society

moieties which induce neoplasia. It is probable that studies of the relationship between chemical structure and carcinogenic activity of nitroso compounds will lead to understanding of these intermediate processes, and might indicate those reactions likely to be related to carcinogenesis.

It seemed at one time that mutagenic activity in bacteria after appropriate metabolic activation might be a measure of carcinogenic activity in animals. This would then enable studies of the metabolic pathways leading to carcinogenesis to be carried out using mutagenic activity of the intermediates, without the need of lengthy chronic toxicity tests in animals. Among approximately 150 N-nitroso compounds which have been examined for both carcinogenicity in animals and mutagenicity in bacteria, most were carcinogenic and mutagenic. However, there was no quantitative relationship between the two activities. While there were only a few N-nitroso compounds which were mutagenic but not carcinogenic, quite a large number were carcinogenic but not mutagenic including some rat liver carcinogens which were not activated to mutagens by rat liver microsomal preparations (Table I).

It appears that considerable metabolism of nitrosamines takes place in the liver, from the finding that most of them are activated by rat liver microsomes to bacterial mutagens. However, relatively few nitrosamines induce liver tumors in rats. The most common target is the esophagus and a wide variety of nitrosamines induce tumors in this organ of rats, some only tumors of the esophagus, others tumors of other organs also. Thus, most nitrosopiperidines induce esophageal tumors in rats (8,9) including 4-methyl-, 4-phenyl- and 4-t-butyl-, but not 4-cyclohexyl-, which is inactive (Figure 1). Most methylated dinitrosopiperazines give rise to esophageal tumors in rats (10). Halogen substitution in nitrosopiperidines produces very potent esophageal carcinogens (11) and 3,4-dichloronitroso-pyrrolidine is an esophageal carcinogen, not a liver carcinogen (12). A variety of asymmetric aliphatic nitrosamines give rise to esophageal tumors in rats, including most derivatives of nitrosomethylethylamine bearing a substituent on the beta carbon atom, even if only deuterium atoms (Figure 2). Nitroso-methylalkylamines from -propyl to -hexyl induce esophageal tumors, while these with longer chains induce tumors of other organs. Many symmetrical aliphatic nitrosamines also induce esophageal tumors in rats, often together with tumors of other sites. They include nitrosodiethylamine, nitrosodi-n-propyl-amine, nitrosobis(2-hydroxypropyl)amine and nitroso-2hydroxy-propyl-2oxopropylamine, while nitrosobis-(2-oxopropyl)amine induces liver tumors, but no esophageal tumors. In contrast, nitrosomethyl-2oxopropylamine induces esophageal tumors in rats. Simple explanations of these differences are made more difficult by the fact that in other species, such as hamsters, the target organs of these carcinogens are often very different,

TABLE I

CARCINOGENIC NITROSAMINES NOT MUTAGENIC TO SALMONELLA
WHEN ACTIVATED WITH RAT LIVER S9

TARGET ORGAN IN RATS

NITROSO-

Di-isopropylamine	Liver, Nasal Cavity
Bis-(oxopropyl)amine	Liver, Colon
Bis-(methoxyethyl)amine	Liver, Esophagus
Bis-(2-hydroxypropyl)amine	Esophagus, Nasal Cavity
Methylethylamine	Liver, Esophagus
Methyl-n-propylamine	Esophagus
Methyl-iso-propylamine	Esophagus
Diethanolamine	Liver
Ethanolisopropanolamine	Liver, Esophagus
Methylneopentylamine	Esophagus
Methylaniline	Esophagus
Methylbenzylamine	Esophagus
Methyltrifluoroethylamine	Esophagus
4-t-Butylpiperidine	Esophagus
2,6-Dimethylpiperazine	Thymus
Trimethylpiperazine	Thymus
4-Benzoyl-3,5-dimethylpiperazine	Forestomach
Methyldiethylurea	Nervous System
Ethyldimethylurea	Nervous System
Diphenylamine	Bladder
Phenylbenzylamine	Esophagus
Dihydroxypropylallylamine	Esophagus
Methyloxopropylamine	Esophagus
Methyl-3-carboxypropylamine	Bladder
Methyl-n-Tetradecylamine	Bladder

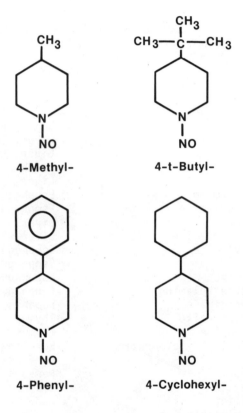

Figure 1.   Structures of 4-substituted nitrosopiperidines.

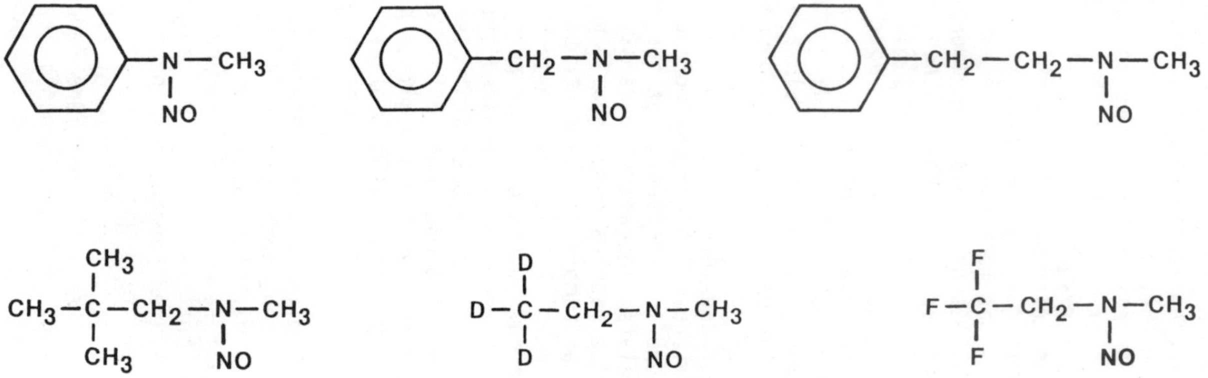

Figure 2. Structures of β-substituted nitrosomethylethylamines: nitrosomethylaniline, nitrosomethylbenzylamine, nitrosomethyl-2-phenylethylamine, nitrosomethylneopentylamine, nitrosomethyltrideuteroethylamine, nitrosomethyltrifluoroethylamine.

including in Syrian hamsters the pancreas, which has not been susceptible in rats to nitrosamine carcinogenesis.

While nitrosodiethanolamine induces liver tumors in rats and nitrosobis(2-hydroxypropyl)amine induces esophageal tumors, nitroso-2-hydroxypropylethanolamine induces both esophageal and liver tumors, suggesting that alternative metabolism of the two sides of the molecule is responsible. The same conclusion is suggested by the results of examining the carcinogenic effect of administering a number of oxygenated cyclic nitrosamines to rats. Nitrosomorpholine, nitroso-1,3-oxazolidine and nitroso-tetrahydro-1,3-oxazine (Figure 3) are all liver carcinogens of almost identical potency, as measured by the rate of death of the animals with tumors. Nitroso-2,6-dimethylmorpholine is a considerably more potent carcinogen than nitrosomorpholine, but induces only esophageal tumors in rats (13). On the other hand, nitroso-2-methylmorpholine, also more potent than nitroso-morpholine, gives rise to both esophageal and liver tumors, possibly determined by oxidation of one side of the molecule or the other. In contrast, nitroso-5-methyloxazolidine is a weaker carcinogen than the unsubstituted compound, and induces only liver tumors.

Because of the profound interspecies differences in carcinogenicity shown by nitroso-2,6-dimethylmorpholine, we have focussed on this compound in our studies of the routes of metabolism that might be related to carcinogenesis. In the rat nitrosodimethylmorpholine induces esophageal tumors, but in the Syrian hamster it induces tumors of the pancreas and of the liver (14). In guinea pigs it gives rise only to angiosarcomas of the liver, which are not tumors of liver parenchyma cells. Nitrosodimethylmorpholine as usually prepared exists as two isomers, cis and trans, which have been separated and purified. In rats, the trans isomer is a much more potent carcinogen than the cis (15), whereas in the hamster and guinea pig the reverse is true, which suggests that the pathways of activation to carcinogenic intermediates in the three species might well be quite different.

Nitrosodimethylmorpholine has been prepared labeled with deuterium in either the alpha positions or in the beta positions. Previous studies with other nitrosamines labeled with deuterium, including nitrosomorpholine (16) and nitrosomethyl-ethylamine (17) showed that the presence of deuterium at a position which led to reduced carcinogenic potency indicated that oxidation at that position was a rate limiting step in activation of the carcinogen. In the case of nitrosodimethyl-morpholine, deuterium in the alpha positions reduced the potency of the compound as an esophageal carcinogen in rats, while deuterium in the beta positions increased carcinogenic potency (18). In hamsters the reverse was true, again indicating that the metabolic pathways leading to carcinogenesis in the two species are different.

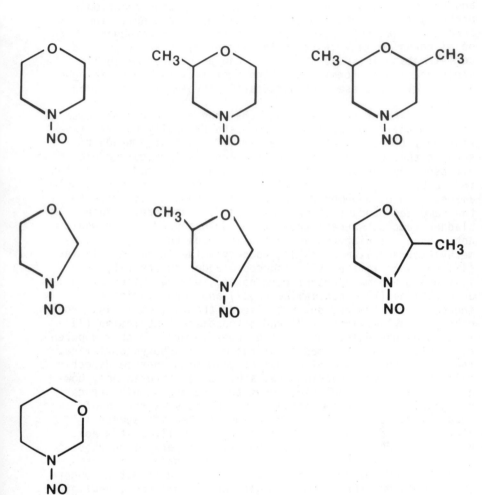

Figure 3.   Structures of oxygen-containing cyclic nitrosamines: nitrosomorpholine, 2-methylnitrosomorpholine,   2,6-dimethylnitrosomorpholine,   nitroso-1,3-oxazolidine, 5-methylnitroso-1,3-oxazolidine, 2-methylnitroso-1,3-oxazolidine, nitrosotetrahydro-1,3-oxazine.

The experiments with deuterium-labeled nitrosamines illustrate two important points. One is that oxidation of nitrosamines takes place at more than one position in the molecule, and the outcome of the balance of such competing reactions probably is the determinant of carcinogenic potency. The second is that the reason for the failure of carcinogenesis to be mirrored in many cases by the microsomally activated bacterial mutagenicity is that there can be several metabolic steps leading to formation of the proximate carcinogenic agent and not all of these need necessarily involve microsomal enzymes.

Our finding some years ago that the asymmetric nitrosamine nitrosomethyl-n-dodecylamine administered orally to rats gave rise only to tumors of the bladder (19), prompted Okada to suggest that beta oxidation of the long chain occurred producing nitrosomethyl-3-carboxypropylamine as a final product excreted in the urine (20). Accordingly, all of the nitrosomethylalkylamines with even-numbered carbon chains should give rise to the same end-product of metabolism and, therefore, induce bladder tumors, whereas those with odd-numbered chains would not. When administered at equimolar doses to rats, nitrosomethyl-tetradecylamine (XII), -dodecylamine (XI), -decylamine (IX) and -octylamine (VII) induced bladder tumors (21) (the last being the most potent carcinogen), whereas nitrosomethylundecylamine (X), -nonylamine (VIII) and -heptylamine (VI) induced liver tumors, but not bladder tumors (Table II). Both nitrosomethylhexylamine (V) and nitrosomethylbutylamine (III) induced esophageal tumors, not bladder tumors, so they apparently could not be oxidized to nitrosomethylcarboxypropylamine. The excretion of this nitrosoamino acid might not be directly related to bladder carcinogenesis by these nitrosamines, however, since it has not been shown to be a potent bladder carcinogen in rats. Examination of the urinary excretion of metabolites following administration of the nitrosomethylalkylamines to rats suggests another possibility. A minor, but significant, component of the urine of rats given those nitrosamines with an even-numbered carbon chain is nitrosomethyl-2-oxopropylamine, and more is produced from the longer-chain compounds than from the short chain compounds; negligible quantities are formed from the odd-numbered chain compounds. Whether nitrosomethyloxopropylamine is a bladder carcinogen in rats remains to be determined, but it is known to be a potent pancreatic carcinogen in hamsters (22).

It seems probable that the shape of a nitrosamine molecule, particularly among cyclic nitrosamines, has a very large bearing on its carcinogenic activity. It is probably the constraints placed on the molecule which makes methylated derivatives of dinitrosopiperazine such potent esophageal carcinogens, compared with the unmethylated compound (10). Similarly, in the case of mononitrosopiperazines, neither 1-nitrosopiperazine, nor 4-methyl-1-nitrosopiperazine is appreciably carcinogenic, although

TABLE II

CARCINOGENESIS IN RATS BY NITROSOMETHYLALKYLAMINES

	Structure	RELATIVE CARCINOGENIC POTENCY	RAT TARGET ORGAN
I	$ON-N \Big\langle \begin{smallmatrix} CH_3 \\ CH_2CH_3 \end{smallmatrix}$	+ + +	LIVER
II	$ON-N \Big\langle \begin{smallmatrix} CH_3 \\ [CH_2]_2\,CH_3 \end{smallmatrix}$	+ + + +	ESOPHAGUS
III	$ON-N \Big\langle \begin{smallmatrix} CH_3 \\ [CH_2]_3\,CH_3 \end{smallmatrix}$	+ + + +	ESOPHAGUS
IV	$ON-N \Big\langle \begin{smallmatrix} CH_3 \\ [CH_2]_4\,CH_3 \end{smallmatrix}$	+ + + +	ESOPHAGUS
V	$ON-N \Big\langle \begin{smallmatrix} CH_3 \\ [CH_2]_5\,CH_3 \end{smallmatrix}$	+ + + +	ESOPHAGUS, LIVER
VI	$ON-N \Big\langle \begin{smallmatrix} CH_3 \\ [CH_2]_6\,CH_3 \end{smallmatrix}$	+ + +	LIVER, LUNG
VII	$ON-N \Big\langle \begin{smallmatrix} CH_3 \\ [CH_2]_7\,CH_3 \end{smallmatrix}$	+ + +	LIVER, BLADDER
VIII	$ON-N \Big\langle \begin{smallmatrix} CH_3 \\ [CH_2]_8\,CH_3 \end{smallmatrix}$	+ + +	LIVER
IX	$ON-N \Big\langle \begin{smallmatrix} CH_3 \\ [CH_2]_9\,CH_3 \end{smallmatrix}$	+ +	BLADDER
X	$ON-N \Big\langle \begin{smallmatrix} CH_3 \\ [CH_2]_{10}\,CH_3 \end{smallmatrix}$	+ +	LIVER, LUNG
XI	$ON-N \Big\langle \begin{smallmatrix} CH_3 \\ [CH_2]_{11}\,CH_3 \end{smallmatrix}$	+ +	BLADDER
XII	$ON-N \Big\langle \begin{smallmatrix} CH_3 \\ [CH_2]_{13}\,CH_3 \end{smallmatrix}$	+ +	BLADDER

the latter induces tumors of the nasal cavity at very high
dose rates; neither compound is significantly mutagenic to
bacteria. On the other hand, methyl substitution in the 3 and
5 positions leads to very potently carcinogenic products.
Both 1-nitroso-3,5-dimethylpiperazine and 1-nitroso-3,4,5-
trimethylpiperazine induced a high incidence of thymic leukemia
in rats within 6 months (23). The 4-acetyl derivative of the
former induced esophageal tumors in the same time, while the
4-benzoyl derivative was a very much weaker carcinogen, giving
rise to only a few tumors of the forestomach after almost 2
years. These contrasting results are difficult to reconcile
with any simple mechanism of carcinogenesis.

    It seems probable that, except for the simple nitrosodi-
alkylamines, the metabolic pathways leading to carcinogenesis
comprise several steps. The relevant pathways might be minor
ones and the major routes of metabolism are probably detoxica-
tion pathways. There seems to be competition between the
various pathways, so that small changes in chemical structure
might make large differences to the pharmacokinetics of the
compound, leading to big changes in carcinogenic potency.
While considerable metabolism of nitrosamines occurs in the
liver, it is likely that specific activating enzymes exist in
various organs and these may be responsible for the pronounced
organ specificity of many nitrosamines. This is strongly suggested by the large difference in carcinogenic potency between
configurational isomers of some cyclic nitrosamines.

LITERATURE CITED

1.   Magee, P.N. and Hultin, T., Biochem. J., 1962, 83, 106.
2.   Magee, P.N. and Farber, E., Biochem. J., 1962, 83, 114.
3.   Druckrey, H., Preussmann, R., Ivankovic, S. and Schmähl,
     D., Z. Krebsforsch., 1967, 69, 103.
4.   McCann, J.E., Choi, E., Yamasaki, E. and Ames, B.N., Proc.
     Natl. Acad. Sci. (USA), 1975, 72, 5135.
5.   Nicoll, J.W., Swann, P.F. and Pegg, A.E., Nature, 1975,
     254, 261.
6.   Singer, B., Bodell, W.J., Cleaver, J.E., Thomas, G.H.,
     Rajewsky, M.F. and Thon, W., Nature, 1978, 276, 85.
7.   Lijinsky, W., Keefer, L., Loo, J. and Ross, A.E., Cancer
     Res., 1973, 33, 1634.
8.   Lijinsky, W. and Taylor, H.W., Internat. J. Cancer, 1975,
     16, 318.
9.   Lijinsky, W. and Taylor, H.W., J. Natl. Cancer Inst.,
     1975, 55, 705.
10. Lijinsky, W. and Taylor, H.W., Cancer Res., 1975, 35,
     1270.
11. Lijinsky, W. and Taylor, H.W., Cancer Res., 1975, 35,
     3209.

12. Lijinsky, W. and Taylor, H.W., Cancer Res., 1976, 36, 1988.
13. Lijinsky, W. and Taylor, H.W., Cancer Res., 1975, 35, 2123.
14. Reznik, G., Mohr, U. and Lijinsky, W., J. Natl. Cancer Inst., 1978, 60, 371.
15. Lijinsky, W. and Reuber, M.D., Carcinogenesis, 1980, 1, 501.
16. Lijinsky, W., Taylor, H.W. and Keefer, L.K., J. Natl. Cancer Inst., 1976, 57, 1311.
17. Lijinsky, W. and Reuber, M.D., Cancer Res., 1980, 40, 19.
18. Lijinsky, W., Saavedra, J.E., Reuber, M.D. and Blackwell, B.N., Cancer Lett., 1978, 5, 215.
19. Lijinsky, W. and Taylor, H.W., Cancer Lett.,1978, 5, 215.
20. Okada, M., Suzuki, E. and Mochizuki, M., Gann, 1976, 67, 771.
21. Lijinsky, W., Saavedra, J.E. and Reuber, M.D., Cancer Res. 1981, 41, 1288.
22. Pour, P., Gingell, R., Langenbach, R., Nagel, D., Grandjean, G., Lawson, T. and Salmasi, S., Cancer Res., 1980, 40, 3585.
23. Singer, S.S., Singer, G.M., Saavedra, J.E., Reuber, M.D. and Lijinsky, W., Cancer Res., 1981, 41, 1034.

RECEIVED July 20, 1981.

Chemistry of Some N-Nitrosamides

CLIVE N. BERRY, BRIAN C. CHALLIS, ANDREW D. GRIBBLE,
and SUSAN P. JONES

Department of Chemistry, Imperial College, London, SW7 2AZ, England

Mechanisms for the thermal and photolytic decomposition of N-nitrosamides are briefly reviewed, and recent results for their decomposition by acidic and basic catalysts are summarised and discussed. In the presence of acids, decomposition proceeds by concurrent pathways involving either deamination (hydrolysis) or denitrosation of conjugate acid intermediates, whose formation is usually rate limiting. Denitrosation predominates at high acidity and is negligible at pH > 2. Deamination, which generates diazohydroxide alkylating agents, is therefore considered to be the more likely transformation under stomach conditions. Decomposition in the presence of bases occurs readily at pH 2-12 by an addition-elimination pathway involving nucleophilic rather than general base catalysis. This reaction, which also generates a diazohydroxide alkylating agent, is considered to be the most important transformation under cellular conditions. It can be induced by nucleotides and nucleic acids in vitro, which are then alkylated to a small extent.

There is much evidence to suggest that carcinogenic N-nitrosamines are metabolised by an oxidative process to produce an alkylating agent (1,2). One potential metabolite is therefore the corresponding N-nitrosamide resulting from 2-electron oxidation at the α-carbon atom, and, indeed, such compounds appear to induce tumours at the site of application without metabolic activation (3). It follows that the chemical properties of N-nitrosamides are relevant to the etiology of cancer.

N-Nitrosamides are much less stable than the parent N-nitrosamines and they can decompose by either thermal, photolytic or acid and base (nucleophilic) catalysed pathways. Thermal decomposition has attracted much attention as a clean method of deamin-

0097-6156/81/0174-0101$05.00/0

ation. A good deal is known about the mechanism of this reaction
and the findings have been critically reviewed ($\underline{4},\underline{5}$). Thermal
decomposition proceeds readily at temperatures from $25-100^{\circ}$C., and
involves initial rearrangement to a diazoester intermediate I,
which then rapidly expels nitrogen to give various products as in

$$R \overset{O}{\underset{\underset{\underset{O}{\overset{\|}{N}}}{\|}}{C}} \overset{R'}{N} \overset{\Delta}{\longrightarrow} \left[R \overset{O}{\overset{\|}{C}} O-N=N-R' \right] \overset{-N_2}{\longrightarrow} RCO_2R', \ RCO_2H, \ \text{olefins} \quad (1)$$

equation 1. Both the ease of rearrangement ($3^{\circ} > 2^{\circ} > 1^{\circ}$) and the
type of product (formed by carbonium ion or free radical substit-
utions and eliminations) are more dependent on the nature of the
R' substituent than R. The diazoester intermediate I is, of
course, a putative alkylating agent, but R' invariably reacts
intramolecularly with the carboxylate ion on the extrusion of N_2.
Photolytic decomposition has also been well-investigated, principally by Chow and his colleagues ($\underline{6}$). These reactions proceed on
irradiation at various wavelengths in both polar and non-polar
solvents. The primary process following photoexcitation is dissociation to a radical pair II, and ensuing chemical events involve
the amidyl and nitric oxide radicals either as a caged pair or in

$$R \overset{O}{\overset{\|}{C}} \overset{R'}{N} \overset{hv}{\longrightarrow} \left[R \overset{O}{\overset{\|}{C}} \overset{\bullet}{NR'} \atop NO^{\bullet} \right] \longrightarrow R \overset{O}{\overset{\|}{C}} NH(CH_2)_3CH(NO)R \ \text{or}$$

II   III

$$R \overset{O}{\overset{\|}{C}} N=CHR + (NOH) \quad (2)$$

IV

the bulk of the solution (equation 2). The amidyl radicals commonly undergo intramolecular H-abstraction followed by coupling
with the nitric oxide to give a $\underline{\delta}$-nitroso product III, or β-elim-
ination to give an N-acylimine IV. In poor H-donor solvents, and
for irradiation at $\overline{\lambda} > 280$ nm, more complex reactions occur for
reasons discussed by Chow ($\underline{6}$) in his review. Neither thermal nor
photolytic decomposition is likely to be involved in biological
transformations of N-nitrosamides, but a different conclusion may
apply to the acid and base catalysed decompositions.

Decomposition by Acid Catalysts

On treatment with gaseous HBr in organic solvents, N-nitrosamides regenerate the parent amide (equation 3) in quantitative
yield ($\underline{7},\underline{8}$). This implies that N-nitrosamide formation is reversible and explains why it is beneficial to add one mole of base

(usually NaOAc) when synthesizing N-nitrosamides from amides and
nitrosyl gases (7). Denitrosation should also occur in aqueous
acid together with hydrolysis reactions characteristic of regular
amides. These expectations are borne out by recent studies (9,10)

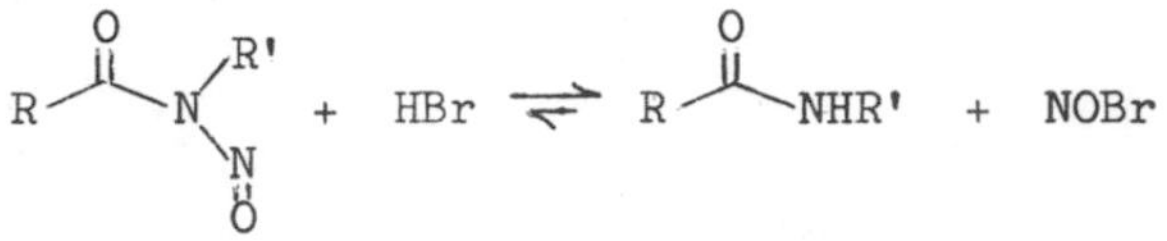

$$R-\overset{\overset{O}{\|}}{C}-\underset{\underset{\overset{\|}{O}}{\overset{N}{|}}}{N}-R' \;+\; HBr \;\rightleftharpoons\; R-\overset{\overset{O}{\|}}{C}-NHR' \;+\; NOBr \qquad (3)$$

which show that typical N-nitrosamides undergo concurrent denitrosation and deamination in aqueous H_2SO_4 and $HClO_4$ (Scheme 1). The
extent of each reaction depends on both the structure of the

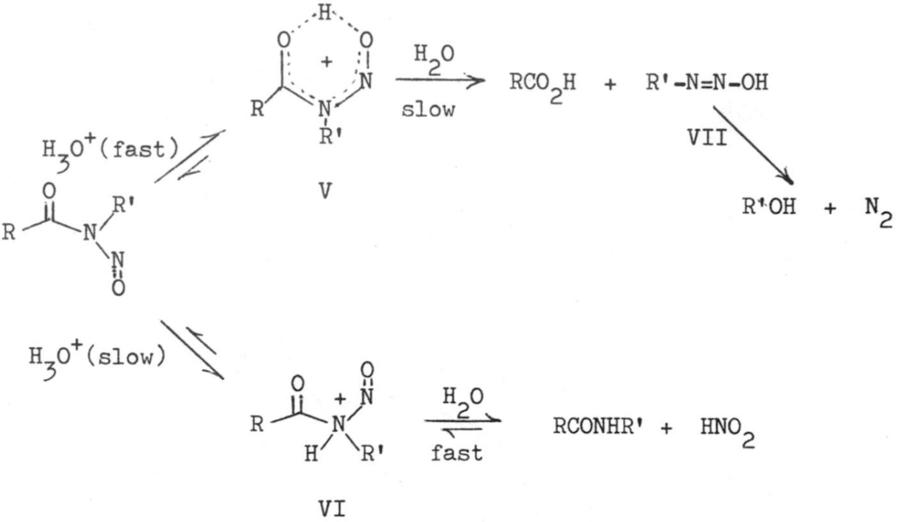

Scheme 1. Concurrent acid catalysed denitrosation and deamination
(hydrolysis) of N-nitrosamides

N-nitrosamide and the solvent acidity, but generally denitrosation
is more strongly acid-catalysed: hence, it is a minor reaction
(<10%) above pH 2 and the major pathway in concentrated acid.
This difference is exemplified by the variation of the pseudo
first order rate coefficients (Rate = k [Substrate]) for the deamination and denitrosation of N-n-butyl-N-nitroso acetamide at $25°C$
with $[H_2SO_4]$ shown in Figure 1. Denitrosation becomes the domin-
ant pathway above 5M. H_2SO_4 (9), but the crossover acidity is much
lower (ca.0.5M. H_2SO_4) for N-nitroso-2-pyrrolidone (10).
The acid catalysed denitrosation of N-nitrosamides is subject
to a substantial solvent deuterium isotope effect $[k_o^{NO}(H)/k_o^{NO}(D)$
ca. 1.9 for N-n-butyl-N-nitroso acetamide in sulphuric acids at
$25°C.(9)$; $k_o^{NO}(H)/k_o^{NO}(D)$ ca. 1.6 for N-nitroso-2-pyrrolidone in

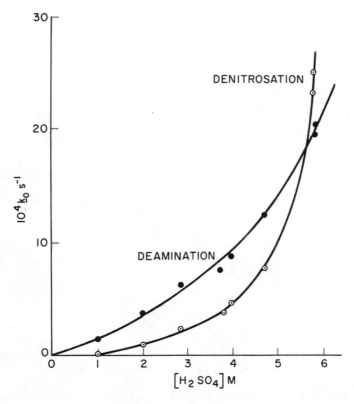

Figure 1.   Variation in the rates of denitrosation and deamination of N-n-butyl-N-nitrosoacetamide with [H_2SO_4] at 25°C (9).

perchloric acids at 25°C.(11)] which implies that the proton
transfer from the solvent to the amino-N-atom is rate limiting as
shown in Scheme 1. This conclusion is supported by the observation of general acid catalysis for the denitrosation of N-nitroso-
2-pyrrolidone (10). With the exception of one or two recent examples (12,13,14,), proton transfer to oxygen and nitrogen bases
is usually rapid, more often than not proceeding on encounter (15).
The exceptional behaviour for N-nitrosamides can be attributed to
the strong electron-withdrawing ability of the nitroso substituent.
The mechanism for the acid catalysed deamination of N-nitrosamides
seems to have a structural dependence not evident for denitrosation. Thus, general acid catalysis is observed (10) for N-nitroso-
2-pyrrolidone, but not for N-n-butyl-N-nitroso acetamide. This
difference has been interpreted as evidence that the deamination
of N-n-butyl-N-nitroso acetamide proceeds via an O-conjugate acid
intermediate V as shown in Scheme 1, whereas both deamination and
denitrosation of N-nitroso-2-pyrrolidone proceed via a common N-
conjugate acid intermediate analogous to VI.

   As far as the biological behaviour of N-nitrosamides is concerned, the decomposition reactions in aqueous acid are representative of those likely to occur in an acidic stomach. Thus N-nitrosamides could act as in vivo transnitrosating and alkylating
agents. Of the two, transnitrosation is likely to be least important, because it must be indirect (ie. proceeding via release of
HNO$_2$) and denitrosation is a minor reaction at pH $>$ 2. Deamination,
however, results in the generation of a diazohydroxide VII, the
putative ultimate carcinogen derived metabolically from N-nitrosamines. Diazohydroxides are powerful alkylating agents which, in
principle, may interact with stomach tissue.

Decomposition by Basic and Nucleophilic Catalysts

   By analogy with regular amides (16), N-nitrosamides should
also undergo hydrolysis by a base catalysed addition-elimination
pathway to yield a diazohydroxide and a carboxylate ion. These
reactions should be particularly favoured by the enhanced leaving
ability of the N-nitrosamino fragment, and they probably explain
the ready generation of diazo alkanes from N-nitrosamides (equation 4) on treatment with strong alkali (17). It is of considerable

$$R-\overset{\overset{O}{\parallel}}{C}-N\overset{CH_2R'}{\underset{\underset{O}{\overset{\parallel}{N}}}{}} \xrightarrow{HO^-} RCO_2^- + R'CH_2N=N-OH \longrightarrow R'CH=\overset{+}{N}=\overset{-}{N} + H_2O \quad (4)$$

interest to establish both the mechanism of these reactions and
their propensity under cellular conditions.

   We have shown (10) that base (or nucleophilic) catalysed
hydrolysis contributes to the decomposition of N-nitroso-2-pyrrol-
idone at 25°C in Cl$_2$HCCO$_2$H and ClH$_2$CCO$_2$H buffers at pH 1-3, and

is the sole decomposition pathway in HCO_2H and CH_3CO_2H buffers at
pH 3.7 - 5.4. Evidence to this effect for CH_3CO_2H buffers is the
correlation of pseudo first order rate coefficients (Rate = k
[Substrate]) with $[CH_3CO_2^-]$ but not $[CH_3CO_2H]$ shown in Figure 2,
and the absence of HNO_2 as a reaction product.

We have also shown (18) that other bases stronger than $CH_3CO_2^-$
(pK_A 4.75) catalyse the decomposition of N-nitroso-2-pyrrolidone
at 25°C. With the exception of imidazole, these reactions follow
uncomplicated second order kinetics (Rate = k_2[Substrate][Base])
and only products of deamination (hydrolysis) are obtained. Generally, k_2 values increase with the base strength of the catalyst
and fit the Brønsted relationship with β = 0.66. However, the absence of significant catalysis by sterically hindered bases (eg.
2,6-lutidine), the strong catalysis by imidazole relative to $HPO_4^=$
$(k_2(\text{Imidazole})/k_2(HPO_4^=)$ = 83) and by hydroxide ion relative to
imidazole $(k_2(HO^-)/k_2(\text{Imidazole})$ = 4,300), and the observation of
second order catalytic terms in imidazole (Rate = [Substrate]
$(k_2[\text{Imidazole}] + k_3[\text{Imidazole}]^2)$ suggests that decomposition by
some (if not all) catalysts involves nucleophilic attack at the
carbonyl C-atom as shown in Scheme 2.

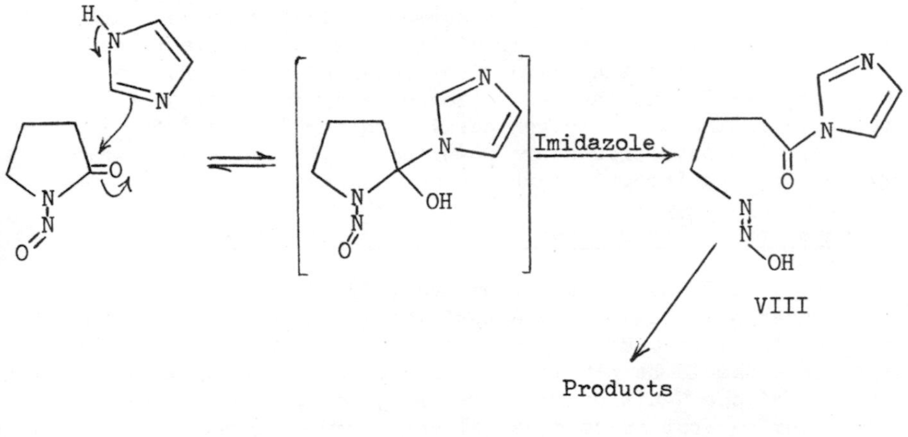

Scheme 2. Imidazole catalysed decomposition of N-nitroso-2-
                                                      pyrrolidone

The tendency for N-nitrosamides to undergo hydrolysis by a
nucleophilic catalysed pathway has been confirmed by studies of
N-alkylnitroso acetamides (19). Results summarised in Table I for
N-n-butyl-N-nitroso acetamide show that its decomposition is also
subject to steric constraints (2,6-lutidine≪pyridine) and is best
effected by strong nucleophiles (eg. imidazole, thiols) irrespective of their base strength (pK_A). Further, the second order dependence on [Imidazole] is more clearly defined for the decomposit-

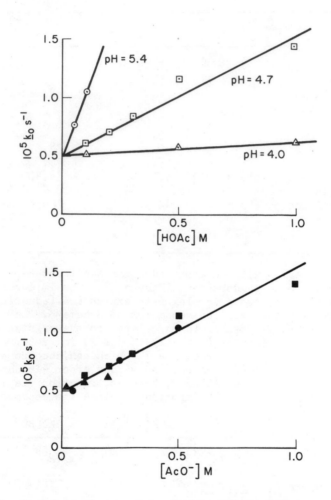

Figure 2. Variation in the rate of deamination of N-n-butyl-N-nitrosoacetamide *with [CH₃CO₂H] and [CH₃CO₂⁻] at 25°C.*

Table I

Rate Coefficients for the Decomposition of N-n-Butyl-N-nitrosoacetamide at 25°C.

Catalyst	pK_A	$10^4 k_2$ $M^{-1} s^{-1}$
$CH_3CO_2^-$	4.75	0.114
Pyridine	5.17	0.14
2,6-Lutidine	6.77	No catalysis
N-Methylimidazole	6.95	15.7
Imidazole	7.05	39.7
$HPO_4^=$	7.21	1.35
$HOCH_2CH_2S^-$	9.72	438,000
HO^-	15.75	378,000
Cysteine thiolate	1.8/8.3/10.8	34,400
Glycine	2.34/9.60	0.14
Methionine	2.12/9.28	No catalysis
Lysine	2.16/9.18/10.79	2.75
Arginine	1.82/8.99/12.48	1.17

ion of N-n-butyl-N-nitroso acetamide than N-nitroso-2-pyrrolidone
and is found to be pH dependent (Figure 3). The imidazole catalysed reaction follows a complex rate expression (equation 5)
consistent with the detailed mechanism of Scheme 3. The occurrence
of second order [Imidazole] terms relates to the existence of two
tetrahedral intermediates (probably IX and X) in prototropic equilibrium. At pH ca.6, intermediate IX which can decompose spontaneously to products predominates, and second order [Imidazole]
terms are not observed. At pH 7, however, reaction via intermediate X becomes increasingly important. Since its decomposition to

$$
\text{Rate} = [\text{Substrate}] \frac{k_1[\text{Im}]\left\{k_3 + K'_x k'_2[\text{Im}] + K_x k_2[\text{HO}^-]\right\}}{k_{-1} + k_3 + K'_x k'_2[\text{Im}] + K_x k_2[\text{HO}^-]}
\tag{5}
$$

where $k_1 = k'_1 + k''_1[\text{HO}^-]$ and $K'_x = K_x[\text{HO}^-][\text{ImH}^+]/[\text{Im}]$

products is facilitated by general acid catalysts (including the
imidazolium ion), second order [Imidazole] terms are observed.
This mechanistic interpretation is supported by the isolation of
N-acetylimidazole as a major product and the observation of only
first order catalysis by N-methylimidazole irrespective of the pH
(19).
    Thus, N-nitrosamides also undergo deamination (hydrolysis) by
an addition-elimination pathway involving nucleophilic rather than

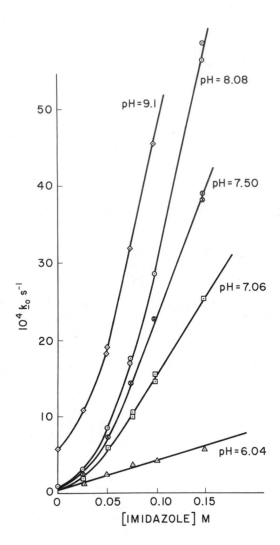

Figure 3. Imidazole catalyzed deamination of N-n-butyl-N-nitrosoacetamide at 25°C.

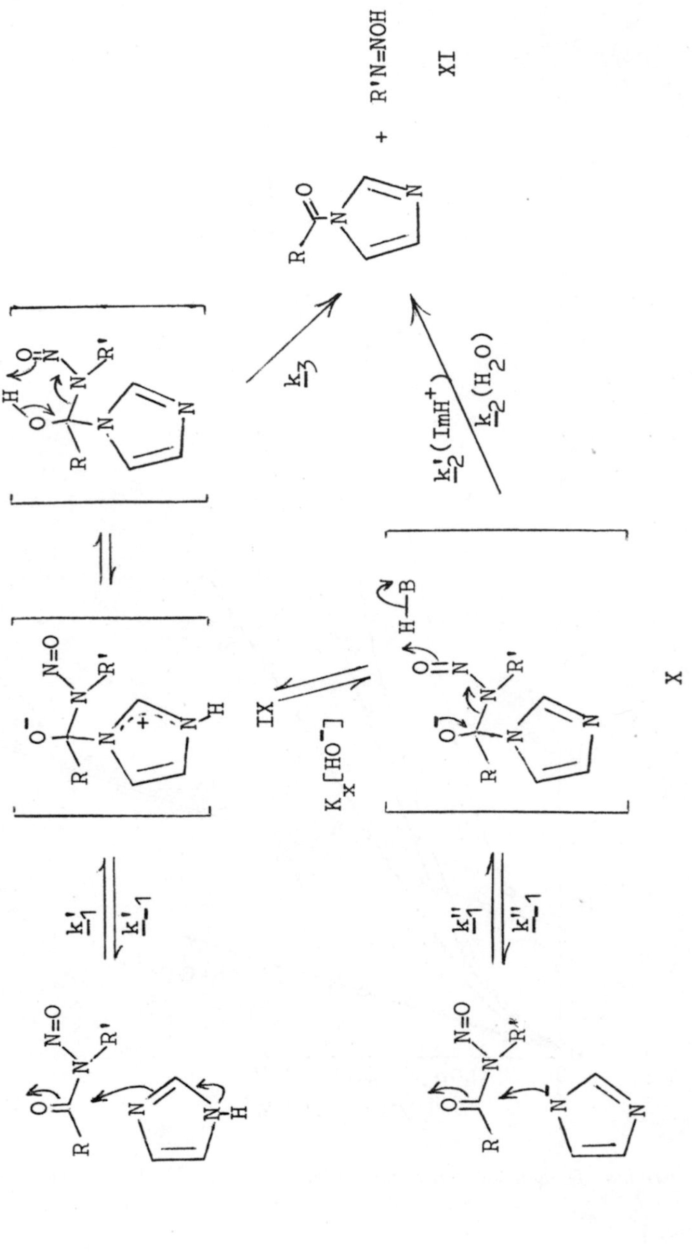

Scheme 3. Imidazole catalysed deamination of N-n-butyl-N-nitroso acetamide

(R = CH$_3$; R' = n-C$_4$H$_9$)

the general base catalysis usually observed for regular amides. This difference can be related to the enhanced leaving ability of the $\underline{N}$-nitrosamino fragment. The reaction proceeds over a wide range of pH ($\underline{ca}$.2-12) and is probably the most important decomposition pathway under cellular conditions. Significantly, it also generates diazohydroxide alkylating agents such as VIII and XI.

Decomposition by Nucleotides and Nucleic Acids

One goal of our current investigation is to show that genetically important cellular constituents can initiate the release of an alkylating agent from the $\underline{N}$-nitrosamide. This might explain how highly reactive diazohydroxide metabolites effect the alkylation of nucleic acids within the cell nucleus. Preliminary evidence (20) summarised in Table II shows that the decomposition of

Table II

Rate Coefficients for the Decomposition of $\underline{N}$-n-Butyl-$\underline{N}$-nitroso acetamide by Nucleotide-5'-monophosphates in O.1M Phosphate Buffers at 25°C.

Catalyst	pH	$10^5 \underline{k}_2$ $M^{-1}s^{-1}$
$HPO_4^=$	-	14
Guanosine	7.23	51
Adenosine	7.26	60
Cytidine	7.34	38.5
Ribose	7.31	35.4
2'-Deoxyguanosine	7.10	14.5
2'-Deoxyadenosine	7.09	18
2'-Deoxycytidine	7.08	13.4
2'-Deoxyribose	7.0	14.5

$\underline{N}$-n-butyl-$\underline{N}$-nitroso acetamide in phosphate buffer at 25°C. is catalysed by nucleotide components. Surprisingly, the strongest catalysis applies to guanosine- and adenosine-5'-monophosphates, but it should be noted that all the $\underline{k}_2$ values cited in Table II are pH dependent. The decomposition of $\underline{N}$-n-butyl-$\underline{N}$-nitroso acetamide in phosphate buffer at 25°C. is also catalysed by calf-liver RNA (Figure 4) to an extent comparable with the catalytic rate coefficients given in Table II (20). Examination of the products from reaction of guanosine- and adenosine-5'-monophosphates with radiolabelled $\underline{N}$-nitroso acetamides in phosphate buffer shows low-level (<2%) alkylation of the purine bases (20). This implies that nucleic acids can decompose $\underline{N}$-nitrosamides and trap the diazohydroxide alkylating agent.

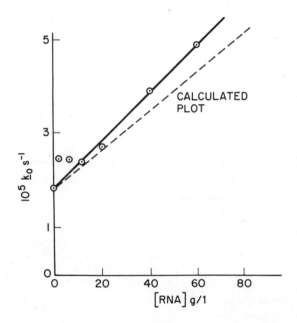

Figure 4.   Decomposition of N-n-butyl-N-nitrosoacetamide *by calf liver RNA in phosphate buffer (pH 7) at 25°C. The dashed line is the calculated rate based on coefficients given in Table II.*

Acknowledgements

   We thank the Cancer Research Campaign and the Science
Research Council for their support of this work.

Literature Cited

1. Magee, P. N.; Montesano, R.; Preussmann, R.; in "Chemical
    Carcinogenesis"; Searle, C. E., Ed.; ACS Monograph 173;
    American Chemical Society: Washington, DC, 1976; p 491.
2. Lawley, P. D.; in "Chemical Carcinogenesis"; Searle, C. E.,
    Ed.; ACS Monograph 173; American Chemical Society: Washington,
    DC, 1976; p 83.
3. Druckrey, H.; Preussmann, R.; Ivankovic, S.; Schmähl, D.
    Z. Krebsforsch. 1967, 69, 103.
4. White, E. H.; Woodcock, D. J.; in "Chemistry of the Amino
    Group"; Patai, S. Ed.; Wiley: New York, 1968; chapter 8.
5. Bieron, J. F.; Dinan, F. J.; in "Chemistry of the Amides";
    Zabicky, J., Ed.; Wiley: New York, 1970; chapter 4.
6. Chow, Y. L.; in "N-Nitrosamines"; Anselme, J-P., Ed.; ACS
    Symposium Series 101; American Chemical Society: Washington,
    DC, 1979; p 18.
7. White, E. H. J. Am. Chem. Soc. 1955, 77, 6008.
8. Huisgen, R. Annalen, 1951, 573, 163.
9. Berry, C. N.; Challis, B. C. J.C.S. Perkin II, 1974, 1638.
10. Challis, B. C.; Jones, S. P. J.C.S. Perkin II, 1975, 153.
11. Challis, B. C.; Jones, S. P.; unpublished results.
12. Hibbert, F.; Awwal, A. J.C.S. Perkin II, 1978, 939.
13. Freeman, K. A.; Hibbert, F.; Hunte, K. P. P. J.C.S. Perkin II,
    1979, 1237.
14. Bade, M. L. J. Am. Chem. Soc. 1971, 93, 949.
15. Eigen, M. Angew. Chem. Intnl. Edn. 1964, 3, 1.
16. Challis, B. C.; Challis, J. A.; in "Comprehensive Organic
    Chemistry, Vol. 2"; Sutherland, I. O. Ed.; Pergamon: Oxford,
    1979; p 957.
17. Lobl, T. J. J. Chem. Educ. 1972, 49, 730.
18. Challis, B. C.; Jones, S. P. J.C.S. Perkin II, 1979, 703.
19. Berry, C. N.; Challis, B. C.; Gribble, A. D. J.C.S. Perkin II,
    to be published.
20. Challis, B. C.; Gribble, A. D.; unpublished results.

RECEIVED August 10, 1981.

CHEMISTRY OF
FORMATION AND BLOCKING

Gas Phase Reactions of N,N-Dimethylhydrazine with Ozone and NO_x in Simulated Atmospheres

Facile Formation of N-Nitrosodimethylamine

WILLIAM P. L. CARTER, ERNESTO C. TUAZON, ARTHUR M. WINER and J. N. PITTS, JR.

Statewide Air Pollution Research Center, University of California, Riverside, CA 92521

The gas phase reactions of unsymmetrical di-methylhydrazine (UDMH) with ozone and with NO have been studied under simulated atmospheric conditions in a 30,000-liter outdoor Teflon chamber using in-situ long path Fourier transform infrared spectros-copy. The reaction of UDMH with O_3 (at ppm levels) in air goes to completion in less than 2 minutes forming N-nitrosodimethylamine as the predominant product ($\gtrsim 60\%$ yield) along with smaller amounts of HCHO, H_2O_2, and HONO. In pure air, UDMH undergoes a slow dark reaction with the only product detected being NH_3 at low yields. The UDMH decay rate is unaffected by the addition of NO, but instead of NH_3, formation of HONO, N_2O, and one or more unidentified products are observed. When a UDMH-NO-air mixture is irradiated, N-nitrosodimethylamine, N-nitrodi-methylamine, HCHO, N_2O, and unknown product(s), believed to be primarily N-nitrosodimethylhydrazine, are formed. The unknown products are formed only during the initial stages, and the nitramine is formed only during the later stages of the irradi-ation. Possible mechanisms which are consistent with the observed data for the UDMH + O_3 and UDMH + NO systems are discussed.

Until recently the atmospheric chemistry of nitrogen-containing compounds such as the hydrazines, which are widely used as fuels in military and space vehicles, has received comparatively little attention. N,N-dimethylhydrazine (also UDMH ≡ unsymmetrical dimethylhydrazine) is used in liquid-fueled rockets, and thus there is a possibility that its use, storage, and handling could result in its release in the atmosphere. There is evidence that N-nitrosodimethylamine can be among its oxidation products (1-3) but, other than studies of the gas phase

0097-6156/81/0174-0117$05.00/0

stability of UDMH at relatively high concentrations (2-4), there
have been relatively few investigations of its atmospheric
reactions.

In view of its potential for nitrosamine formation, a more
detailed knowledge of the atmospheric reactions and products of
UDMH is clearly desirable. In order to provide such data for
UDMH and other hydrazines we have studied their dark reactions in
air, with and without added O_3 or NO, and have investigated their
atmospheric photooxidation in the presence of NO (5,6). In this
paper, we report the results we have obtained to date for UDMH.

Experimental

The facility, methods of procedure, and materials employed
in this study have been discussed in detail elsewhere (5,6), and
are only briefly described here. The outdoor reaction chamber
employed in this study consists of a 30,000-liter FEP-type Teflon
bag of triangular cross section held semi-rigidly by a framework
of steel pipes. The chamber houses a set of multiple-reflection
optics (capable of pathlengths in excess of 1 km) which is interfaced to a Midac interferometer and associated data system.

Measured amounts of the reactants in glass bulbs (for UDMH
and O_3) or glass syringes (for NO) were flushed into the chamber
by a stream of N_2 through a disperser tube which runs the length
of the chamber. Fans attached to the Teflon-coated aluminum end
panels provide rapid mixing of reactants. For dark experiments,
the chamber was covered with an opaque tarpaulin; the latter
could be removed readily for sunlight irradiations. The chamber
was thoroughly flushed with clean ambient air after each run, and
was additionally purged and filled with a total of five volumes
of purified air (7) prior to each experiment.

Nitric oxide (NO) and nitrogen dioxide (NO_2) were monitored
by a Bendix chemiluminescence instrument. However, since the
"NO_2" readings of this type of instrument are known to include
the contribution of HNO_3, HONO, PAN, and other organic nitrates
(8), we report these readings as "gaseous nitrate."

The growth and decay of all other species (including O_3)
were monitored by Fourier transform infrared (FT-IR) spectroscopy
at a total pathlength of 460 meters and a spectral resolution of
1 cm^{-1}. At this pathlength, the intense absorptions of H_2O and
CO limit the usable IR spectral windows to the approximate
regions 750-1300, 2000-2300, and 2400-3000 cm^{-1}. Each spectrum
(700-3000 cm^{-1}) was adequately covered by the response of the
Cu:Ge detector. Approximately 40 seconds were required to
collect the 32 interferograms co-added for each spectrum.

Reactant and product analyses were obtained from the intensities of infrared absorption bands by successive subtraction of
absorptions by known species. Low noise reference spectra for
UDMH and several reaction products were generated for this purpose in order to minimize the increase in the noise level of the
residual spectrum with each stage of subtraction.

Results and Discussion

   Dark Decay of UDMH in Air. UDMH was observed to undergo a
gradual dark decay in the 30,000-liter Teflon chamber at a rate
which depended on humidity. Specifically, at 41°C and 4% RH the
observed UDMH half-life was ~9 hours (initial UDMH ≈ 4.4 ppm) and
at 40°C and 15% RH, the half-life was ~6 hours (initial UDMH
≈2.5 ppm). The only observed product of the UDMH dark decay was
NH_3, which accounted for only ~5-10% of the UDMH lost. In particular, no nitrosamine, nitramine, or hydrazone were observed.
Formaldehyde dimethylhydrazone was observed in previous studies
which employed higher UDMH concentrations and reaction vessels
with relatively high surface/volume ratios (2,4).
   The mechanism of the UDMH dark decay is unknown, but it is
presumed to be heterogeneous in nature. It is probably not wall
adsorption, since much slower decay rates were observed previously in the absence of O_2 (4).

   Dark Reaction of UDMH with O_3. When O_3 was injected into
UDMH-air mixtures, consumption of UDMH and O_3 was "instantaneous"
and formation of N-nitrosodimethylamine was immediately observed.
The reaction was complete within 2 minutes, by which time either
the hydrazine or the O_3 was totally consumed. Figure 1 shows IR
spectra before and 2 minutes after ~2 ppm of O_3 was injected into
air containing ~2 ppm of UDMH. The nitrosamine is positively
identified by its IR absorptions at 1296, 1016, and 848 cm^{-1}.
Also formed in the UDMH-O_3 system, but in lesser yields, were
HCHO, H_2O_2, and HONO.
   The experimental conditions and products observed at
selected times in the UDMH + O_3 experiments are shown in Table I.
In general, the nitrosamine yields ranged from ~60% when the
reaction was carried out in a slight excess of UDMH to ~100% when
O_3 was in excess. The HCHO, H_2O_2, and HONO yields were ~13%,
10%, and 3% of the reacted UDMH when near-stoichiometric mixtures
were employed. NO_2 was probably also formed to some extent, but
the interference by other nitrogenous compounds on the NO_2
analyzer (8) made the determination of its exact yield uncertain.
The observed O_3/UDMH stoichiometry was ~1.5:1.
   The data are consistent with the following mechanisms for
the UDMH-O_3 reaction:

$$(CH_3)_2N-NH_2 + O_3 \longrightarrow (CH_3)_2N-\overset{\bullet}{N}H + OH + O_2 \qquad (1)$$

$$(CH_3)_2N-NH_2 + OH \longrightarrow (CH_3)_2N-\overset{\bullet}{N}H + H_2O \qquad (2)$$

$$(CH_3)_2N-\overset{\bullet}{N}H + O_3 \longrightarrow (CH_3)_2N-N\overset{O\bullet}{\underset{H}{\diagup}} + O_2 \qquad (3)$$

Table I. Experimental Conditions and Concentrations of Reactants and Products at Selected Times in the UDMH + O₃ Dark Experiments

Conditions			Elapsed	Concentrations (ppm)^a								
Exp. No.	T (avg)	RH (avg) (%)	Time (min)	UDMH	O_3	$(CH_3)_2NNO$	H_2O_2	HCHO	NH_3	HONO	"Gaseous Nitrate"^{b}	$(CH_3)_2NNO_2$
1	21	10	-7	1.71					0.116		0.02	
			0		~2.0^c ← INJECTED							
			2	0.386	--	0.81	0.14	0.19	0.106	0.023		--
			61	0.170	--	0.78	0.08	0.08	0.079	0.062	0.92	--
			69	← START OF SUNLIGHT IRRADIATION (1401 PDT, May 9, 1980)								
			79	--	--	0.41	0.07	0.34	0.077	0.19	0.86	0.15
			128^d	--	0.51			0.50	--	--	0.68	0.48
2	21	10	-9	1.93					0.116		0.106	
			0		~2.0^c ← INJECTED							
			3	0.479	--	0.82	0.13	0.17	0.099	--		--
			73	0.293	--	0.77	0.045	0.051	0.069	0.052	0.83	--
			84		~2.0^c ← INJECTED							
			88	--	1.47	1.09	0.13	0.21	--	0.066	1.42	--
			120^e	--	1.09	1.11	0.15	0.22	--	0.058	1.22	--

^{a}Dash means below FT-IR detection sensitivity; blank means no measurement was made.

^{b}"Gaseous nitrate" is total amount of species converted to NO by the molybdenum catalyst of a commercial NO-NO$_x$ analyzer ($\underline{8}$). This includes NO$_2$, HONO and organic nitrates, and possibly nitrosamines and nitramines.

^{c}Calculated amount injected.

^{d}Additional products observed at end of irradiation: CO (0.9 ppm), CH$_3$ONO$_2$ (0.05 ppm), HCOOH (0.043 ppm), N$_2$O (0.035 ppm).

^{e}Additional products detected at end of run: CO (0.3 ppm), N$_2$O (0.033 ppm).

$$(CH_3)_2N\overset{O\cdot}{\underset{H}{\diagdown}} + O_2 \longrightarrow (CH_3)_2N-NO + HO_2 \tag{4}$$

$$HO_2 + HO_2 \longrightarrow H_2O_2 + O_2 \tag{5}$$

Other possible mechanisms have been considered (5), but they either predict formation of products which are not observed, do not explain the observed O_3/UDMH stoichiometry, or are inconsistent with the results of the UDMH-NO stoichiometry and the formation of nitrosamine and H_2O_2 in this system. The other products observed, and the fact that the nitrosamine and H_2O_2 yields are somewhat less than the predicted 100% and 50% of the UDMH consumed, can be attributed to possible secondary reactions of the nitrosamine with the OH radical.

It should be noted that the UDMH + O_3 mechanism is probably quite different from that appropriate for hydrazines with hydrogens on both nitrogen atoms. In our study of the reactions of O_3 with N_2H_4 and monomethyl hydrazine (MMH) (5,6), the data were best explained by assuming the initial hydrazine consumption reactions to be analogous to reactions (1) and (2), but with the N-amino radical formed reacting rapidly with O_2, e.g.,

$$CH_3NH-NH + O_2 \longrightarrow CH_3N=NH + HO_2 \tag{6}$$

to give rise to products other than nitrosamines. Since a reaction analogous to (6) is not possible for UDMH, its mechanism and products are significantly different.

Irradiation of the UDMH + O_3 Reaction Products. One experiment was conducted in which the UDMH + O_3 reaction products (with UDMH in slight excess) were irradiated by sunlight. The results are shown in Table I and Figure 1. It can be seen that rapid consumption of UDMH, the nitrosamine, and HONO occurred, with N-nitrodimethylamine (also dimethylnitramine) and additional formaldehyde being formed. The formation of nitramine upon irradiation of the nitrosamine is consistent with results of previous studies in our laboratories (9,10), and probably occurs as shown:

$$(CH_3)_2N-NO + h\nu \longrightarrow (CH_3)_2N\cdot + NO \tag{7}$$

$$(CH_3)_2N\cdot + NO_2 \xrightarrow{M} (CH_3)_2N-NO_2 \tag{8}$$

It should be noted that the immediate formation of the nitramine in the photolysis of the UDMH + O_3 products indicates the presence of NO_2 in that mixture.

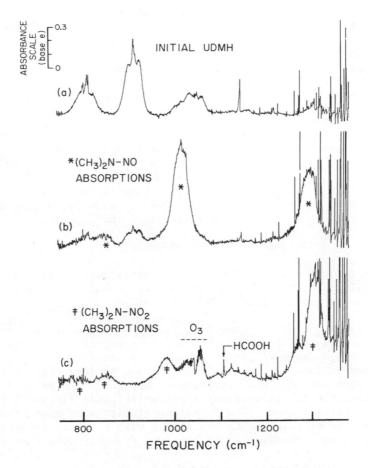

Figure 1.   Selected IR spectra from the UDMH + O₃ experiments (NH₃ absorptions subtracted): (a) UDMH prior to O₃ injection; (b) 2 min after O₃ injection; (c) approximately 1 h into sunlight irradiation of reaction mixture.

Dark Decay of UDMH in the Presence of NO. When 1.3 ppm of
UDMH in air was reacted in the dark with an approximately equal
amount of NO, ~0.25 ppm of UDMH was consumed and formation of
~0.16 ppm HONO and ~0.07 ppm N_2O was observed after ~3 hours.
Throughout the reaction, a broad infrared absorption at ~988 cm^{-1}
corresponding to an unidentified product(s), progressively grew
in intensity. The residual infrared spectrum of the unknown
product(s) is shown in Figure 2a. It is possible that a very
small amount (≲0.03 ppm) of N-nitrosodimethylamine could also
have been formed but the interference by the absorptions of the
unknown product(s) made nitrosamine (as well as nitramine) detection difficult. No significant increase in NH_3 levels was
observed, in contrast to the UDMH dark decay in the absence of
NO. Approximately 70% of the UDMH remained at the end of the 3-
hour reaction period; this corresponds to a half-life of ~9 hours
which is essentially the same decay rate as that observed in the
absence of NO.

As with the UDMH-air dark decay, the UDMH-NO-air decay reactions are unknown and probably heterogeneous in nature. The fact
that the presence of NO does not significantly change the UDMH
decay rate but changes the products formed suggests that the
initiation reaction is the same in both cases, but that the NO
reacts with the intermediates formed. Additional investigation
is required to characterize this process.

Irradiation of UDMH-NO-Air Mixtures. One experiment was
performed in which approximately equimolar amounts of NO and
UDMH were irradiated for 2.5 hours. The reactant and product
concentrations are summarized in Table II and are shown graphi-
cally for selected species in Figure 3. The UDMH-NO irradiation
experiment can be considered to have three stages as summarized
below:

(1) The first ~30 minutes of the irradiation is characterized by relatively high NO levels and relatively slow decay of
UDMH. During this period, formation of both the nitrosamine and
the unknown product(s) is significant, but only minor formation
of the nitramine occurs. Although NO was consumed, relatively
little NO_2 was formed, indicating that (unlike most photochemical
smog systems) NO to NO_2 conversion was not occurring to a
significant extent. HONO levels remained remarkably high,
despite its known rapid photolysis in sunlight (11), thus
suggesting that HONO formation was also rapid. A typical residual infrared spectrum taken during this period of the irradiation
is shown in Figure 2b.

(2) Following the above initial period, the remaining UDMH
suddenly dropped to concentrations below the FT-IR detection
limit, formation of the unknown product(s) stopped, while formation of the nitrosamine continued and formation of the nitramine
began. The rate of NO consumption also increased, although not
as much as that of UDMH.

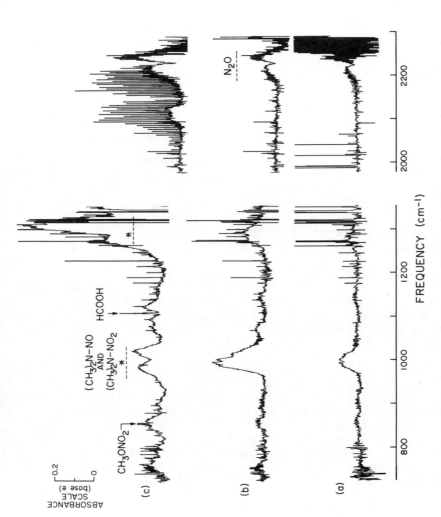

Figure 2. Selected residual IR spectra from the UDMH + NO experiments: (a) spectrum of un-known(s) formed in approximately 3 h from UDMH + NO in the dark; (b) residual spectrum at 19 min into the UDMH + NO irradiation; (c) residual spectrum at 200 min into the irradiation.

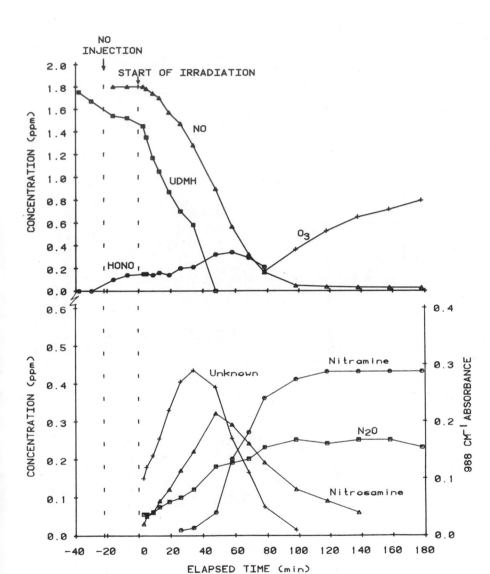

Figure 3.   Concentration–time profiles for reactants and selected products in the UDMH–NO irradiation.

Table II. Reactant and Product[a] Concentrations for the UDMH-NO Irradiation[b]

Irrad. Time (min)	CONCENTRATION (ppm)[c]											~988 cm^{-1} absorbance (base e)
	UDMH	NO	Me$_2$NNO	Me$_2$NNO$_2$	HONO	N$_2$O	"Gaseous Nitrate"[d]	HCHO	NH$_3$	NHO$_3$	O$_3$	
-7	1.52	1.80	--	--	0.14	--	0.14	--	0.143	--	--	--
3	1.45	1.80	0.03	--	0.15	0.055	0.15	--	0.147	--	--	0.10
5	1.35	1.78	0.05	--	0.15	0.055	0.14	--	0.141	--	--	0.12
9	1.17	1.74	0.06	--	0.14	0.061	0.15	0.070	0.136	--	--	0.14
13	1.05	1.70	0.09	--	0.16	0.074	0.15	0.093	0.131	--	--	0.17
19	0.87	1.57	0.12	--	0.14	0.088	0.17	0.15	0.126	--	--	0.22
26	0.70	1.47	0.17	0.014	0.20	0.099	0.22	0.20	0.119	--	--	0.27
34	0.58	1.28	0.22	0.02	0.21	0.12	0.30	0.45	0.113	--	--	0.29
48	--	0.895	0.32	0.06	0.32	0.18	e	0.67	0.074	--	--	0.26
58	--	0.565	0.29	0.20	0.34	0.19	0.82	0.77	0.035	--	--	0.17
68	--	0.320	0.24	0.27	0.29	0.20	0.95	0.82	--	--	--	0.11
78	--	0.162	0.19	0.36	0.21	0.23	1.05	0.74	--	0.041	0.170	0.05
98	--	0.045	0.12	0.41	--	0.25	0.96	0.60	--	0.140	0.364	0.01
118	--	0.035	0.09	0.43	--	0.24	0.84	0.55	--	0.215	0.528	--
138	--	0.030	0.06	0.43	--	0.25	0.73	0.51	--	0.265	0.651	--
158	--	0.028	--	0.43	--	0.25	0.60	0.48	--	0.277	0.717	--
178	--	0.026	--	0.43	--	0.23	0.60	--	--	0.290	0.796	--

[a] Other products observed at 178 min. CH$_3$ONO$_2$ = 0.05 ppm, HCOOH = 0.07 ppm.

[b] T(avg.) = 38°C, RH(avg.) = 4%, irradiation began at 1202 PDT on May 13, 1980.

[c] Dash means below FT-IR detection sensitivity.

[d] Species converted to NO by the molybdenum catalyst of the commercial NO-NO$_2$ analyzer (8). These include NO$_2$, HONO, organic nitrates, and possibly nitrosamines and nitramines.

[e] Data not available.

(3) After the UDMH was consumed, the nitrosamine was converted to the nitramine, the unknown product(s) was consumed, the NO and HONO levels dropped, and eventually O_3 formation took place. The reactions occurring during this stage are probably similar to those which occur during the photolysis of the UDMH-O_3 products discussed earlier. A typical residual infrared spectrum taken during this period of the irradiation is shown in Figure 2c.

In the initial stage of the irradiation, the O_3 level is suppressed by NO, and thus reaction with OH (reaction 2) is probably the process which consumes UDMH. The N-amino radical formed could react with O_2, NO, NO_2, or O_3:

$$(CH_3)_2N\text{-}\overset{\bullet}{N}H + O_2 \xrightarrow{M} (CH_3)_2N\text{-}N\overset{\overset{O_2^{\bullet}}{\diagup}}{\diagdown}_H \qquad (9)$$

$$(CH_3)_2N\text{-}\overset{\bullet}{N}H + NO \xrightarrow{M} (CH_3)_2N\text{-}N\overset{\overset{NO}{\diagup}}{\diagdown}_H \qquad (10)$$

$$(CH_3)_2N\text{-}\overset{\bullet}{N}H + NO_2 \xrightarrow{M} (CH_3)_2N\text{-}N\overset{\overset{NO_2}{\diagup}}{\diagdown}_H \qquad (11)$$

$$(CH_3)_2N\text{-}\overset{\bullet}{N}H + O_3 \rightarrow \rightarrow (CH_3)_2N\text{-}NO + HO_2 \qquad (3,4)$$

Reaction (9) is unlikely because reaction of amino radicals with O_2 has been shown to be slow (12,13); if it occurred, NO to NO_2 conversion via

$$(CH_3)_2N\text{-}N\overset{\overset{O_2^{\bullet}}{\diagup}}{\diagdown}_H + NO \longrightarrow (CH_3)_2N\text{-}N\overset{\overset{O^{\bullet}}{\diagup}}{\diagdown}_H + NO_2 \qquad (12)$$

would be expected, which is contrary to what is observed. On the other hand, even though O_3 levels are suppressed by NO, reaction (3) still may be occurring since O_3 is regenerated by rapid NO_2 photolysis

$$NO_2 + h\nu \longrightarrow O + NO \qquad (13)$$

$$O + O_2 + M \longrightarrow O_3 + M \qquad (14)$$

and is thus in a photostationary state governed by $[NO_2]/[NO]$. The formation of nitrosamine in the initial stages of the run suggests that reaction (3) is indeed occurring to some extent.

The above mechanism suggests that the unknown product(s) is nitrosohydrazine and/or nitrohydrazine. The position of the unknown IR absorption at ~988 cm^{-1} is not inconsistent with the expected N-N stretching frequency for these compounds. However, the positions and contours of other weaker absorptions, which

may provide further information as to the identity of the unknown
compound(s), are less well-defined due to the number of successive subtraction of overlapping known bands and the strong interferences of background H_2O absorptions. The ~988 cm^{-1} band itself is likely to be a composite, with the high frequency shoul-
der seen to be somewhat variable in content at various stages of
the reactions.

Like nitrosamines, nitrosohydrazines probably undergo rapid
photolysis in sunlight,

$$(CH_3)_2N-N\overset{NO}{\underset{H}{\diagup}} + h\nu \longrightarrow (CH_3)_2N-\overset{\bullet}{N}H + NO \qquad (15)$$

which can account for the rapid disappearance of the unknown once
the UDMH is consumed. On the other hand, the nitrohydrazine,
like nitramines, should be more stable under irradiation than is
the unknown product observed here. Thus the unknown product is
probably primarily nitrosohydrazine. Consistent with this conclusion is the fact that the NO_2 levels are quite low compared to
that of NO during the initial stage, as indicated by the low
"gaseous nitrate" readings (Table II) and the low amount of
nitramine formed.

The relatively high concentrations of HONO observed during
the initial period can be attributed to high OH and NO levels,
with HONO being in photostationary state due to its rapid
photolysis (11).

$$OH + NO \overset{M}{\longrightarrow} HONO \qquad (16)$$

$$HONO + h\nu \longrightarrow OH + NO \qquad (17)$$

From the prevailing NO and HONO levels occurring during this
period of the irradiation, the HONO photolysis rate (11,14), and
the rate constant for the OH + NO reaction (15), we estimate that
steady state OH levels of ~2 x 10^7 molecule cm^{-3} were present.
From this OH radical concentration and assuming an UDMH + OH rate
constant similar to those observed for N_2H_4 and MMH (5,16), we
calculate a UDMH decay rate which is in reasonable agreement with
what is observed. Thus, the HONO level measured during the
initial period is entirely consistent with our assumed mechanism.

The sudden consumption of the remaining UDMH, and the
increased relative importance of N-nitrosamine formation at ~30
minutes into the photolysis can be rationalized by assuming that
at that time the $[NO_2]/[NO]$ ratio, and thus the photostationary
state $[O_3]$, has become sufficiently high that O_3 may be reacting
with the hydrazine directly, and that reaction (3) begins to
dominate over reaction (10). This results in higher rates of
UDMH consumption by the OH radicals formed in the UDMH + O_3 reaction (1), and by the OH radicals generated by the reaction of NO

with the HO_2 radical (resulting from reactions 3 and 4):

$$HO_2 + NO \longrightarrow OH + NO_2 \qquad (18)$$

Reaction (18) also converts NO to NO_2 which further increases the O_3 levels, thus accelerating the overall process.

The formation of the other products observed during the irradiation can be attributed to secondary reactions. For example, formation of N_2O, HCHO, and traces of CH_3ONO_2 (Table II) may result from the reaction of the nitrosamine with OH:

$$(CH_3)_2N\text{-}NO + OH \longrightarrow \overset{\displaystyle \cdot CH_2}{\underset{\displaystyle CH_3}{\diagdown}}N\text{-}NO + H_2O \qquad (19)$$

$$\overset{\displaystyle \cdot CH_2}{\underset{\displaystyle CH_3}{\diagdown}}N\text{-}NO + O_2 \xrightarrow{\ M\ } \overset{\displaystyle \cdot O_2CH_2}{\underset{\displaystyle CH_3}{\diagdown}}N\text{-}NO \qquad (20)$$

$$\overset{\displaystyle \cdot O_2CH_2}{\underset{\displaystyle CH_3}{\diagdown}}N\text{-}NO + NO \longrightarrow \overset{\displaystyle \dot{O}CH_2}{\underset{\displaystyle CH_3}{\diagdown}}N\text{-}NO + NO_2 \qquad (21)$$

$$\overset{\displaystyle \cdot OCH_2}{\underset{\displaystyle CH_3}{\diagdown}}N\text{-}NO \longrightarrow HCHO + CH_3\dot{N}\text{-}NO \qquad (22)$$

$$CH_3\dot{N}\text{-}NO \longrightarrow CH_3\cdot + N_2O \qquad (23)$$

$$CH_3\cdot \xrightarrow{\ O_2\ } \overset{\displaystyle NO}{\underset{\displaystyle NO_2}{+}} \xrightarrow{\ O_2\ } HCHO + HO_2 \qquad (24)$$

$$CH_3\cdot \xrightarrow{\ O_2\ } \overset{\displaystyle NO}{\underset{\displaystyle NO_2}{+}} \xrightarrow{\ NO_2\ } CH_3ONO_2 \qquad (25)$$

Another probable source of the relatively high yield of N_2O observed in this experiment is reaction of OH with the nitroso-hydrazine:

$$(CH_3)_2N\text{-}N\overset{\displaystyle NO}{\underset{\displaystyle H}{\diagdown}} + OH \longrightarrow (CH_3)_2N\text{-}\dot{N}\text{-}NO + H_2O \qquad (26)$$

$$(CH_3)_2N-\overset{\bullet}{N}-NO \longrightarrow (CH_3)_2N\bullet + N_2O \qquad (27)$$

Secondary product formation is also expected to result from the reaction of OH with the nitramine, but the mechanism and products formed in this case is more uncertain.

Conclusions

The results of the study reported here show clearly that, upon release into the atmosphere, N,N-dimethylhydrazine (UDMH) can be rapidly converted to N-nitrosodimethylamine by its reaction with atmospheric ozone. A similar conclusion can be reached concerning nitrosamine formation from other unsymmetrically disubstituted hydrazines.

Although nitrosamine and nitramine formation from UDMH (and similarly substituted hydrazines) will probably be the major reaction pathway as long as the atmosphere into which the hydrazines are emitted contains some ozone (as does the "natural" troposphere), our results indicate that different products would potentially be formed if these compounds are emitted into polluted atmospheres where O_3 is suppressed by high levels of NO. Under these conditions, although nitrosamine formation appears to occur to some extent, formation of an unknown product (or set of products) is also observed. On the basis of mechanistic considerations we believe this product to be primarily a nitrosohydrazine. Upon photolysis, this compound may give rise to an N-nitrohydrazine, or, when O_3 is present, to the nitrosamine. The toxicity of nitrosohydrazines and nitrohydrazines are unknown, but it seems unlikely that they are innocuous compounds.

Although the results obtained have been useful in indicating the major atmospheric fate of UDMH and similar hydrazines, this study must be considered to be largely exploratory in nature. Additional work is required to obtain more quantitative data under a wider variety of reaction conditions in order to provide a firmer basis upon which to establish the assumed mechanism. Such studies are now underway in our laboratory.

Acknowledgments

The authors wish to gratefully acknowledge the valuable assistance of Mr. Richard Brown in conducting the experiments. Helpful discussions with Dr. Daniel Stone, project officer, and Lt. Col. Michael MacNaughton, project manager, are also acknowledged. This study was supported by funds from the U.S. Air Force Contract No. AF-F08635-78-C-0307.

Literature Cited

1. International Agency for Research on Cancer, Monographs on the Evaluation of Carcinogenic Risk to Man, Vol. 4, p. 137-144, 1974.
2. Urry, W. H.; Olsen, A. L.; Bers, E. M.; Krause, H. W.; Ikoku, C.; Gaibuez, A. "Autooxidation of 1,1-Dimethylhydrazine," NAVEWEPS Report 8798, NOTS Technical Publication 3903, September 1965.
3. Loper, G. L. "Gas Phase Kinetic Study of Air Oxidation of UDMH," in Proceedings of the Conference on Environmental Chemistry of Hydrazine Fuels, Tyndall AFB, 13 September 1977, Report No. CEEDO-TR-78-14, 1970, p. 129.
4. Stone, D. A. "The Vapor Phase Autooxidation of Unsymmetrical Dimethylhydrazine and 50 Percent Unsymmetrical Dimethyl-hydrazine-50 Percent Hydrazine Mixtures," Report No. ESL-TR-80-21, April 1980.
5. Pitts, Jr., J. N.; Tuazon, E. C.; Carter, W. P. L.; Winer, A. M.; Harris, G. W.; Atkinson, R.; Graham, R. A. "Atmospheric Chemistry of Hydrazines: Gas Phase Kinetics and Mechanistic Studies," Final Report, U.S. Air Force Contract No. F08635-78-C-0307, July 31, 1980.
6. Tuazon, E. C.; Carter, W. P. L.; Winer, A. M.; Pitts, Jr., J. N. "Reactions of Hydrazines with Ozone under Simulated Atmospheric Conditions," Environ. Sci. Technol., in press, 1981.
7. Doyle, G. J.; Bekowies, P. J.; Winer, A. M.; Pitts, Jr., J. N. Environ. Sci. Technol., 1977, 11, 45.
8. Winer, A. M.; Peters, J. W.; Smith, J. P.; Pitts, Jr., J. N. Environ. Sci. Technol., 1974, 8, 1118.
9. Tuazon, E. C.; Winer, A. M.; Graham, R. A.; Schmid, J. P.; Pitts, Jr., J. N. Environ. Sci. Technol., 1978, 12, 954.
10. Pitts, Jr., J. N.; Grosjean, D.; Van Cauwenberghe, K.; Schmid, J. P.; Fitz, D. R. Environ. Sci. Technol., 1978, 12, 946.
11. Stockwell, W. R.; Calvert, J. G. J. Photochem., 1978, 8, 193.
12. Lesclaux, R.; Demissy, M. Nov. J. Chemie., 1977, 1, 443.
13. Lindley, C. R. C.; Calvert, J. G.; Shaw, J. H. Chem. Phys. Lett., 1979, 67, 57.
14. Peterson, J. T. "Calculated Actinic Fluxes (290-700 nm) for Air Pollution Photochemistry Application," GVA 600/4-76-025, 1976.
15. NASA Panel for Data Evaluation, "Chemical Kinetic and Hydro-chemical Data for Use in Stratospheric Modeling," Evaluation Number 2, JPL Publication 79-27, April 1979.
16. Harris, G. W.; Atkinson, R.; Pitts, Jr., J. N. J. Phys. Chem., 1979, 83, 2557.

RECEIVED August 10, 1981.

Possible Mechanisms of Nitrosamine Formation in Pesticides

LARRY K. KEEFER

Analytical Chemistry Section, Division of Cancer Cause and Prevention,
National Cancer Institute, Bethesda, MD 20205

The chemists' role in civilization's attempts to prevent cancer by minimizing human exposure to carcinogens can be considered pivotal, because exposure prevention strategies are most simply and systematically
developed from a knowledge of the molecular pathways
by which carcinogens are formed as unwanted environmental contaminants. Examples in which mechanistic
studies have led to substantial reductions of N-
nitroso compound concentrations in commercial herbicides are summarized from the literature. However,
the diversity in plants of possible nitrosating agents,
as well as of nitrosatable substrates, suggests that
additional pesticide-nitrosamine problems might well
remain to be discovered.

In 1976, Ross et al. first reported that the powerful carcinogens, N-nitrosodimethylamine (NDMA) and N-nitrosodi-n-propylamine (NDPA) were found at parts per million (ppm) concentrations
in certain commercial herbicide formulations (1,2). Since then,
scientists have learned that carcinogenic N-nitroso compounds
can be formed in a wide variety of media of interest to those
who manufacture, use, or study pesticide products. Some of these
nitrosation-supporting media are listed in Table I.
These findings posed a question of urgent importance to society: How can we keep these unwanted impurities out of the products of commerce? To answer this question, chemists began studying the mechanisms of nitrosamine formation in the products and
processes involved. If the mechanisms could be understood in sufficient detail, it was reasoned, it should be possible to develop
a strategy for controlling nitrosamine formation and perhaps sup-
pressing it altogether.

Table I. Media in Which Pesticide Nitrosation has been Observed

Medium	Example
Manufacturers' reaction mixtures	Ross et al., 1977 (1,2)
Formulations during storage	Ross et al., 1977 (1,2)
Laboratory model systems	Kearney, 1980 (3)
Soil	Oliver and Kontson, 1978 (4)
In vivo Plants	Schmeltz et al., 1977 (5)
Animals	Eisenbrand et al., 1974 (6)
Microbial culture	Alexander, 1981 (7)
GC Injection port	Fan and Fine, 1978 (8)

These studies led to some important insights regarding the pathways by which nitrosamine contamination can occur in pesticides, many of which have been described in the literature since the problem was discovered. Let us look at selected papers which have provided this information so we can understand the mechanisms involved, and then at some of the strategies for contamination control which have grown out of these principles.

Mechanisms of Nitrosamine Formation

The literature concerning mechanisms of nitrosamine formation in general has been the subject of several excellent reviews, e.g. that of Douglass et al. (9). However, the basic principles of nitrosamine formation will be briefly stated here by way of introduction.

Almost any class of reduced nitrogen compound can serve as the nitrosatable precursor of an N-nitroso compound. To make matters worse, every nitrogen coordination state from primary to quaternary has been converted to a nitrosamine. A glance at the list of some of the known nitrosatable substrates given in Table II illustrates what a wide variety of N-nitroso compounds analysts might expect to encounter in our complex environment.

Table III shows that any of the higher oxidation states of nitrogen can serve as a nitrosating agent. To form a nitrosamine, all that need happen is for a nitrosating agent to encounter a nitrosatable substrate under favorable conditions, which might (but need not) involve acceleration of the reaction rate by one of the chemical or physical agents indicated in Table IV.

Table II.   Nitrosatable Substrates

	Amides
	Amines
Primary	Ammonium compounds
Secondary	Carbamates
Tertiary	Cyanamides
or	Guanidines
Quaternary	Hydrazines
	Hydroxylamines
	Ureas

Table III.   Nitrosating Agents

Example	Nitrogen Oxidation State	Reference
Coordinated NO₃	+5	Croisy et al., 1980 (10)
Nitrogen dioxide	+4	Challis et al., 1978 (11)
Nitrites, nitrosamines, nitro compounds	+3	Douglass et al., 1978 (9)
Nitric oxide	+2	Ragsdale et al., 1965 (12)

Table IV.   Possible Accelerators of Nitrosation

Nucleophiles	Fan and Tannenbaum, 1973 (13)
Electrophiles	Croisy et al., 1980 (10)
Physical stimuli	Challis et al., 1980 (14)
Microbes (dead or alive)	Yang et al., 1977 (15)

With such a diversity of N-nitrosation pathways theoretically possible, it is comforting to note that only a few combinations of circumstances have been implicated in environmental nitrosamine formation thus far. Two of these are so facile and prevalent that, as of 20 years ago, they were the only recognized mechanisms of N-nitrosation. They involve the interaction of di- or trisubstituted ammonia derivatives with a nitrite ion, as illustrated in Figure 1 for the secondary amines, under the catalytic influence of acid. Note the important special cases of nucleophilic displacement of water from the nitrous acidium ion, H_2O-NO^+, by a second nitrite ion to yield N_2O_3 (as in the reaction at the top of Figure 1), and by nitrate (bottom of Figure 1, $X^-=NO_3^-$) to yield N_2O_4. Both of these excellent nitrosating agents have been studied in considerable detail in recent years by Challis and coworkers (11). A closer look at the reactions of N_2O_4 with amines to form nitramines as well as nitrosamines can be found in Figure 2. Several examples are now known (10, 17) in which environmentally important nitrosation apparently occurs by mechanisms other than those shown in Figures 1 and 2, but the acid/nitrite and NO_x mechanisms still appear to be the principal ones.

Established Pathways of Herbicide Nitrosation

How does this understanding of molecular mechanisms of nitrosamine formation help us solve the practical problem of controlling any contamination associated with the production, storage, or use of pesticides? Let us look at two excellent examples of relevant chemical problem-solving.

Nitrosamines in Aromatic Carboxylic Acids. Most of the herbicides originally found by Ross et al. (1,2) to be contaminated were formulations whose active ingredients were carboxylic acids containing no nitrogen at all. A typical active ingredient was 2,4-D, which has the structure shown in Figure 3. How could a compound like this contain NDMA? It turns out that the guilty mixtures were formulated as their dimethylamine salts. Early research showed that the formulations most heavily contaminated were placed in metal cans together with nitrite to keep the containers from corroding (1,2). Since not much acid is required to induce reaction in an amine-nitrite mixture, [none if certain other electrophilic species, possibly even complexes of the transition metals (10) used to make the cans, are present], the formation of NDMA in a mixture like this is not surprising. The postulated reaction involved is shown in Figure 4.

To prevent this contamination, there were several approaches which this mechanistic information suggested. But the one which the manufacturers proposed (1) was to stop using nitrite to control corrosion of the container. Most dimethylamine phenoxyalkanoic acid products now contain non-detectable NDMA (18).

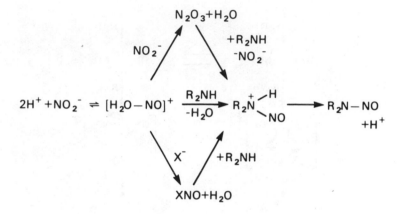

Figure 1. N-Nitrosation mechanisms known in 1961, according to Ridd (16).

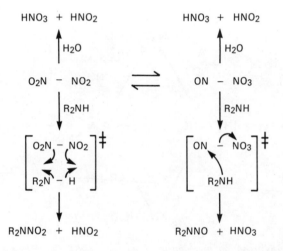

Figure 2.   Mechanisms of N-nitration and N-nitrosation by N₂O₄ according to Challis et al. (11).

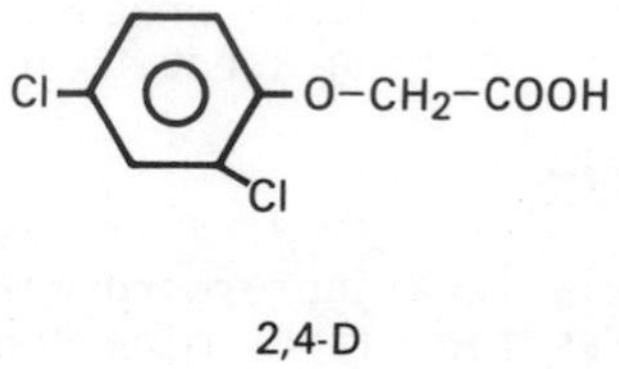

2,4-D

Figure 3. Structure of 2,4-D (2,4-dichlorophenoxyacetic acid).

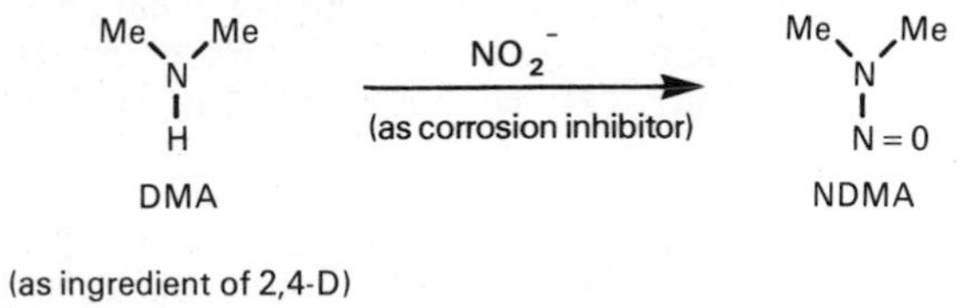

DMA

(as ingredient of 2,4-D)

NDMA

Figure 4. Postulated pathway by which 2,4-D became contaminated with NDMA, as suggested by Ross et al. (2).

   Dinitroaniline Herbicides. Another major contamination
problem was that of Treflan, which was contaminated with over a
hundred ppm of NDPA (2). Mechanistic sleuthing led to the postulate (1,2,3) that the nitrosamine was produced during Treflan
manufacture, as shown in Figure 5. According to this mechanism,
a small amount of nitrite (produced, perhaps, when nitric acid
acted as an oxidizer instead of as a nitrating agent) could
combine under nitration conditions with the HNO_3 to produce
N_2O_4, which then could react with dipropylamine when it was added in the next step. This hypothesis led to an experiment (19)
in which the newly synthesized chloro intermediate shown at the
left of Figure 5a was scrubbed with sodium carbonate under aeration at 70° for one hour (Figure 6). This step successfully reduced NDPA contamination (19), thanks presumably to the oxidative
conversion of residual nitrosating agent to nitrate.

Environmental Significance

   It is now a relatively straightforward matter to analyze commercial pesticide mixtures for nitrosamine contamination as part
of the normal quality control process (2,18), and, as illustrated
above, some of the major contamination problems have begun to be
solved by means of technological improvements. If further research brings about the circumstance in which there is no detectable N-nitroso compound in any pesticide sold anywhere, will
that mean that pesticide use is safe from nitrosamine-formation
problems? I believe that considerably more research must be done
before a responsible answer to this question can be given. To
elaborate on this position, let me focus on the herbicides for
a moment.

   Nitrosatable Substrates in Herbicides. Herbicides contain
many ingredients which are nitrosatable substrates, and they are
are used on plants which contain many more. If any of these compounds contacts a nitrosating milieu, a potentially carcinogenic
N-nitroso compound could result. Several studies have been reported (20-23) which were designed to test this possibility, as
well as the extent to which plants ingested by humans could take
up and store potentially carcinogenic N-nitroso compounds formed
in the plants' environment. Most of this work yielded negative
results.
   For several reasons, this was not surprising. First of all,
most of the data on the disposition of nitrosamines in plant
parts relies on the use of compounds such as NDMA, which, by analogy to its behavior in animals (24), might be expected to have a
short biological half-life in plants; it could be especially difficult to detect such compounds when the analyses are performed
several weeks after applying the precursors to the plants. In
addition, one might expect to find relatively little nitrosation
of dimethylamine in a plant, as such relatively basic amines are

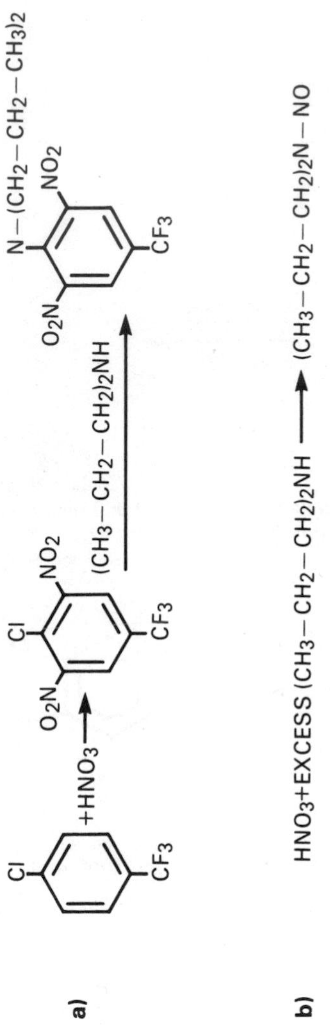

a)

b) $HNO_3 + EXCESS\ (CH_3-CH_2-CH_2)_2NH \longrightarrow (CH_3-CH_2-CH_2)_2N-NO$

Figure 5.   Synthesis of Treflan (a), and the probable source of contamination thereof by NDPA (b), from Kearney (3).

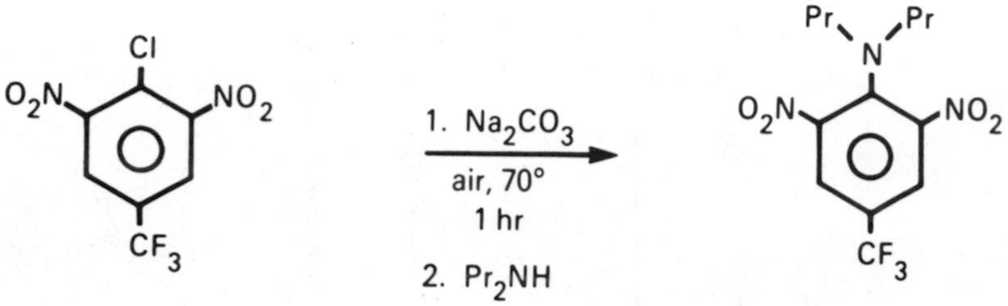

Figure 6.   A troubleshooting step now patented for Treflan manufacture, devised by Cannon and Eizember (19).

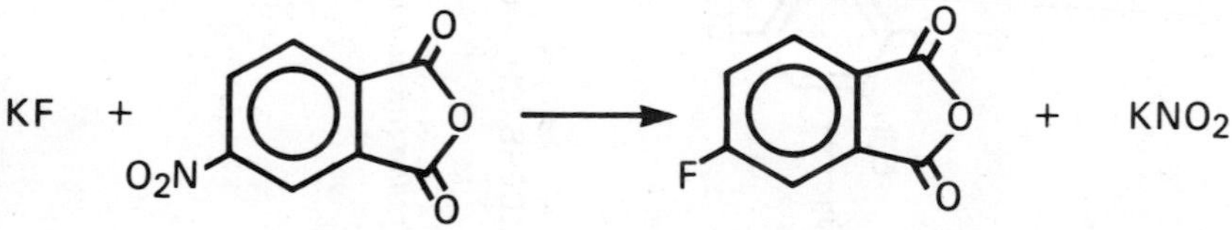

Journal of Organic Chemistry

Figure 7.   Nucleophilic displacement of nitrite from a nitro compound, from Markezich et al. (26).

not rapidly nitrosatable in animals (24). Finally, detectable
nitrosamine formation would not be expected unless at least one
specific nitrosatable substrate is present in relatively high
concentrations; positive results might have been more likely if
these experiments had been conducted on plants which are richer
in nitrosatable alkaloids, or which were more heavily treated
with nitrosatable exogenous amines.

Nitrosating Agents in Plants and Herbicides. As to the
other essential precursor for nitrosamine formation, I believe
that the diversity of nitrosating agent sources in plants may
also prove to be much greater than is currently recognized.
Consider the fact that many organic nitro compounds are used as
herbicide ingredients (18). Looking again to the literature of
mechanistic organic chemistry, it is now established that certain
nitro compounds can convert secondary amines to their carcinogenic
N-nitroso derivatives (25); e.g. Bronopol (2-nitro-2-bromo-1,3-di-
hydroxypropane) has been shown to be an active nitrosating agent
for the alkanolamines used in cosmetic formulations (17). Mech-
anistically, nitro compounds can give rise to free nitrosating a-
gents in several ways: by heterolytic cleavage to nitrite (17,26),
as illustrated in Figure 7; by photolysis, as shown in Figure 8
(27), to nitrogen dioxide, which can dimerize to N_2O_4 or potentially even recombine with the aryl radical from which it was homo-
lytically generated to form an aryl nitrite ester, an as yet
apparently uncharacterized compound type which should be a powerful nitrosating agent; or by autoxidation (28) or microsomal
oxidation (29) to nitrous acid, as illustrated in Figure 9. It
has also been suggested that one fundamental mode of herbicide
action is to inhibit the normal process of nitrite reduction,
so that nitrite builds up in the plant system, leading to toxicity and death (30); while herbicide-treated plants are not normally harvested or used, the possibility of nitrosamine formation
therein should be borne in mind as having potential significance.

Prediction

What if this nitrite buildup, whether by biological or chemical action, occurred in the amine-rich plants mentioned previously? The prediction that significant nitrosamine concentrations
could then be found is supported by the one set of experiments
performed in such a system so far. Schmeltz et al. (5) showed
that tobacco grown in fields on which the antisuckering agent,
MH-30 (which is maleic hydrazide compounded as the diethanolamine
salt) was used had relatively high concentrations of nitrosodiethanolamine (NDElA), while those crops grown without MH-30 application contained much less NDElA. It was concluded, therefore,
that the NDElA arose by in vivo nitrosation of the diethanolamine
in the plant. This finding is illustrated in Figure 10.

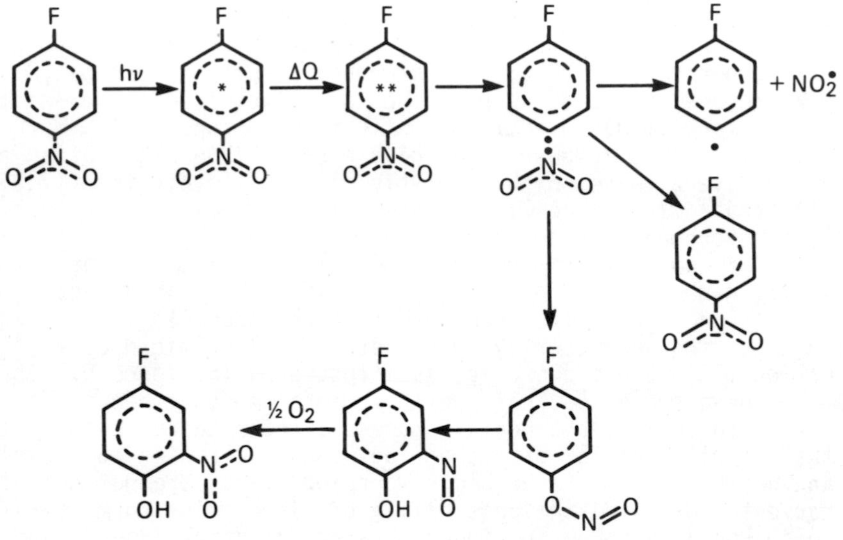

Helvetica Chimica Acta

Figure 8.   Photolytic generation of nitrogen dioxide and an aryl nitrite ester from a nitro compound, as formulated by Lippert and Kelm (27).

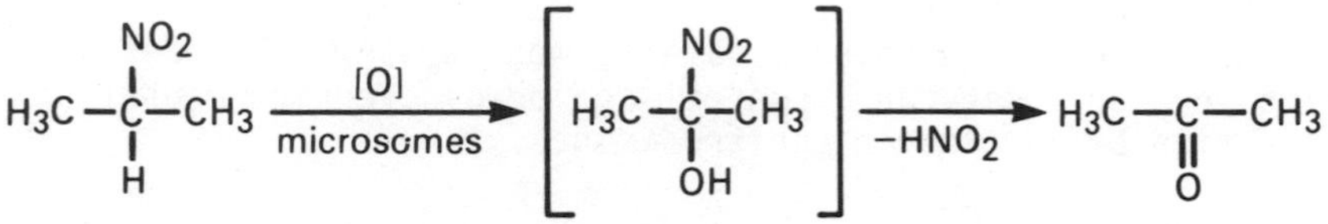

Biochemical Pharmacology

Figure 9. Metabolic generation of nitrite from a nitro compound in vitro, from Ullrich et al. (29).

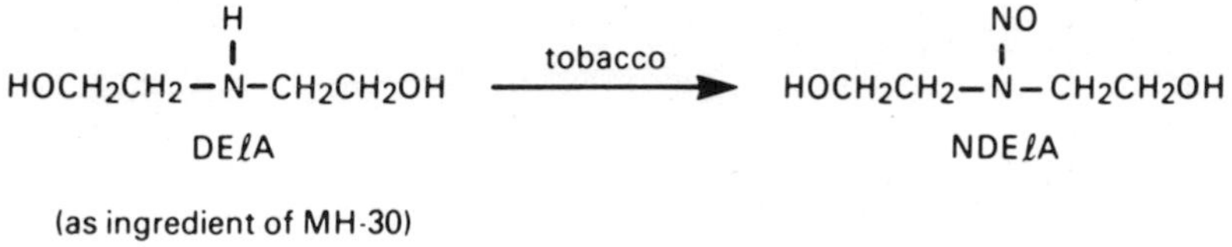

(as ingredient of MH-30)

Figure 10. Mechanism of contamination of tobacco by NDELA, from Schmeltz et al. (5).

Conclusions

Of course, the "prediction" made in the previous paragraph was not really a prediction, because Schmeltz et al. (5) had already published the data which answered the "question" before it was asked. But it did seem to emphasize appropriately one of this paper's two major points -- that no reliable negative generalizations about the extent of environmental nitrosamine formation associated with pesticide use can be made on the basis of data available currently.

As to the second chief point of this paper, let me finish by restating it. If you want to solve a nitrosamine contamination problem, learn everything you can about the chemistry of your product and the process by which it was made, then learn everything you can about the mechanisms of nitrosamine formation. Sifting through the two sets of information for common factors should provide the simplest, safest, most systematic means by which you can understand, control and predict environmental contamination by carcinogenic nitrosamines.

Literature Cited

1. Rawls, R. L. Chem. & Eng. News Sept. 20, 1976, 33-34.
2. Ross, R. D.; Morrison, J.; Rounbehler, D. P.; Fan, S.; Fine, D. H. J. Agric. Food Chem. 1977, 25, 1416-1418.
3. Kearney, P. C. Pure & Appl. Chem. 1980, 52, 499-526.
4. Oliver, J. E.; Kontson, A. Bull. Environm. Contam. Toxicol. 1978, 20, 170-173.
5. Schmeltz, I.; Abidi, S.; Hoffmann, D. Cancer Letters 1977, 2, 125-132.
6. Eisenbrand, G.; Ungerer, O.; Preussmann, R. Fd. Cosmet. Toxicol. 1974, 12, 229-232.
7. Alexander, M. Science 1981, 211, 132-138.
8. Fan, T. Y.; Fine, D. H. J. Agric. Food Chem. 1978, 26, 1471-1472.
9. Douglass, M. L.; Kabacoff, B. L.; Anderson, G. A.; Cheng, M. C. J. Soc. Cosmet. Chem. 1978, 29, 581-606.
10. Croisy, A. F.; Fanning, J. C.; Keefer, L. K.; Slavin, B. W.; Uhm, S.-J., in E. A. Walker, L. Griciute, M. Castegnaro, and M. Börzsönyi, Eds., "N-Nitroso Compounds: Analysis, Formation, and Occurrence"; IARC Scientific Publications No. 31, International Agency for Research on Cancer: Lyon, 1980, pp. 83-92.
11. Challis, B. C.; Edwards, A.; Hunma, R. R.; Kyrtopoulos, S. A.; Outram, J. R., in E. A. Walker, L. Griciute, M. Castegnaro, and R. E. Lyle, Eds., "Environmental Aspects of N-Nitroso Compounds"; IARC Scientific Publications No. 19, International Agency for Research on Cancer: Lyon, 1978, pp. 127-142.

12. Ragsdale, R. O.; Karstetter, B. R.; Drago, R. S. Inorg. Chem. 1965, 4, 420-422.
13. Fan, T-Y.; Tannenbaum, S. R. J. Agric. Food Chem. 1973, 21, 237-240.
14. Challis, B. C.; Outram, J. R.; Shuker, D. E. G., in E. A. Walker, L. Griciute, M. Castegnaro, and M. Börzsönyi, Eds., "N-Nitroso Compounds: Analysis, Formation, and Occurrence"; IARC Scientific Publications No. 31, International Agency for Research on Cancer: Lyon, 1980, pp. 43-56.
15. Yang, H. S.; Okun, J. D.; Archer, M. C. J. Agric. Food Chem. 1977, 25, 1181-1183.
16. Ridd, J. H. Quart. Revs. 1961, 15, 418-441.
17. Schmeltz, I.; Wenger, A. Fd. Cosmet. Toxicol. 1979, 17, 105-109.
18. Zweig, G.; Selim, S.; Hummel, R.; Mittelman, A.; Wright, D. P., Jr.; Law, C., Jr.; Regelman, E., in E. A. Walker, L. Griciute, M. Castegnaro, and M. Börzsönyi, Eds., "N-Nitroso Compounds: Analysis, Formation, and Occurrence"; IARC Scientific Publications No. 31, International Agency for Research on Cancer: Lyon, 1980, pp. 555-562.
19. Cannon, W. N.; Eizember, R. F. U.S. Patent 4,120,905, October 17, 1978; cf. Chem. Abstr. 1979, 90, 54647.
20. Sander, J.; Schweinsberg, F.; LaBar, J.; Bürkle, G.; Schweinsberg, E. GANN Monograph on Cancer Research 1975, 17, 145-160.
21. Dressel, J. Z. Lebensm. Unters.-Forsch. 1977, 163, 11-13.
22. Dressel, J. Qual. Plant.-Plant Foods Hum. Nutr. 1976, 25, 381-390; cf. Chem. Abstr. 1976, 85, 191315.
23. Dean-Raymond, D; Alexander, M. Nature 1976, 262, 394-396.
24. Magee, P. N.; Montesano, R.; Preussmann, R., in Searle, C. E., Ed., "Chemical Carcinogens"; American Chemical Society: Washington, D. C., 1976; pp. 491-625.
25. Fan, T. Y.; Vita, R.; Fine, D. H. Toxicol. Letters 1978, 2, 5-10.
26. Markezich, R. L.; Zamek, O. S.; Donahue, P. E.; Williams, F. J. J. Org. Chem. 1977, 42, 3435-3436.
27. Lippert, E.; Kelm, J. Helv. Chim. Acta 1978, 61, 279-285.
28. Franck, B; Conrad, J.; Misbach, P. Angew. Chem. Intl. Ed. Engl. 1970, 9, 892.
29. Ullrich, V.; Hermann, G.; Weber, P. Biochem. Pharmacol. 1978, 27, 2301-2304.
30. Klepper, L. " A Mode of Action of Herbicides: Inhibition of the Normal Process of Nitrite Reduction", Research Bulletin 259. The Agriculture Experiment Station, University of Nebraska-Lincoln College of Agriculture: Lincoln, 1974.

RECEIVED July 31, 1981.

Formation and Inhibition of
N-Nitrosodiethanolamine in an Anionic
Oil–Water Emulsion

B. L. KABACOFF[1]—Revlon Research Center, Incorporated, 945 Zerega Avenue,
Bronx, NY 10473

R. J. LECHNIR[1] and S. F. VIELHUBER[1]—Raltech Scientific Services,
P.O. Box 7545, Madison, WI 53707

M. L. DOUGLASS[1]—Colgate Palmolive Company, 909 River Road,
Piscataway, NJ 08854

The effect of various types of inhibitors with respect to structure and solubility on the formation of
N-Nitrosodiethanolamine was studied in a prototype
oil in water anionic emulsion. Nitrosation
resulted from the action of nitrite on diethanolamine at pH 5.2-5.4. Among the water soluble inhibitors incorporated into the aqueous phase, sodium bisulfite and ascorbic acid were effective. Potassium
sorbate was much less so. The oil soluble inhibitors
were incorporated into the oil phase of the emulsion.
Of these, only ascorbyl palmitate was effective.
Butylated hydroxyanisole and α-tocopherol were not.
It was found that when α-tocopherol was dispersed in
the aqueous phase in absence of the oil, nitrite was
readily destroyed. It was concluded that the
reduced inhibition obtained with butylated hydroxyanisole and α-tocopherol was at least in part due to
their isolation from the nitrite in water phase. On
the other hand, the reducing portion of the ascorbyl
palmitate molecule could be in the water phase at
the oil/water interface.

During the three years since trace levels of N-Nitrosodiethanolamine (NDElA) were reported (1) in some cosmetic products,
the Cosmetic, Toiletry and Fragrance Association's Nitrosamine
Task Force has conducted several research programs to determine
its sources. In one, analytical methods were developed for determination of NDElA and nitrite in many widely used cosmetic
ingredients. No significant NDElA contamination of the major
product ingredients was found (2,3). Such a result suggests that
the nitrosamine contamination of some products results from in
situ formation during manufacture or storage.

[1] The authors represent the Nitrosamine Task Force of the Cosmetic, Toiletry, and
Fragrance Association, 1110 Vermont Avenue NW, Washington, DC 20005

A research program in progress at Raltech Scientific Services is designed to find inhibitors which will prevent nitrosamine formation in cosmetic products. A review of the literature (4) indicated that the oil phase of emulsions may play an important role in nitrosation chemistry. Thus, results from studies in water alone could be misleading when reduced to practice.

Evidence exists that the relative solubility of amines and inhibitors in heterogeneous oil-water systems could be decisive in formation of nitrosamines and blocking these reactions. Nitrosopyrrolidine formation in bacon predominates in the adipose tissue despite the fact that its precursor, proline, predominates in the lean tissue (5,6,7). Mottram and Patterson (8) partly attribute this phenomenon to the fact that the adipose tissue furnishes a medium in which nitrosation is favored. Massey, et al. (9) found that the presence of decane in a model heterogeneous system caused a 20-fold increase in rate of nitrosamine formation from lipophilic dihexylamine, but had no effect on nitrosation of hydrophilic pyrrolidine. Ascorbic acid in the presence of decane enhanced the synthesis of nitrosamines from lipophilic amines, but had no effect on nitrosation of pyrrolidine. The oil-soluble inhibitor ascorbyl palmitate had little influence on the formation of nitrosamines in the presence or absence of decane.

As in the case of foods and other consumer products, nitrites have been found in cosmetic raw materials (3). Other potential nitrosating species have not been reported in these substances.

In view of these observations, the reaction media chosen for the present study are cosmetic-related emulsion systems containing both oil-soluble and water-soluble secondary amines and blocking agents. Nitrite was chosen as the nitrosating agent. This paper reports initial results on the nitrosation of the water-soluble amine diethanolamine (DEA), chosen because of its ubiquitous presence in cosmetic products. Inhibition of NDElA formation by a select group of water- and oil-soluble inhibitors in an oil-in-water emulsion stabilized by an anionic surfactant is described. The nitrosation of an oil-soluble amine and the effects of nitrosation blocking agents in emulsion systems will be the subject of a future paper.

Experiment Procedures

Composition of Emulsion. The prototype oil/water emulsion described in Table I contained ingredients typical of a large number of cosmetic products, although simplified somewhat to avoid analytical problems. The aqueous phase contained sodium lauryl sulfate (SLS) as emulsifier, 0.2% (19 m$\underline{M}$) DEA as precursor to NDElA and 0.1% benzoic acid as preservative.

The oil phase included a fatty acid, a fatty alcohol and hydrocarbons. The emulsion had a pH of 5.1 to 5.4, exhibited emulsion and pH stability at 37° for at least 21 days, and was shown to be an oil-in-water type by being readily dispersible in water and by its uptake of a water-soluble dye. In inhibition and positive control experiments sodium nitrite was added at a level of 10 ppm (0.217 m$\underline{M}$) NO_2. Thus, nitrite was the limiting reagent giving a theoretical NDElA yield of 29 ppm.

Table I

O/W Emulsion Composition

	Aqueous Phase	Oil Phase
1	1% Sodium Lauryl Sulfate	2% Cetyl Alcohol
2	2% DEA	1% Paraffin Oil
3	1% Benzoic Acid	1% Stearic Acid
4	94.7% Water	

Preparation of Emulsions. The entire aqueous phase was stirred until all solids were dissolved. Sufficient water was withheld from the formulation so small volumes of experimental and control components could be added to emulsion subsamples. Sulfuric acid (1 $\underline{N}$) was added to the aqueous phase to decrease the pH to 5.7. The two phases in separate containers were blanketed with nitrogen, sealed, and heated to 75° in an 80° water bath (about 30 minutes). The hot oil phase was stirred slowly and blanketed with nitrogen, then the hot aqueous phase was quickly added while stirring. The emulsion was blanketed with nitrogen and slowly stirred (about 2 hours) in the stoppered container until ambient temperature ($\sim$25°) was reached. Subsamples of the master batch were removed for the addition of experimental components and stored in 1-oz containers. The containers had been washed with hot tap water, deionized water, and methanol, then dried at 120°.

In experiments with water-soluble inhibitors, the subsample was stirred under nitrogen during post-addition of an aqueous solution of the inhibitor followed by an aqueous sodium nitrite solution. Aliquots were weighed into 1-oz ointment jars, covered with nitrogen, sealed, and stored at 37° for later replicate analyses. Preparation of the positive control subsample was identical except that water was added in place of inhibitor.

Since post-addition of oil-soluble inhibitors would not assure their presence in the oil phase of the emulsion, separate emulsions were prepared.

Each inhibitor was added to the oil phase of each emulsion before heating and combining the water and oil phases. Therefore, positive controls containing no inhibitor could not be made from the same base emulsion. Instead, a separate emulsion batch run simultaneously was used as a positive control.

In the first set of experiments, it was found that two out of the three oil soluble inhibitors had little effect on the amount of NDElA formed. It was considered possible that these had broken down during heating at 75°. To rule out this possibility, the oil-soluble inhibitors were added to the oil phase when the temperature had been lowered to about 52°. The two phases were then mixed as described above. No appreciable difference was observed in the degree of inhibition and the results are pooled in Table III.

Analytical Method. Three grams of emulsion were weighed into a tared test tube containing 300 μL of 10% ammonium sulfamate, and the sample was immediately mixed with a vortex-type shaker. Ten percent calcium chloride (300 μL) was added and the sample was mixed, heated to 50° for 5 minutes, and mixed again. After being cooled to room temperature, the sample was filtered through glass wool to remove calcium sulfate and calcium stearate. Anion exchange was used to remove benzoic acid from the extracts. A 1.5-cm Pasteur pipette column to Biorad X-1 (hydroxide form, 100 - 200 mesh) was slurry packed. One milliliter of extract was passed through the column followed by two 0.5-mL washed with Milli-Q water. The NDE1A content of these extracts was determined by HPLC using a Zorbax (DuPont) octadecyl silane column and a UV detector at 237 nm and using Milli-Q water as carrier solvent at 1.0 mL/min.

Validation assays for the extraction and quantitation of added NDE1A from the emulsion were conducted. The range of recoveries at NDE1A levels of 0.3, 1.0, 3.0, 10.0, and 30.0 ppm (4-6 replicates at each level) was 94%-102% with a mean of 98%. The coefficient of variation at each level fell between 2% and 7% with a mean of 4%.

Experimental Results. The inhibition of NDE1A formation after 16 days at 37° for three water-soluble and three oil-soluble inhibitors is displayed in Tables II and III, respectively. So that relative effectiveness could be compared, the concentration of each inhibitor was chosen to provide an inhibitor/nitrite mole ratio = 10. Each inhibitor was tested in two separate emulsion batches with two or more aliquots analyzed. The data shown for positive controls are from similar replicate experiments.

Table II

Water Soluble Inhibitors (10 Molar Excess*)

Inhibitor	NDE1A Produced in 16 days at 37°C						Inhibition	
	Without Inhibitor			With Inhibitor				
	ppm	SD	N	ppm	SD	N	%	SD
Ascorbic Acid	3.00	.04	2	0.57	.30	2	79	3
	3.42	.14	2	0.81	.04	4		
K. Sorbate	3.91	.10	3	2.23	.10	3	32	16
	3.00	.04	2	2.39	.03	3		
Na Bisulfite	3.91	.01	3	0	0	3	100	--
	3.00	.04	2	0.01	.01	3		

* Based on nitrite present

Table III

Oil Soluble Inhibitors (10 Molar Excess*)

| Inhibitor | NDElA Produced in 16 Days at 37°C | | | | | | Inhibition | |
| | Without Inhibitor | | | With Inhibitor | | | | |
	ppm	SD	N	ppm	SD	N	%	SD
BHA	3.27	.12	5	3.00	.28	5	16	11
	2.89	.13	5	2.19	.14	5		
Alpha Tocopherol	3.37	.19	5	3.17	.11	5	0**	--
	3.00**	.04	2	2.74	.79	4		
Ascorbyl	2.74	.17	5	0.47	.16	5	90	10
Palmitate	3.00*	.04	2	0.09	.02	3		

* Emulsion without inhibitor run just previously.
** Calculated value is 8% inhibition but value obtained is not statistically significant.

Inhibition of NDElA formation after 16 days at 37° by ascorbic acid was examined more extensively at ascorbic acid/nitrite mole ratios = 0.1, 0.25, 0.5, 1.0, 5.0, and 10.0. Figure 1 shows that a linear relationship exists between log NDElA formed and ascorbic acid/nitrite mole ratios over the range 0-5. However, the trend was reversed at mole ratio = 10, where inhibition was significantly less than at ratio = 5. This is shown by the following data obtained from experiments where each ascorbic acid level was tested in two emulsion batches:

Ascorbic Acid/ Nitrite Mole Ratio	Batch	Mean % Inhibition	N	SEM
5	1	90.5	4	0.61
5	2	92.4	2	1.55
10	1	76.3	3	0.57
10	2	81.2	2	8.05

Discussion

The criteria for choosing inhibitors in this study were the ability to compete with diethanolamine for the nitrite and lack of toxicity. An attempt was made to cover as broad a group as possible within the limits of feasibility. Ascorbic acid in its water soluble form and its oil soluble form, the palmitate, represent the enediols. Sorbate is a diene fatty acid which has been shown to inhibit nitrosation (10). Since the pK of sorbic acid is 4.76, at the pH of these experiments, both water soluble sorbate ion and oil soluble sorbic acid are present in significant amounts. Sodium bisulfite is a strong inorganic reducing agent which has an acceptable lack of toxicity at the concentration

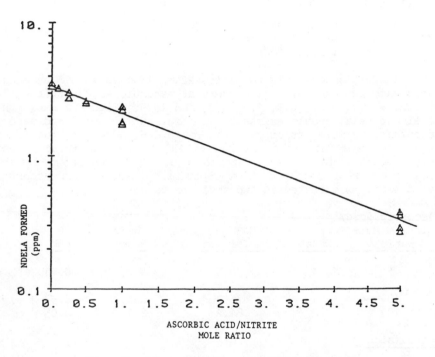

Figure 1.   Effect of ascorbic acid on NDELA formation.

used here. Two phenols which are frequently used as antioxidants are included, BHA and α-tocopherol. Neither of these have open para positions, the presence of which can lead to catalysis (4).

Studies have shown that one of the inhibitors, ascrobic acid, may catalyze nitrosation (11,12) especially at high concentrations (13). In the present study, the inhibitory effect of a range of concentrations was investigated. Between a mole ratio of ascorbate to nitrite of zero to 5:1, there was an inverse linear relationship between the ascrobate concentration and the log of NDElA formed (Figure 1). At a mole ratio of 10:1, the degree of inhibition decreased significantly. Conceivably at still higher concentrations, catalysis may occur. The ascorbate effect on nitrosation is still quite unclear and should be investigated further, because of the frequency of its use in blocking this reaction.

Sodium bisulfite gave essentially complete protection against NDElA formation and thus was superior to the other inhibitors in this regard. However, its strong reducing properties as well as its reactivity with carbonyl compounds found in fragrances may militate against its use in some cosmetic products. Three of the blocking agents showed poor inhibition, potassium sorbate and the lipophilic inhibitors, BHA and α-tocopherol.

The poor results with the two phenols were unexpected in view of their effectiveness as antioxidants in emulsions. R. Stadnick and A. Govil investigated the effect of α-tocopherol in a system identical to that used in the present study except for the absence of the oil phase (14). They found that after incubation for 90 hours, the recovery of nitrite in the presence of α-tocopherol was half that observed in its absence.

Since α-tocopherol destroys nitrite in the system in absence of the oil phase, we may postulate that the ineffectiveness of these two oil soluble inhibitors resulted from their absence from the aqueous phase. Diethanolamine is miscible with water and presumably its nitrosation occurs in the aqueous phase. There is a significant difference in the solubility characteristics of ascorbyl palmitate. The reducing portion of the molecule is water soluble. Thus the ascorbate moiety may be in the aqueous phase while the fatty acid tails may lie within the oil globules. The α-tocopherol and the BHA may well be effective if they are dispersed in the aqueous phase after preparation of the emulsion. This will be investigated in future experiments.

Acknowledgements. The authors express their gratitude for the technical contributions of Drs. M. Goodman, A. Cooper, and I.E. Rosenberg.

Literature Cited

1.   Fan, T.Y.; Goff, U.; Song, L.; Fine, D.H.; Arsenault, G.P.; Biemann, K. Food Cosmet. Toxicol. 1977, 15, 423–430.

2. CTFA Final Report on Midwest Research Institute Program for
    Development of Analytical Methods for NDElA Using HPLC-TEA,
    Determination of "Nitrites", Determination of NDElA by a
    non-TEA Method. July 10, 1980. Cosmetic, Toiletry and
    Fragrance Association, Inc., 1110 Vermont Avenue, N.W.,
    Washington, D.C. 20005.
3. "Analysis of Cosmetic Ingredients for Nitrites," by Raltech
    Scientific Services. Available from the Cosmetic, Toiletry
    and Fragrance Association, Inc.
4. Douglass, M.L.; Kabacoff, B.L.; Anderson, G.A.; Cheng, M.C.
    J. Soc. Cosmet. Chem. 1978, 29, 581-606.
5. Gough, T.A.; Goodhard, K.; Walter, C.L. Sci. Fd. Agric.
    1976, 27, 181.
6. Fiddler, W.; Pensabene, J.W.; Fagan, J.C.; Thorne, E.J.;
    Protrowski, E.G.; Wasserman, A.E. J. Food Sci. 1974, 39,
    1070-1071.
7. Coleman, M.H. J. Food Technology, 1978, 13, 55-69.
8. Mottram. D.S.; Patterson, R.L.S. J. Sci. Food Agric., 1977,
    28, 352-354.
9. Massey, R.C.; Dennis, M.J.; Crows, C.; McWeeny, D.J.; "N-
    Nitroso Compounds; Analysis, Formation and Occurance,"
    IARC Scientific Publication No. 31, International Agency
    for Research on Cancer, Lyon, France, 1980; p. 291.
10. Tanaka, K.; Chung, K.C.; Hayatsu, H.; Kada, T. Fd. Cosmet.
    Toxicol., 1978, 16, 209-215.
11. Chang, S.K.; Harrington, G.W.; Rothstein, M.; Shergalis,
    W.A.; Swern, D.; Vohra, S.K. Cancer Res., 1979, 39, 3871-
    3874.
12. Kawabata, T.; Shazuk, H.; Ishibashi, T. Nippon Suisan
    Gakkaishi, 1974, 40, 1251.
13. Kim, Y.K.; Tannenbaum, S.R.; Wishnok, J.S. "N-Nitroso
    Compounds: Analysis, Formation and Occurance", IARC Scien-
    tific Publication No. 31, International Agency for Research
    on Cancer, Lyon, France, 1980, 207-214.
14. Stadnick, R.; Govil, A. personal communication, 1981,
    March 27.

RECEIVED July 30, 1981.

The Role of Bacteria in Nitrosamine Formation

D. RALT and S. R. TANNENBAUM

Department of Nutrition & Food Science, Massachusetts Institute of Technology, Cambridge, MA 02139

Bacteria have been implicated in the formation of N-nitroso compounds under a wide variety of conditions representing both in vitro and in vivo situations. Mechanisms of participation and/or catalysis include a) decrease of the pH of the system, b) reduction of nitrate to nitrite, c) adsorption of amine onto the cell surface or cytoplasmic membrane, d) actual enzymatic formation. The literature of the field will be reviewed and experimental evidence which tests the above mechanisms will be presented.

Recently nitrosamines have attracted attention because of their marked carcinogenic activity in a wide variety of animal species (1, 2). Nitrosamines are likely to be carcinogens in man as well: human exposure to these compounds is by ingestion, inhalation, dermal contact and in vivo formation from nitrite and amines. Nitrite and amines react most rapidly at an acidic pH. A variety of factors, however, make nitrosation a potentially important reaction above pH 7: these include the presence of microorganisms, and the possibilities of catalysis by thiocyanate, metals and phenols, and of transnitrosation by other nitroso compounds.

J. Sander (3) and M. J. Hill and coworkers (4) were among the first to recognize the role that bacteria might play in endogenous nitrosamine formation. Since then several other workers have suggested that microorganisms play a role in N-nitrosation. Hicks et al. (5) and Radomski and coworkers (6) have shown an association of nitrosamines in human urine with chronic urinary tract infections and possibly with bladder cancer. We have shown (7) that oral microorganisms reduce nitrate to nitrite, and that salivary nitrite reacts with secondary amines added to saliva to form the corresponding nitrosamines: the reaction rate is increased in the presence of bacteria. Crissey et al. (8) have

shown that contact with the fecal stream is necessary in rats for
the experimental induction of colon tumors following uretero-
sigmoidostomy. It is conceivable that nitrate introduced into
the fecal stream from the urine is reduced by bacteria to nitrite,
which can react to form N-nitroso compounds in the colon.

Only a few studies have been undertaken on the role of bacteria in nitrosation at neutral pH in the intestinal tract itself.
Hashimoto et al. (9) demonstrated that by changing the equilibrium
between different species of microorganisms in the gut, (by
administration of antibiotics and inoculation with nitrate-reducing strains) dimethylnitrosamine (NDMA) accumulated in the stomach
and cecum of rats fed dimethylamine (DMA) and nitrate. The
authors describe only one experimental group, however, and the
mechanism of NDMA formation is not clear.

Other investigators have carried out in vitro experiments
using intestinal contents or isolated bacteria in order to assess
the potential role of microorganisms in the formation of N-nitroso
compounds (3, 4, 10-13). The format of these experiments is
essentially as follows.

Bacteria are added to complex medium supplemented with DMA,
glucose and nitrate (or nitrite); the utilization of DMA and the
formation of NDMA are then followed and compared to the level
of NDMA in the uninoculated control medium. Three problems arise
under those conditions: i) The bacterial reduction of nitrate to
nitrite is very rapid; therefore an appropriate control medium
should contain nitrite, not nitrate; ii) Bacterial growth (utilizing glucose) causes a decrease in pH even in buffered media, so
control media should be adjusted to the final pH of the growing
culture; and iii) In some experiments, the method of detection
of NDMA has not been specific (11, 13); therefore differences
between culture and medium could have been due to other metabolites accumulating in the growing culture. We felt that these
complications made it impossible to isolate the role of microorganisms in N-nitrosation. We therefore initiated carefully
controlled experiments designed to determine whether microorganisms can specifically catalyze formation of NDMA from nitrite and
DMA.

In order to define the conditions of the growing cultures,
buffered medium (VL) inoculated with E. coli ATCC 11775 and supplemented with nitrate, glucose and DMA was incubated at 37°C,
and pH, nitrite concentration, nitrate concentration, cell growth
and nitrosamine formation were followed (Fig. 1). Within 2 hrs,
>90% of the nitrate is converted to nitrite (some of the nitrite
is further reduced) and over 8 hrs the pH drops from 7.3 to 6.0.
This would indicate that in experiments carried out for 20 hrs or
more the control medium should be adjusted to pH 6.0 to 6.5 and
nitrite should be added rather than nitrate. Such a control
medium (VL) was supplemented with nitrite and DMA; and NDMA formation was followed (Fig. 2). It can be seen that even without the
addition of cells the rate of nitrosation is 4 fold greater than

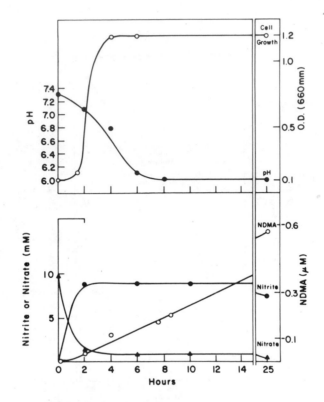

Figure 1. pH, nitrite, nitrate, and NDMA concentration as a function of cell growth in VL medium. E. coli ATCC 11775 was inoculated into sealed tubes containing VL medium (described in the legend to Table I) supplemented with 0.2% glucose, 0.05% dimethylamine, and 0.08% sodium nitrate. The tubes were incubated at 37°C and cell growth was followed with spectronic 20 (Bausch & Lomb) at 660 nm. Nitrite and nitrate were measured as described in (14) and NDMA detection as described in the legend to Table I.

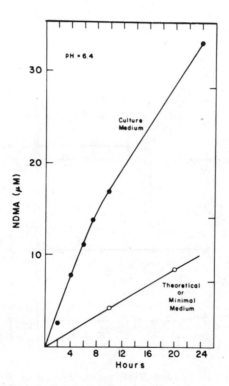

Figure 2.   Accumulation of NDMA in VL medium at pH 6.4. The pH of VL medium was adjusted to 6.4 with HCl and NDMA formation was followed at 37°C. Nitrite concentration in the medium was 0.2% and dimethylamine 0.05%. Theoretical curve was calculated according to Fan and Tannenbaum (15).

the rate predicted for chemical nitrosation under these conditions. This suggests that the complex medium itself contains
catalysts for N-nitrosation. None of the three strains of $\underline{E}$.
$\underline{coli}$ or two strains of proteus that we examined, enhanced the
rate of NDMA formation above this rate in the medium alone. When
minimal salt medium was used instead of the complex VL, no en-
hancement of reaction rate was observed, and the NDMA formed was
as expected from chemical nitrosation.

It seemed possible that a catalytic effect of the bacteria
upon nitrosation might be masked by the high level - 0.2% - of
nitrite used in these experiments, so we repeated the work with
lower concentrations of nitrite (Table I). The catalysis by
the medium alone is considerably greater at these limiting concentrations of nitrite, and again the bacteria do not further
enhance the rate of nitrosation.

Table I
Formation of Dimethylnitrosamine (NDMA) in VL medium

	NaNO$_2$ μM	NDMA theoretical μM	NDMA actual μM	actual/theoretical
medium	5.79	0.16	4.46	27.9
	2.89	0.04	0.81	20.3
medium + cells	5.79	0.16	3.11	19.4
	2.89	0.04	0.54	13.5

VL medium, pH 6.4, with or without cells was incubated in sealed
tubes for 10 hrs at 37°C. NDMA was determined by gas chromatography with the Thermal Energy Analyzer as a detector (Thermo
Electron Corp., Waltham, MA). The identity of NDMA was confirmed
by GC mass spectrometry.

VL medium includes: Trypticase (BBL) 1%, Yeast Extract
(Difco) 0.5%, Meat Extract (Difco 0.2%, Hemin 0.0007%, Na$_2$CO$_3$
0.4%, NaCl 0.05%, NH$_4$Cl 0.1%, MgSO$_4$·7H$_2$O 0.02% and phosphate
buffer 1%. The medium was supplemented with 0.05% dimethylamine
(DMA) and 0.2% glucose.

These results led us to reexamine the published work on
nitrosation in the presence of bacteria (Table II). In two of
the studies (4, 10) the amount of NDMA formed is actually less
than the amount that can be predicted from theoretical consideration of the uncatalyzed chemical reaction alone (15). In another
study (12), the yield of NDMA is again slightly lower than the
predicted chemical yield at the final pH (6.0) of the growing
culture. The work by Kunisaki and Hayashi (13), on the other
hand, does indicate that resting cells of $\underline{E. \; coli}$ B catalyze

Table II

Summary of some published studies on the formation of dimethylnitrosamine (NDMA) in the presence of bacteria

	micro-organism	pH	nitrite/nitrate conc.	amine conc.	Incubation (hrs)	NDMA (μM) theoretical	NDMA (μM) actual
Hawksworth & Hill, 1971	Intestinal E. coli	6.5	0.2%	0.1%	18	1	<0.2
Klubes et al., 1972	Intestinal flora	Initial pH = 7	1.4%	0.3%	20	108	95
Hashimoto et al., 1975	E. coli (feces)	Initial pH = 7.5*	0.2%	0.05%	72	68	41
Kunisaki & Hayashi, 1979	E. coli B	8.0	0.7%	0.2%	1	0.7	95
Ralt and Tannenbaum, this report	------	6.4	0.04%	0.05%	10	.2	5

*drops to ~6.0, as in Figure 1.

nitrosation, although their method of detection - u.v. photolysis of the putative NDMA and measurement of the released nitrite - is less than perfectly selective.

Other studies on the involvement of microorganisms with N-nitrosation have either been insufficiently controlled or have been only incompletely reported. Radomski et al. (6) found NDMA at a level of 2 ppm in bacterially contaminated urine which contained 25 ppm nitrite. Kinetic data is absent, however, and the actual concentration of urinary nitrite in the bladder is not clear; no control experiments were done in the absence of the bacteria to clarify their contribution to the formation of NDMA. Thacker and Brooks (11) reported the formation of NDMA by proteus strains in urine, but detection of the nitrosamine was carried out only by gas chromatography (equipped with EC detector) - an insufficiently selective method - and quantitative data are not presented. It has been shown (16) that longer chain secondary amines are catalytically nitrosated in the presence of cells - living or not - but the catalysis presumably results from micellar effects which would not involve a hydrophilic substrate such as DMA.

We conclude that the major role of bacteria in the nitrosation of dimethylamine is the reduction of nitrate to nitrite and the lowering of the pH of the medium. Furthermore, the complex medium itself catalyzes nitrosation. The nature of this catalysis is not known, although it could be due to the presence of carbonyl compounds, cysteine, or a variety of other compounds which are known to catalyze nitrosation (17).

It has recently been shown (14) that nitrate from food can be detected in the feces of germfree animals but not in the feces of conventional animals. This result and the results of Witter et al. (18-20) suggest that nitrate is available in the lower gastrointestinal tract and is metabolized by the microbial flora. The intestine may thus be a site for the endogenous formation of N-nitroso compounds.

Acknowledgement

This work was supported by the National Cancer Institute Grant No. 1-P01-CA26731-02.

Literature Cited

1. Magee, P.N.; Montesano, R.; Preussmann, R. "Chemical Carcinogens"; American Chemical Society Monograph 173: Washington, DC, 1976, pp. 491-625.
2. Druckrey, H.; Preussmann, R.; Ivankovic, S.; Schmahl, D. Z. Krebsforsch. 1967, 69, 103-201.
3. Sander, J. Arch. Hyg. Bakt. 1967, 151, 22-28.
4. Hawksworth, G.M.; Hill, M.J. Brit. J. Cancer 1971. 25, 520-526.

5. Hicks, R.M.; Gough, T.A.; Walters, C.T. "Environmental
    Aspects of N-nitroso Compounds", Walker, E.A.; Castegnaro,
    M.; Griciute, L.; Lyle, R.E., Eds.; IARC No. 19: Lyon,
    1978, pp. 465-475.
6. Radomski, J.L.; Greenwald, D.; Hearne, W.L.; Block, N.L.;
    Woods, F.M. J. Urol. 1978, 120, 48-50.
7. Tannenbaum, S.R., Archer, M.C.; Wishnok, J.S.; Bishop, W.W.
    J. Natl. Cancer Inst. 1978, 60, 251-253.
8. Crissey, M.M.; Glenn, D.S.; Gittes, R.F. Science 1979,
    207, 1079-1080.
9. Hashimoto, S.; Yokokura, T.; Kawai, Y.; Mutai, M. Fd.
    Cosmet. Toxicol. 1976, 14, 553-556.
10. Klubes, P.; Cerna, I.; Rabinowitz, A.D.; Londorf, W.R. Fd.
    Cosmet. Toxicol. 1972, 10, 757-767.
11. Thacker, L.; Brooks, J.B. Infect. Immun. 1974, 9, 648-653.
12. Hashimoto, S.; Kawai, Y.; Mutai, M. Infect. Immun. 1975,
    11, 1405-1406.
13. Kunisaki, N.; Hayashi, M. Appl. Envir. Microbiol. 1979,
    37, 279-282.
14. Green, L.C.; Tannenbaum, S.R.; Goldman, P. Science 1981,
    212, 56-58.
15. Fan, T.Y.; Tannenbaum, S.R. Agric. and Food Chem. 1973,
    21, 237-240.
16. Yang, S.R.; Okun, J.D.; Archer, M.C. J. Agric. Food Chem.
    1977, 25, 1181-1183.
17. Kunisaki, N.; Hayashi, M. J. Natl. Cancer Inst. 1980,
    65, 791-794.
18. Witter, J.P.; Balish, E. Appl. Environ. Microbiol. 1979,
    38 861-869.
19. Witter, J.P.; Balish, E.; Gatley, S.J. Appl. Environ.
    Microbiol. 1979, 38, 870-878.
20. Witter, J.P.; Gatley, S.J.; Balish, E. Science 1979, 204,
    411-413.

RECEIVED August 10, 1981.

Formation of *N*-Nitroso Compounds in Foods

J. IAN GRAY

Department of Food Science and Human Nutrition, Michigan State University,
East Lansing, MI 48824

N-Nitrosamines, formed principally from the reaction
of naturally occurring secondary amines with nitrites
that may be added to foods or produced by bacterial
reduction of nitrates, have been identified in many
food systems including cured meat products, nonfat
dried milk, dried malt and beer. In addition, the
presence of less volatile and non-volatile N-nitroso
compounds or their precursors in foods have been
suggested from a number of model system studies.
This paper reviews the possible formation of these
compounds in selected food products as well as
identifying several future research areas.

There has probably been no topic in the past decade that
has generated as much discussion and research as the presence of
N-nitroso compounds in food systems. Many of these compounds
are carcinogenic and, in addition, some exhibit mutagenic,
embryopathic and teratogenic properties. Thus, it is under-
standable that the possible occurrence of these compounds in
food has caused considerable concern. N-Nitrosamines are formed
principally from the reaction of naturally occurring secondary
amines with nitrites that may be added to foods or produced by
the bacterial reduction of nitrates (1). These compounds have
been reported in various foods including wheat flour, mushrooms,
alcoholic beverages, cheese, milk and soybean oil as well as in
meat and fish products. The authenticity of some of these
reports is questionable, however, because many of the analytical
procedures used lacked the necessary specificity for the
unequivocal identification of N-nitrosamines at the very low
μg/kg levels normally found in foods. Many reviews on
N-nitrosamines have been published in the past few years,
dealing with their formation and occurrence in foods, and their
toxicology and human health hazards (2-9).
    In recent years, the development of the highly specific and
sensitive **Thermal Energy Analyzer** (10) has greatly facilitated

the detection of N nitrosamines in other food systems including nonfat dried milk, dried malt and beer, as well as in experimental animal diets (11). In this paper, the main focus will be on the formation of volatile N-nitrosamines in cured meats, and to a lesser extent, in dairy products and beer. Results of some model system studies, from which the presence of less volatile and non-volatile N-nitroso compounds or their precursors in foods have been suggested, will also be discussed.

N-Nitrosamines in Cured Meat Products

The food items of major importance as far as the formation of N-nitrosamines are concerned are the cured meat products, especially bacon. N-Nitrosopyrrolidine (NPYR), and to a lesser extent, N-nitrosodimethylamine (NDMA) have been isolated consistently from cooked bacon (Table 1). Although NPYR is not detected in raw bacon, it is found almost invariably after cooking, the levels depending on cooking conditions and other less well defined factors (12). Interestingly, the amounts of N-nitrosamine being detected in the cooked bacon or rendered fat constitute only a portion of the total quantity of N-nitrosamine formed. During frying, a substantial portion of these compounds is volatilized in the fumes. This phenomenon has been investigated by several workers who reported a wide range of values for the percentages of N-nitrosamines found in the vapor (Table II). Obviously, the mode of cooking, as well as the moisture content and ratio of lean to adipose tissue in the bacon samples, influences the amount of N-nitrosamines in the vapor.

Table 1
N-Nitrosamine Formation (μg/kg) in Fried Bacon

Investigators	N-Nitrosopyrrolidine		N-Nitrosodimethylamine	
	Bacon	Cooked-out fat	Bacon	Cooked-out fat
Crosby et al.(13)	tr 40	----	tr	----
Sen et al.(14)	4–25	----	2–30	----
Fiddler et al.(15)	2–28	6–24	----	----
Pensabene et al.(16)	11–38	16–39	----	----
Gray et al.(17)	tr–23	tr–41	----	---
Pensabene et al.(18)	2–45	5–55	2–9	2–34
Sen et al.(19)	2–22	15–34	tr–17	3–12
Pensabene et al.(20)	2–6	11–34	----	----

Mechanism of NPYR Formation The consistent occurrence of NPYR in fried bacon and cooked-out fat has led to an intensive search for both the precursors and mechanism that could account for its formation. Although model system studies have implicated a number of compounds including proline, collagen, putrescine, spermidine, pyrrolidine and glycyl-L-glycine as possible precursors of NPYR (30), the most probable precursor of NPYR in bacon appears to be proline. Free proline is present in pork belly at a concentration of approximately 20 mg/kg (29,30,32).

Table II

Percentage of N-Nitrosamines in the Fumes Produced During the Frying of Bacon or Ham

Investigators	N-Nitrosamine (%)		Sample
	NPYR	NDMA	
Gough et al.(21)	60-95	75-100	bacon
Hwang and Rosen (22)	14-37	----	bacon
Warthesen et al (23)	20-40	----	pork belly[a]
Sen et al. (24)	28-82	28-92	bacon
Gray and Collins (25)	27-49	----	pork belly[a]
Mottram et al. (26)	57-75	73-80	bacon
Gray et al. (27)	----	56-80	pork belly
Janzowski et al. (28)	45-52	74-83	ham[b]
Bharucha et al. (29)	up to 32	up to 62	bacon

[a]Contained added nitrite; [b]contained added N-nitrosoamino acids

How proline is converted to NPYR has not yet been fully elucidated and could conceivably occur by either of two pathways (29,30). One pathway involves the initial N-nitrosation of proline, followed by decarboxylation, while in the other, proline is first decarboxylated to pyrrolidine followed by N-nitrosation to NPYR. Since the conversion of N-nitrosoproline (NPRO) to NPYR occurs at a much lower temperature than the transformation of proline to pyrrolidine, the pathway involving intermediacy of NPRO is thus the more likely route (29). It has been reported that preformed NPRO in raw bacon is not the primary precursor of NPYR in cooked bacon (29,33-5), as shown by the fact that ascorbyl palmitate, when added to bacon, inhibits the formation of NPYR (33). However, this by no means rules out the intermediacy of NPRO which could be formed at the higher temperatures attained during the frying process (29,36).

The mechanism of NPYR formation has been studied by Coleman (37) and Bharucha et al. (29). Coleman (37) reported that the requirement for a high temperature, the inhibitory effects of water and antioxidants, and the catalytic effect of a lipid hydroperoxide are consistent with the involvement of a free radical in the formation of NPYR. Similarly, Bharucha et al. (29) suggested that, since both NPYR and NDMA increase substantially towards the end of the frying process, N-nitros-amine formation during frying of bacon occurs essentially, if not entirely, in the fat phase after the bulk of the water is removed and therefore by a radical rather than an ionic mechanism. These authors speculated that, during the frying of

bacon, nitrous acid is converted essentially into N_2O_3 by
continuous removal of water, and N_2O_3, in turn, undergoes
dissociation at higher temperatures (>100°C) to nitric oxide
and $NO_2^.$. Since, nitric oxide is relatively stable, it was
concluded that the $NO_2.$ radical can act as the chain initiator
and abstract the amino proton from proline to give a radical
which combines with the NO· radical to give NPRO.

Factors Influencing NPYR Formation. The major factors which
influence the formation of NPYR in cooked bacon have been well
documented (2,30) and include the method of cooking, frying
temperature and time, nitrite concentration, ascorbate
concentration, preprocessing procedures, presence of lipophilic
inhibitors, and possibly smoking.
Cooking methods. It has been well established that pan-frying
of bacon results in more NPYR formation than other cooking
procedures such as microwave cooking (16,38) and grilling (29).
Bharucha et al.(29) explained the reduced yields of N-nitrosamines during grilling as being due to the cooked-out fat
running out of the heated area. Consequently, the bacon slices
never reach the same temperature as during pan frying. It has
also been demonstrated that both frying temperature and time
clearly influence the levels of NPYR in cooked bacon. Pensabene
et al.(16) showed that bacon samples from one belly formed no
NPYR when fried for 105 minutes at 99°C, while samples from
the same belly, fried to the same "doneness" at 204°C for 4
minutes, produced 17 μg/kg of NPYR. Bharucha et al.(29)
reported that the maximum amount of N-nitrosamine was produced
when the bacon was fried for 12 minutes at 360°F, after
starting with a cold frying pan. Very little N-nitrosamine was
found in the rendered fat after 4 minutes of heating; however,
the N-nitrosamine level increased sharply with time and reached
a maximum at around 12 minutes and then began to decline. Two
explanations were offered to explain the initial low formation
of N-nitrosamine: (i) the N-nitrosamines were actually formed at
about 100°C, but being steam-volatile, were removed with the
water vapor; or (ii) the N-nitrosation occurred at temperatures
greater than 100°C, after the major portion of the water was
removed.
Nitrite concentration The kinetics of N-nitrosamine formation
in vitro has been studied at length (39,40) and, in moderately
acidic media, the reaction rate is directly proportional to the
concentration of the free amine (non-protonated) and to the
square of the concentration of the undissociated nitrous acid.
Therefore, it is not surprising that the amount of nitrite
permitted in bacon has received considerable attention. Although, there have been suggestions that it is the initial and
not the residual nitrite that influences N-nitrosamine formation
in bacon (41), recent evidence seems to indicate that the lowest
residual nitrite gives the least probability of N-nitrosamines

being formed (8,42). Consequently, it has been recommended that
the in-going nitrite levels for bacon be reduced from 156 to 120
mg/kg, with the simultaneous inclusion of 550 mg/kg of sodium
ascorbate (43). Similarly, in Canada, the amount of nitrite to
be used in the preparation of side bacon has been reduced from
200 to 150 mg/kg, calculated before any smoking, cooking or
fermentation (30). Period surveillance of the levels of
N-nitrosamines in cooked bacon over the past few years suggests
that in both the U.S. and Canada, the concentration of these
compounds has been decreasing steadily (2,44,45).

     Recently, Robach et al. (46) investigated the effects of
various concentrations of sodium nitrite and potassium sorbate
on N-nitrosamine formation in commercially prepared bacon.
Bacon, processed with 40 mg/kg of nitrite and 0.26 sorbate
contained an average of 8.7 µg/kg of NPYR, whereas samples
prepared with 120 mg/kg of nitrite contained an average of 28.1
µg/kg of NPYR. This marked reduction in NPYR levels is clearly
due to the reduced levels of nitrite, although it has been
reported that sorbic acid also possesses anti-N-nitrosamine
activity (47).

N-Nitrosamine inhibitors   Ascorbic acid and its derivatives,
and ∝-tocopherol have been widely studied as inhibitors of the
N-nitrosation reactions in bacon (33,48-51). The effect of
sodium ascorbate on NPYR formation is variable, complete
inhibition is not achieved, and although results indicate lower
levels of NPYR in ascorbate-containing bacon, there are examples
of increases (52). Recently, it has been concluded (29) that
the essential but probably not the only requirement for a
potential anti-N-nitrosamine agent in bacon are its (a) ability
to trap NO· radicals, (b) lipophilicity, (c) non-steam
volatility and (d) heat stability up to 174°C (maximum frying
temperature). These appear important requirements since the
precursors of NPYR have been associated with bacon adipose
tissue (15). Consequently, ascorbyl palmitate has been found to
be more effective than sodium ascorbate in reducing N-nitrosamine formation (33), while long chain acetals of ascorbic acid,
when used at the 500 and 1000 mg/kg levels have been reported to
be capable of reducing the formation of N-nitrosamines in the
cooked-out fat by 92 and 97%, respectively (49).

     The inhibition of formation of NPYR and NDMA in fried
bacon by the use of cure-solubilized α-tocopherol (500 mg/kg)
has been demonstrated by Fiddler et al. (50). Walters et al.
(53) also reported reduced levels of N-nitrosamines in the
vapors during the frying of bacon in fat containing
α-tocopherol. It has also been shown that α-tocopherol is
dispersed quite effectively during frying of bacon slices;
therefore, application to bacon may be made by spray or dip to
overcome the problem of water insolubility (51). Controlled
addition of this antioxidant may be an effective and practical
way of reducing the concentration of N-nitrosamines in cooked

bacon and minimizing the remote possibility of any health
hazards arising from the consumption of such foods (54).
Preprocessing. Storage of pork bellies also has a definite
effect on NPYR formation in fried bacon (20). Bacon, made from
fresh bellies produced significantly less (p < 0.05) NPYR than
that made from bellies that had been either stored for 1 week in
a refrigerator or frozen for 3 months and then thawed before
using. It was suggested that the higher levels of NPYR results
from the increase in both amines and amino acids that occurs
during extended storage (20). Several investigators (31,32)
have shown that the free proline contents in whole and lean
tissue of green pork bellies increased approximately 50% after
storage at 2°C for 1 week. Over the same period, free proline
in the adipose tissue increased approximately 90%.
Smoking. The effects of smoking on the formation of N-nitrosamines in bacon has been investigated recently by Bharucha et
al. (49). They reported that unsmoked bacon samples generally
tended to contain more N-nitrosamines, presumably because of
their higher nitrite content at the time of frying. Sink and
Hsu (55) showed a lowering of residual nitrite in a liquid smoke
dip process for frankfurters when the pH also was lowered. The
effects of smoke seem to be a combination of pH decrease and
direct C-nitrosation of phenolic compounds to lower the residual
nitrite in the product (56). This is an area which requires
further study since certain C-nitrosophenols have been shown to
catalytically transnitrosate amines in model systems (57).
       To date, the majority of bacon studies have centered on
the formation and inhibition of N-nitrosamines in brine-cured
bacon. Two recent investigations, however, have indicated the
presence of high concentrations of NPYR in dry-cured bacon after
frying (35,58). NPYR levels ranging from 39-89 μg/kg were
reported by Pensabene et al. (35), while those cited by the
Nitrite Safety Council (58) ranged from traces-320 μg/kg. These
findings have identified dry-cured bacon as one cured meat
product category requiring further study. Evaluation of cure
formulation changes or process control adjustments which may
reduce or eliminate N-nitrosamine formation in dry-cured bacon
obviously are necessary (58).

N-Nitrosamines in Cured Meats Other Than Bacon. The presence of
N-nitrosamines in cured meats other than bacon has been the
subject of several recent surveys (19,58,59). In general, the
majority of the positive samples contained extremely low levels
of N-nitrosamine, usually less than 1 μg/kg (19,58). This low
level could, in part, be attributed to the discontinuation of
the use of nitrite-spice premixes in the mid-1970's (19). In
the Holland study (59), the predominant N-nitrosamines detected
were NDMA and N-nitrosomorpholine (NMOR) and, generally, values
of 4 μg/kg were obtained for each N-nitrosamine. More
correctly, these are presumptive N-nitrsoamine levels since mass
spectral confirmation of their identities was not achieved.

In a similar study, Gray et al. (60) investigated the possible formation of N-nitrosamines in heated chicken frankfurters which been prepared with various levels of nitrite (0-156 mg/kg). As expected, apparent N-nitrosamine levels increased with increasing concentrations of nitrite, but did not exceed 4 µg/kg except for two samples which contained 8 and 11 µg/kg of NMOR. The presence of these relatively high levels of NMOR was confirmed by mass spectrometry and raised the question as to its mode of formation. It was shown to be due to the morpholine present in the steam entering the smokehouse, as this amine is commonly used as a corrosion inhibitor in steam process equipment (61). The detectable levels of NMOR in the Canadian study (59) were also attributed in part to the use of morpholine as an anti-corrosion agent in the steam supply (62).

N-Nitrosamines in Dairy Products.

Dairy products have also been extensively scrutinized for the possible presence of volatile N-nitrosamines (Table III). Cheeses of the Gouda and Edam types as produced in certain European countries are very susceptible to late blowing as a result of the development of clostridia in the cheese (1). One of the most successful methods of preventing this defect is the addition of potassium or sodium nitrate. The germinating clostridia spores are very susceptible to nitrite which is produced from the added nitrate, primarily by the action of the milk enzyme, xanthine oxidase. The practice of adding nitrate to cheese milk is sometimes questioned since it may lead to the formation of N-nitrosamines. Gough et al. (63) examined 21 different varieties of cheese commonly available in the United Kingdom including cheeses to which nitrate had been added during manufacture. There was no greater occurrence of NDMA in these samples than in cheese made without added nitrate. Levels of NDMA were similar for all cheeses (1-5 µg/kg) except for one sample of Stilton which contained 13 µg/kg. A similar range of values were obtained by Sen et al. (64) for 31 samples of cheese imported into Canada, many of which were known to have been prepared with the addition of nitrate. Havery et al. (67) failed to detect any of 14 N-nitrosamines in 17 samples of cheese, 10 of which were the imported variety and which had been processed with nitrate as an additive.

Commercially available nonfat dried milk and dried buttermilk have also been shown to contain small but detectable levels of NDMA (65,66,68). It has been suggested that N-nitrosamine formation is possible in foods that are dried in a direct-fired dryer (65). In such a dryer, the products of combustion come into direct contact with the food being dried, and N-nitrosamine formation is probably due to the reaction between secondary and/or tertiary amines in the food and the oxides of nitrogen that are produced during fuel combusion (65).

Table III
N-Nitrosamine Content of Selected Dairy Products

Type of Product	N-Nitrosodimethylamine ($\mu g/kg$)	Investigators
Cheese		
Gouda, Havarti, Provolone Camembert, Cheddar, Edam	trace-19	Sen et al. (64)
Cheddar, Gouda, Cheshire Edam, St. Pauline, Stilton	1-13	Gough et al.(63)
Yogurt	not detected	Gough et al.(63)
	not detected	Lakritz and Pensabene (68)
Nonfat dried milk	trace-4.5	Libbey et al.(65)
	0.3 - 0.7	Sen and Seaman (66)
Dried buttermilk	0.9 - 1.8	Libbey et al.(65)
Whole milk (pasteurized)	0.05 - 0.17	Lakritz and Pensabene (68)

N-Nitrosamine in Beers and Malt

In the past two years, considerable attention has also focused on the presence of N-nitrosamines in beer and other alcoholic beverages. In 1978, it became known that minute amounts of NDMA were present in many domestic and foreign beers, and in most brands of Scotch whisky (Table IV). Spiegelhalder et al. (69) analyzed 158 samples of different types of beer available in West Germany and reported 70% of these to contain NDMA, with a mean concentration 2.7 $\mu g/kg$. Goff and Fine (72) reported NDMA levels ranging from 0.4-7.0 $\mu g/kg$ in 18 brands of domestic (USA) and imported beers. In addition, six out of seven brands of Scotch whisky contained NDMA at levels between 0.3 and 2.3 $\mu g/kg$. Scanlan et al. (71) reported NDMA in 23 of 25 beer samples, levels of N-nitrosamine ranging from 0-14 $\mu g/kg$ with an average of 5.9 $\mu g/kg$.

It has been shown in these studies that the principal, and probably only significant source of NDMA, is malt which had been dried by direct-fired drying (71,73). It is well known that malts kilned by indirect firing have either low or non-detectable levels of NDMA (74). Consequently, changes in malting procedures have been implemented in both the U.S. and Canada which have resulted in marked reductions in N-nitrosamine levels in both malts and beer (70,74). For example, sulfur dioxide or products of sulfur combustion are now used routinely by all maltsters in the U.S. to minimize N-nitrosamine formation (70). The Canadian malting industry, on the other hand, has

also made a basic decision to change the kilning process from a
direct to an indirect firing technique. As a result, as of
January 1981, malts produced in Canada are no longer exposed to
kiln gases during drying (74).
    Research on the formation of NDMA in beer has centered
on three possible sources of amine precursors (74). N-Nitrosamine formation from amines such as hordenine, gramine and
methyltyramine which are formed endogenously in the germination
of barley has been discussed elsewhere in this symposium (75).
The Brewers Association of Canada (74) investigated the

Table IV.
Occurrence of N-Nitrosodimethylamine in Beers and Malt

Product	N-Nitrosodimethylamine (μg/kg)	Investigators
Beer		
Pilsen lager	0-6.5	Spiegelhalder et al.(69)
	0-9	Havery et al.(70)
	0-14	Scanlan et al.(71)
Dark lager	0.5-47	Spiegelhalder et al.(69)
	0-3	Havery et al.(70)
	0.5-0.8	Scanlan et al.(71)
Ale	0-0.5	Spiegelhalder et al.(69)
	0-3	Havery et al.(70)
	3-7	Scanlan et al.(71)
Malt liquor	0-3	Havery et al.(70)
	0.5-5	Scanlan et al.(71)
Malt	0-86	Havery et al.(70)
	115	Hotchkiss et al.(73)
	3.1-67[a]	Sen and Seaman(66)
	1.3-6.6[b]	Sen and Seaman(66)

[a]Samples analyzed in 1979, [b]Samples analyzed in 1980.

possibility that amines such as dimethylamine might be present
in malts as a result of infection of barley by smut, or from the
treatment of barley with herbicides. Results indicate that the
degree of smut infection in barley does not materially affect
either the dimethylamine content in barley or the amount of NDMA
which is formed when the barley is malted. Similarly, this
study also revealed that no relationship exists between
herbicide treatment of barley and the NDMA content of malts or
beers.

Formation of Non-Volatile N-Nitroso Compounds in Food Systems

    The functions of nitrite and the formation of volatile
N-nitrosamines in food systems and their toxicity have been well

reviewed. Many questions, however, still remain unresolved.
Our knowledge of the distribution of N-nitroso compounds in
foods is limited to the formation, isolation and identification
of the volatile N-nitrosamines. This group is readily separated
for analytical purposes, but may constitute only a small
proportion of the total N-nitroso compounds to which man is
exposed (5). The presence of less volatile and non-volatile
N-nitroso compounds in foods must be investigated, especially
since their presence has been suggested recently by several
model system studies. Conditions whereby these compounds may
arise include the (i) interaction of amino acids and amines with
fatty esters, (ii) interaction between oxidized lipids and free
amino acids, (iii) reaction between reducing sugars and amino
acids, and (iv) reaction between cysteamine and acetaldehyde,
glucose and glyoxal.

Formation of N-substituted amides in food systems. Fatty esters
react readily with many α-amino acids at temperatures as low as
150°C to give N-substituted amides as the major reaction
products (76). This reaction involves decarboxylation of the
amino acid and displacement of the alcohol moiety of the fatty
ester by the amine which is formed. The presence of a secondary
amino group in these compounds makes them very susceptible to
N-nitrosation. Recent studies, however, indicate that the
formation of primary amines via the decarboxylation of α-amino
acids appears to be unlikely under normal cooking conditions,
e.g. oven roasting of pork or pan-frying of bacon, due to
insufficient energy for the decarboxylation reaction (77). High
temperatures (minimum 150°C for 45 minutes) are required for
the decarboxylation of norleucine. It was concluded that under
the conditions encountered in the processing and cooking of
foods, only amines would react with fatty acids (or esters) to
yield substantial quantities of secondary amides.

   The N-nitrosation of these amides was subsequently studied
as a function of pH and temperature (78). There was no apparent
pH maximum for the reaction, N-nitrosamide formation increasing
with increasing hydrogen ion concentration. The rates of
N-nitrosation decreased rapidly as the pH increased and little
reaction occurred above pH 3. A unit drop in pH from 2 to 1
increased the rate of N-nitrosation by a factor of 5-8 times.
The rate constants for the reaction remained relatively constant
over the pH range 1-3.5, supporting the nitrous acidium ion
mechanism.

   Decomposition studies confirmed that N-nitrosamides are
much less stable than volatile N-nitrosamines such as NPYR and
NDMA (79). Thermal studies utilizing heating conditions
commonly encountered in the cooking of bacon and pork roasts
indicated that N nitrosomethylpropionamide (NOMP) was degraded
to the extent of 74-97% compared to 3-14% for NPYR and NDMA
(Table V). It was tentatively concluded that the major

contribution of N-substituted amides, if present in foods, may
be as precursors of N-nitroso compounds formed by in vivo
N-nitrosation reactions.

Table V

A Comparison of the Stabilities of N-Nitroso Compounds
in Sealed Ampules during Heating (79)

Cooking Condition	% Decomposition
Bacon frying (8 min,)	
NOMP (top of bacon, 100°C)	74
NOMP (on pan, 130°C)	91
NPYR (on pan, 130°C)	3
NDMA (on pan, 130°C)	14
Pork roasting (180°C, 2h)	
NOMP	97
NPYR	10
NDMA	8

Journal of Agricultural and Food Chemistry

In a continuation of this study, Fooladi (80) rehydrated
freeze-dried pork belly slices with water containing pentylamine. After equilibrating for 24 hours at 4°C, the slices
were fried for 4 minutes on each side at 175°C in a preheated
pan. Analysis revealed no amide formation in either the fried
slices or cooked-out fat. However, when the latter was heated
for an additional 8 minutes, N-substituted amides were formed in
proportion to the fatty acid composition of pork belly adipose
tissue. When the study was repeated with norleucine, no amide
formation took place in the cooked-out fat, even after the
additional heating period. Furthermore, when pork bellies were
stitch pumped with pentylamine (or norleucine) and trilaurin,
smoked, sliced and fried, amide formation was not evident in
either the bacon or in the cooked-out fat. The stability of
N-nitrosopentylpalmitamide was also studied by injecting a
solution of the N-nitrosamide in oil into the pork belly,
storing overnight at 4°C and frying as before. The overall
decomposition was approximately 93%, and only trace amounts were
present in both the fried bacon and cooked-out fat. The major
conclusion from this investigation was that N-nitrosamides are
unlikely to be present in heat-processed foods.

Interaction of lipid oxidation products and amino compounds.
Amino acids and primary amines may be involved in other
reactions which could lead to the formation of compounds having
the potential to undergo N-nitrosation. Malonaldehyde, produced
as a result of oxidation of lipids, particularly polyunsaturated
fatty acids, has been shown to react with amino acids to produce

conjugated Schiff bases of the general formula, RN=CH-CH=CH-NH-R
(81). Similarly, a diene conjugated enamine-imine compound has
been identified in a freeze-dried cellulose system containing
malonaldehyde and phosphatidyl ethanolamine, and which had been
stored at ambient temperature and a relative humidity of 14% for
30 days (82). Preliminary studies in our laboratory indicate
that these compounds can be N-nitrosated under appropriate
conditions (83). N-Nitrosoenamines (α,β-unsaturated N-nitrosamines) have been synthesized and proven to be very useful
intermediates in the synthesis of a number of compounds (84).
        Products of lipid oxidation may also influence the
formation of N-nitroso compounds in other ways. Malonaldehyde,
prepared by the acid hydrolysis of its tetraacetal, has been
shown to greatly influence the rate of N-nitrosation of
dimethylamine, decreasing the formation of NDMA at pH 3, and
increasing it over the pH range 4 to 7 (85). Glucose, furfural,
benzaldehyde and glyoxal, when tested under the same conditions
had little influence on NDMA formation. It was suggested that
malonaldehyde formed in the fat of adipose tissue, might promote
the formation of N-nitrosamines in fried bacon. Further studies
are obviously required in this area, since lipid hydroperoxides
have also been reported to have a catalytic effect on N-nitrosamine formation in model systems (37).
        Two other model system studies involving interaction of
amino and carbonyl compounds require mentioning. Amadori
compounds (N-substituted 1-amino-1-deoxy-2-ketoses), produced in
the Maillard reaction, are weakly basic, easily N-nitrosated
secondary amines that occur widely in most heat-processed foods
(86). Fructose-L-tryptophan, when treated with excess sodium
nitrite, gave positive dose responses for the Ames mutagen
assays on Salmonella typhimurium TA 98 and TA 100 in the absence
of microsomal activation. Sakaguchi and Shibamoto (87) have
reported that many heterocyclic compounds containing a nitrosatable nitrogen, including thiazolidines, are produced from the
reaction of cysteamine and glucose, acetaldehyde, or glyoxal.
Thiazolidine and its 2-alkyl derivatives readily react with
nitrite to produce N-nitroso compounds, some of which display
mutagenic responses toward S. typhimurium TA 100 (88). At
present, thiazolidines have not been identified in food systems
(88).

Conclusion

        Results of a systematic study (89) of food from the German
market indicate the average daily intake for male persons
amounts to 1.1 μg for NDMA and 0.1-0.15 μg for NPYR.
Approximately 64% of this total daily intake for NDMA is found
in beer, while another 10% comes from cured meat products. It
is also important to note that food is not only the only source
of N-nitrosamine or N-nitrosatable amines to which we are

exposed. A recent paper by Fine (90) indicates that the largest
human exposure to these compounds is in the industrial section.
For example, leather tanners are exposed to NDMA at levels up to
47 µg/m^3, corresponding to 440 µg per day. Workers in the
curing areas of tire factories are exposed to NMOR at levels up
to 27 µg/m^3, corresponding to a daily intake of 250 µg per
person. These industrial exposures are considerably greater
than exposure via foodstuffs, beer, cosmetics and herbicides.

Acknowledgments.
        Part of the work reported in this paper was supported by
Grant No. 1 RO1CA26576-01, awarded by the National Cancer
Institute, DHEW. Michigan Agricultural Experiment Station
Journal Article No.   10029   .

Literature Cited

1. Gray, J.I., Irvine, D.M., Kakuda, Y.   J. Food Protect.
    1979, 42, 263-272
2. Gray, J.I., Randall, C.J.   J. Food Protect. 1979, 42, 168-179
3. Crosby, N.T. Residue Rev. 1976, 64, 77-135
4. Fiddler, W. Toxicol. Appl. Pharmacol. 1975, 31, 352-360
5. Issenberg, P. Fed. Proc. 1975, 36, 1322-6
6. Scanlan, R.A. C.R.C. Critical Rev. Food Technol. 1975, 5,
    363-402
7. Shank, R.C. Toxicol. Appl. Pharmacol. 1975, 31, 361-8
8. Sebranek, J.G. Food Technol. 1979, 33,(7), 58-63
9. Wogan, G.N., Tannenbaum, S.R. Toxicol. Appl. Pharmacol.,
    1975, 31, 375-383
10. Fine, D.H., Rufeh, F., Lieb, D., Rounbehler, D.P. Anal.
    Chem. 1975, 47, 1188-1191
11. Kann, J., Spiegelhalder, B., Eisenbrand, G., Preussmann, R.
    Z. Krebsforsch. 1977, 40, 321-3
12. Rubin, L.J. Can. Inst. Food Sci. Technol. J. 1977, 10, A11-13
13. Crosby, N.T., Foreman, J.K., Palframan, J.F., Sawyer, R.
    Nature (London) 1972, 238, 342-3
14. Sen, N.P., Donaldson, B., Iyengar, J.R., Panalaks, T. Nature
    (London) 1973, 241, 473-4
15. Fiddler, W., Pensabene, J.W., Fagan, J.F., Thorne, E.J.,
    Piotrowski, E.G., Wasserman, A.E. J. Food Sci. 1974, 39,
    1070-1
16. Pensabene, J.W., Fiddler, W., Gates, R.A., Fagan, J.C.,
    Wasserman, A.E. J. Food Sci. 1974, 39, 314-6
17. Gray, J.I. Collins, M.E., Russell, L.F. Can Inst. Food Sci.
    Technol. J. 1977, 10, 36-9
18. Pensabene, J.W., Feinberg, J.I., Dooley, C.J., Phillips,
    J.G., Fiddler, W. J. Agric. Food Chem. 1979, 27, 842-5
19. Sen, N.P., Seaman, S., Miles, W.F. J. Agric. Food Chem. 1979
    27, 1354-7

20. Pensabene, J.W., Fiddler, W., Miller, A.J., Phillips, J.G.
    J. Agric. Food Chem. 1980, 28, 966-970
21. Gough, T.A., Goodhead, K., Walters, C.L. J. Sci. Food Agric.
    1976, 27, 181-5
22. Hwang, L.S., Rosen, J.D., J. Agric. Food Chem. 1976, 24,
    1152-4
23. Warthesen, J.J., Bills, D.D., Scanlan, R.A., Libbey, L.M.
    J. Agric. Food Chem. 1976, 24, 892-4
24. Sen, N.P., Seaman, S., Miles, W.F. Food Cosmet. Toxicol. 1976
    14, 167-170
25. Gray, J.I., Collins, M.E. J. Food Sci. 1977, 42, 1034-7
26. Mottram, D.S., Patterson, R.L.S., Edwards, R.A., Gough, T.A.
    J. Sci. Food Agric. 1977, 28, 1025-9
27. Gray, J.I., Collins, M.E., MacDonald, B. J. Food Protect.
    1978, 43, 31-5
28. Janzowski, C., Eisenbrand, G., Preussmann, R. Food Cosmet.
    Toxicol. 1978, 16, 343-8
29. Bharucha, K.R., Cross, C.K., Rubin, L.J. J. Agric. Food
    Chem. 1979, 27, 63-9
30. Gray, J.I. J. Milk Food Technol. 1976, 39, 686-692
31. Lakritz, L., Spinelli, A.M., Wasserman, A.E. J. Food Sci.
    1976, 41, 879-881
32. Gray, J.I., Collins, M.E. Can. Inst. Food Sci. Technol. J.
    1977, 10, 97-9
33. Sen, N.P., Donaldson, B., Seaman, S., Iyengar, J.R., Miles
    W. J. Agric. Food Chem. 1976, 24, 397-401
34. Hansen, T., Iwaoka, W., Green, L., Tannenbaum, S.R. J. Agric.
    Food Chem. 1977, 25, 1423-6
35. Pensabene, J.W., Feinberg, J.I., Piotrowski, E.G., Fiddler,
    W. J. Food Sci. 1979, 44, 1700-2
36. Lee, M-L. 1981. M.S. Thesis, Michigan State University
37. Coleman, M.H. J. Food Technol. 1978, 13, 55-69
38. Herring, H.K. Proc. Meat Industry Research Conf. 1973, p.47
39. Mirvish, S.S. J. Natl. Cancer Inst. 1970, 44, 633-9
40. Mirvish, S.S. Toxicol. Appl. Pharmacol. 1975, 31, 325-351
41. Sen, N.P., Iyengar, J.R., Donaldson, B.A., Panalaks, T. J.
    Agric. Food Chem. 1974, 22 540-1
42. Dudley, R. Meat Ind. 1979, 26(2),24
43. Federal Register, 1975, 40, 52614
44. Coffin, D.E., Proc. 58th Annual Meeting, Meat Packers Council
    of Canada, 1978, Montreal, Canada
45. Havery, D.C., Fazio, T., Howard, J.W. J. Assoc. Off. Anal.
    Chem. 1978, 61, 1379-1382
46. Robach, M.C., Owens, J.L., Paquette, M.W., Sofos, J.N., Busta
    F.F. J. Food Sci. 1980, 45, 1280-4
47. Tanaka, K., Chung, K.C., Hyatsu, H., Kada, T. Food Cosmet.
    Toxicol. 1978, 16, 209-215
48. Greenberg, R.A. Proc. Int. Symp. Nitrite Meat Prod., 1973,
    Zeist, p. 179
49. Bharucha, K.R., Cross, C.K., Rubin, L.J. J. Agric. Food Chem.
    1980, 28, 1274-1281

50. Fiddler, W., Pensabene, J.W., Piotrowski, E.G., Phillips, J.G., Keating, J., Mergens, W.J., Newmark, H.L. J. Agric. Food Chem. 1978, 26, 653-6
51. Mergens, W.J., Newmark, H.L. Proc. Meat Industry Research Conf. 1979, p.79
52. Mottram, D.S., Patterson, R.L.S. J. Sci. Food Agric. 1977, 28, 1025-9
53. Walters, C.L., Edwards, M.W., Elsey, T.S., Martin, M. Z. Lebensm. Unters.-Forsch. 1976, 162, 377-385
54. Sen, N.P., in "Safety of Foods," p. 319-349, ed. Graham, H.D. AVI Publishing Co., Westport, Conn., 1980
55. Sink, J.D., Hsu, L.A. J. Food Sci. 1977, 42, 1489-1493, 1503
56. Knowles, M.E. Nature (London), 1974, 249, 672-3
57. Davies, R., Massey, R.C., McWeeny, D.J. Food Chem. 1980, 6, 115-122
58. Nitrite Safety Council. Food Technol. 1980, 34,(7),45-53,103
59. Holland, G., Wood, D.F., Randall, C.J. Can. Inst. Food Sci. Technol. J. 1981, in press.
60. Gray, J.I., Bussey, D.M., Dawson, L.E., Price, J.F., Stevenson, K.E., Owens, J.L., Robach, M.C. J. Food Sci. 1981, in press
61. Fajen, J.M., Carson, C.A., Rounbehler, D.P., Fan, T.Y., Vita, R., Goff, V.E., Wolf, M.H., Edwards, G.S., Fine, D.H., Reinhold, V., Bieman, K. Science 1979, 205, 1262-4
62. Randall, C.J. 1980. Personal Communication.
63. Gough, T.A., McPhail, M., Webb, K.S., Wood, B.J., Coleman, R.F. J. Sci. Food Agric. 1977, 28, 345-351
64. Sen, N.P., Donaldson, B.A., Seaman, S., Iyengar, J.R., Miles, W.F. Paper presented at Symp. Nitrosamines in Cheese, 1977, Ottawa, Canada
65. Libbey, L.M., Scanlan, R.A., Barbour, J.F. Food Cosmet. Toxicol. 1980, 18, 459-461
66. Sen, N.P., Seaman, S. J. Assoc. Off. Anal. Chem. 1981, in press.
67. Havery, D.C., Kline, D.A., Miletta, E.M., Joe, F.L., Fazio, T. J. Assoc. Off. Anal. Chem. 1976, 59, 540-6
68. Lakritz, L., Pensabene, J.W. J. Dairy Sci. 1981, 64, 371-4
69. Spiegelhalder, B., Eisenbrand, G., Preussmann, R. Food Cosmet Toxicol. 1979, 17, 29-31
70. Havery, D.C., Hotchkiss, J.H., Fazio, T. J. Food Sci. 1981, 46, 501-5
71. Scanlan, R.A., Barbour, J.F., Hotchkiss, J.H., Libbey, L.M. Food Cosmet. Toxicol. 1980, 18, 27-9
72. Goff, E.V., Fine, D.H. Food Cosmet. Toxicol. 1979, 17, 569-573
73. Hotchkiss, J.H., Barbour, J.F., Scanlan, R.A. J. Agric. Food Chem. 1980, 28, 678-680
74. Brewers Association of Canada, Final Report, Dec. 15, 1980, PDR Contract with Agriculture Canada

75. Mangino, M.M., Scanlan, R.A., O'Brien, T.J. 1981, Paper
    presented at ACS National Meeting, Atlanta, Georgia
76. Sims, R.J., Fioriti, J.A. J. Am. Oil Chem. Soc. 1975, 52,
    144-7
77. Kakuda, Y., Gray, J.I. J. Agric. Food Chem. 1980, 28, 580-4
78. Kakuda, Y., Gray, J.I. J. Agric. Food Chem. 1980, 28, 584-7
79. Kakuda, Y., Gray, J.I. J. Agric. Food Chem. 1980, 28, 584-591
80. Fooladi, M. 1981, Ph.D. Thesis, Michigan State University
81. Dillard, C.J., Tappel, A.L. Lipids, 1973, 8, 183-9
82. Dugan, L.R., Rao, G.V. Technical Report 72-27-FL, 1972,
    United States Army Laboratories, Natick, Massachusetts
83. Mandagere, A., Gray, J.I. 1981, Unpublished data
84. Kupper, R., Michejda, C.J. J. Org. Chem. 1980, 45, 2919-2921
85. Kurechi, T., Kikugawa, K., Ozawa, M. Food Cosmet. Toxicol.
    1980, 18, 119-122
86. Coughlin, J.R., Wei, C.I., Hsieh, D.P.H. Russell, G.F. 1979,
    Paper presented at ACS/CS J. Chem. Cong., Honolulu, Hawaii
87. Sakaguchi, M., Shibamoto, T. J. Agric. Food Chem. 1978, 26,
    1179-1183
88. Sekizawa, J., Shibamoto, T. J. Agric. Food Chem. 1980, 28,
    781-3
89. Spiegelhalder, B., Eisenbrand, G., Preussmann, R. Oncology
    1980, 37, 211-6
90. Fine, D.H., Oncology 1980, 37, 199-202

RECEIVED August 10, 1981.

N-Nitrosomorpholine Synthesis in Rodents Exposed to Nitrogen Dioxide and Morpholine

SIDNEY S. MIRVISH, PHILLIP ISSENBERG, and JAMES P. SAMS

Eppley Institute for Research Cancer, University of Nebraska Medical Center, Omaha, NE 68105

We examined the possibility that N-nitroso-morpholine (NMOR) was produced in vivo when mice were gavaged with morpholine and exposed to atmospheric nitrogen dioxide (NO_2). Using a "stopping solution" adjusted to pH 1 that contained ascorbate and sulfamate to prevent artifactual nitrosation, and cis-2,6-dimethylmorpholine to detect any such nitrosation, we demonstrated that no significant NMOR was formed in rat stomach contents or blood, or the whole mouse, after exposure to morpholine and NO_2. We verified the results of Iqbal et al. (Science 207:1475, 1980) that NMOR was produced when mice were treated with morpholine and NO_2, and the whole mouse homogenate in aqueous methanol was worked up without using a stopping solution. We showed that the NMOR was produced in the homogenate by a nitrosating agent present in several organs of NO_2-exposed mice.

Nitrogen dioxide (NO_2) is an important atmospheric pollutant, arising mainly from motor vehicles or stationary installations burning fossil fuels (1). The main nitrogen oxide species are NO_2 (the more toxic agent) and nitric oxide (NO). Peak 1-hour NO_2 concentration may exceed 0.4 ppm in Los Angeles and 0.14 ppm in many U.S. cities (1). Fresh cigarette smoke contains 0.33 mg NO/cigarette and traces of NO_2 (2), but the NO_2 content of stale smoke presumably rises as the NO is oxidized.

Nitrosamines are readily produced from the reaction of amines with nitrous acid, i.e., acidified nitrite (3), and are also produced in vivo when amines and nitrite are administered, as reflected by tumor induction [reviewed in (3)], and the in vivo appearance of nitrosamines (3,4). Challis and Kyrtopoulos (5) showed that gaseous NO_2 reacts directly with amines in neutral or alkaline aqueous solutions to produce nitrosamines and nitramines. The kinetics of the extremely rapid reaction of

nitrogen oxides with amides in organic solvents was also recently
studied (6). Nitrosamines in certain foods, e.g., beer, probably
arise from reactions of NO_2. These findings raised the possibility that nitrosamines are produced in vivo by the reaction of
amines and inhaled NO_2. To test this possibility, we gavaged
rats and mice with morpholine, exposed them to NO_2, and estimated
the in vivo formation of N-nitrosomorpholine (NMOR).
    Iqbal et al. (7) reported in 1980 that NMOR was produced
in vivo under circumstances similar to those that we were using.
Accordingly, our results represent in part a check on this finding, using both the method of Iqbal et al. (7) and one devised
by us.

Materials and Methods

    Animal treatment. Male 100-g MRC-Wistar or Sprague-Dawley
rats were fasted from the evening of the previous day but were
given water. Non-fasted 30-35 g Swiss mice were used. Except
where specified otherwise, the animals received a standard commercial diet (Wayne Lab Blox, Allied Mills). In the standard
experiments, animals were gavaged with 25 mg morpholine (Fisher
Scientific Co.) in 10 mL water/rat or 2 mg morpholine in 0.2 mL
water/mouse. A stream of NO_2 was prepared by suitable mixing
(using Teflon tubing) of gas from a cylinder of 106 ppm NO_2 in
N_2 (Linde Division, Union Carbide Corp.) and compressed air.
The mixture was passed at 2 L/min through a glass cylinder (9 cm
diam. x 35 cm) containing 3 rats or 4 mice. The effluent gas
was analyzed for NO_2 at hourly intervals by the Griess-Saltzmann
reaction (8). Our results agreed closely with the value of
106 ppm for the NO_2 cylinder, specified by the supplier. Atmospheric NO_2 concentration varied in different experiments, partly
on purpose and partly because the valve settings were not entirely reproducible.
    Our method for NMOR determination. The following procedure
(carried out in a single day) was developed for blood, stomach
contents, the homogenized whole mouse, and diet [(a standard
semisynthetic diet prepared as in (9)]. After the rats were
killed with CO_2, we collected the blood (with a heparinized
syringe from the heart) and the entire stomach contents. The
whole mouse was frozen in liquid N_2 and homogenized as in (7).
One to 5 g material was mixed by a glass rod with 1 mL/g material
of "stopping solution", that contained 25 mg/mL of ammonium sulfamate, 25 mg/mL of sodium ascorbate, and 2.0-2.5 mg/mL of
cis-2,6-dimethylmorpholine (cis-DMM), and was adjusted to pH
1 with H_2SO_4. The pH of the slurry was readjusted to 1 with concentrated H_2SO_4. The mixture was stirred with 10 g Celite 560
(from Johns Manville Corp.; previously sifted to remove < 60 mesh
particles), and packed dry in a "Monoject" plastic 50 mL syringe
barrel (Shermwood Industries) prepacked with 8 g Celite. The
column was eluted (without flow-rate control) with 100 mL

n-hexane and then 100 mL CH_2Cl_2. The CH_2Cl_2 eluate was evaporated
to 4-6 mL in a Kuderna-Danish concentrator and then to 1.0-1.5 mL
with a N_2 stream. A 1-5 μL aliquot was subjected to gas chromatography (GC) on a Bendix model 2200 apparatus, using a 2-m x
2-mm column of 20% Carbowax-20 M-TPA on 100/120-mesh Chromo-
sorb-GHP (Supelco Inc.) at 190°C. For detection we used a Thermal
Energy Analyzer (TEA; Thermo-Electron Corp., model 502), linked
with a Spectraphysics System I integrator. Typical GC-TEA tracings are shown in Figure 1. An "analysis" signifies the full
procedure, applied to blood or stomach contents of a single rat
or 5 g of a whole-mouse homogenate prepared from 1 or 2 mice.

DMM was chosen for detecting nitrosation during the workup
because of its similarity to morpholine. DMM is a good model
because, in a kinetic experiment [10 mM DMM from Aldrich Chem.
Co. and 20 mM nitrite in water, pH 3, 25°C, 60 min reaction],
it was nitrosated to yield 3.62 mM 2,6-dimethyl-N-nitrosomorpho-
line (DMNM), which was a 10% greater yield than that for a similar nitrosation of morpholine to give NMOR. This indicated a
slightly larger rate constant for DMNM than for NMOR formation
(3). Crude DMM is a 2:1 mixture of the cis and trans isomers
(10). GC analysis of the product of the kinetic run showed that
the 2 isomers were nitrosated at similar rates. Cis-DMNM [reten-
tion time (RT), 320 sec] was well separated from NMOR (RT, 430
sec), but trans-DMNM (RT, 405 sec) was not. Accordingly, we prepared pure cis-DMM, b.p. 133°C, by spinning-band fractional distillation of crude DMM and used it in the analytical procedure.
The RT of N-nitrosopyrrolidine (NPYR) was 390 sec.

The stopping solution composition was based on experiments
showing that lowering the pH from 9 to 5 to 1 reduced formation
of cis-DMNM, perhaps because this prevented elution of the amine
from the column, and that addition of ammonium sulfamate lowered
cis-DMNM formation, relative to the situation where ascorbate
alone was used as a nitrite trap. The hexane wash of the column
was introduced to remove a large peak near the solvent front
in the GC-TEA, perhaps due to neutral fats. Using the described
procedure, the recovery of 70-160 ng NMOR added to 2 g semisynthetic diet was 92 ± 19% (mean ± S.D. for 13 measurements). The
recovery of 227 ng NMOR from 5-8 g whole mouse homogenate using
the Iqbal method was 101 ± 57% for 7 measurements.

Method of Iqbal et al. (7) for NMOR determination. The published method (7) was used, with slight modifications. In brief,
the frozen mouse (stored for 1-7 days in liquid N_2) was blended
to give a powder and 5-g samples were reblended with 75 mL
ice-cold 35% aqueous methanol. The mixture (which was at pH 7.6)
was centrifuged at 5°C and the pellets were re-extracted with
aqueous methanol. The combined aqueous method extract was extracted twice with CH_2Cl_2. Our method was used from then on,
i.e., the extract was passed through a Celite column, concentrated, and subjected to GC-TEA as before.

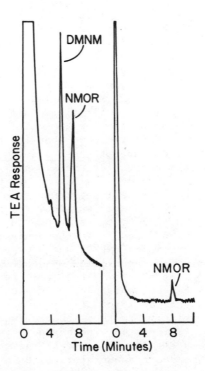

Figure 1. Two GC-TEA tracings of experiments with analysis by our method. The left tracing shows the results of an experiment, similar to that in Table I, Experiment 4, in which nitrite was added to the diet and the hexane wash of the column (which reduced the solvent-front peak) was omitted. The right tracing shows one of the positive experiments (including the hexane wash) on the homogenate of a mouse exposed to morpholine and NO_2 (Table II, Experiment 3).

Results

Rats were used initially because we have studied the intra-gastric fate of nitrite (11) and intragastric methylnitrosurea (MNU) formation (12) in this species. The exposure routes were the atmosphere for NO_2 and either gavage of aqueous solutions or feeding in the diet for morpholine. In experiment (exp.) 1 of Table I, rats were gavaged with morpholine, left 30 min, exposed to NO_2 (12–21 ppm) for 30 min, left in air for 30 min (to give time for nitrosation to proceed), and killed. The waiting period after morpholine administration was designed to allow time for the amine to diffuse from the stomach to sites where NO_2 could react with it. The waiting period after NO_2 exposure was inserted to allow time for nitrosation to proceed. The observed NMOR level was <15 ng/g blood or stomach contents, much less than the cis–DMNM level, and was therefore considered to be artifactual, i.e., not produced in vivo. "Exp. 1" is the composite result of 11 experiments carried out on 8 different days. In exp. 2 of Table I, rats were fasted overnight, given 2 g semi-synthetic diet [as in (9)] containing 10 mg morpholine over 1 hour [as in (11) and (12)], exposed to NO_2 for 30 min, left in air for 30 min, and killed. Here too, no significant NMOR was observed.

Table I

NMOR production in rats, with analysis by our method.[a]

Exp. no.	Organ	No. of analyses	NMOR, ng/g tissue	DMNM, ng/g tissue	NMOR real?
1.	MOR (1 g/kg gavaged, NO_2 (12–21 ppm) in air[b]				
	Blood	4	6 + 3	56 + 23	no
	Stomach	16	14 + 3	118 + 59	no
2.	MOR (5 g/kg) in diet, NO_2 (26 ppm) in air[c]				
	Stomach	3	5 + 2	15 + 8	no
3.	MOR (5 g/kg) and $NaNO_2$ (4 g/kg) in diet[c]				
	Stomach	3	47,000 + 6,000	1,600 + 800	yes
4.	MOR (0.5 g/kg) and $NaNO_2$ (0.4 g/kg) in diet[c]				
	Stomach	3	37 + 11	32 + 17	yes

[a] Results are given as mean + S.E.

[b] Rats received commercial diet.

[c] Semisynthetic diet was used.

Exps. 3 and 4 represented positive controls, in which morpholine and sodium nitrite ($NaNO_2$) were administered in the diet.
This system is known to induce tumors attributed to in vivo NMOR
production (3,13). The rats were fasted overnight, presented
with 2 g diet containing freshly added morpholine and $NaNO_2$,
and killed 2 h later. At the higher doses, the large NMOR yield
was clearly produced for the most part in vivo. The DMNM yield
in this group indicated that some NMOR was also produced during
the workup. The NMOR yield in exp. 4 was 1/1,270 of that in exp.
3, approximately in accord with the 1/1,000 ratio derived from
the third-order nitrosation kinetics (3). (Both the morpholine
and the $NaNO_2$ doses in exp. 4 were 1/10 for those in exp. 3.
Reaction rate is proportional to amine concentration and to nitrite concentration squared, and hence should be reduced 1/10
because of the drop in morpholine concentration and a further
1/100 because of the drop in nitrite concentration.)

We then repeated the experiments by Iqbal et al. (7) on
Swiss mice. These were gavaged with morpholine, immediately exposed to NO_2 for 4 h, and then killed. Unlike the experiments
with rats, an entire mouse was homogenized and the homogenate
was stirred with aqueous methanol (in the absence of stopping
solution) and analyzed for NMOR. In exp. 1 of Table II, analysis
of the homogenate by the Iqbal method gave 140 ng NMOR/g tissue.
Two similar experiments were performed, in which 10 mg cis-DMM
was added to the aqueous methanol together with the mouse homogenate (exp. 2). Large amounts of cis-DMNM were detected, but
little NMOR, indicating that the NMOR in exp. 1 was formed during
the workup. In exp. 3, mice were exposed to morpholine and NO_2,
and homogenized as in exps. 1 and 2, but the whole mouse homogenate was worked up by our method. Very little cis-DMNM was observed, indicating that there was no artifactual nitrosation
under these conditions. The NMOR yield was zero in 11 analyses
and 71 and 88 ng/g tissue in 2 analyses carried out on the same
day, indicating that, overall, there was no significant in vivo
NMOR production.

Most of the results in exp. 1 of Table II were obtained
with mice stored frozen in liquid N_2 for at least 1 day after
the NO_2 exposure. In one case (included in exp. 1), the entire
experiment was performed on a single day, with results similar
to those for the remaining mice. In another check, our method
was applied to a mouse homogenate, but this was mixed with stopping solution using a Waring blender (and not just by stirring).
As in exp. 3 of Table II, the results were negative, indicating
that the method of mixing was not responsible for the low results
with our method.

The results in Table II indicated that a nitrosating agent,
produced in vivo, nitrosated morpholine or DMM in the homogenate,
but did not nitrosate morpholine in vivo. To demonstrate more
clearly the presence of this agent, we used the Iqbal method
to prepare homogenate from mice exposed to NO_2, but not gavaged

with morpholine, and then studied the effect of adding morpholine, DMM and pyrrolidine to this homogenate (Table III). In exps. A1–A3, 5 g homogenate was mixed with 105 μmol amine in 5 ml H_2O, rehomogenized immediately with the aqueous methanol, and worked up by the Iqbal method. In exp. A4, 105 μmol of each of the 3 amines was added to a single homogenate. All 3 amines were nitrosated, demonstrating again that the homogenate of NO_2-treated mice contained a nitrosating agent. Each amine yielded a GC-TEA peak at the correct position for the corresponding nitrosamine, confirming that the nitrosamines were indeed produced. (This was useful, since the structures were not confirmed by mass spectrometry.) The nitrosamine yield followed the order DMNM > NMOR > NPYR, suggesting that nitrosation was related to lipophilicity, which followed the same order for the nitrosamines (14) and hence, probably, for the non-ionized amine, and to the ease of formation from nitrite, which followed the order DMNM $\doteq$ NMOR > NPYR [see Materials and Methods and (3)].

Table II

In vivo production of NMOR from morpholine and 22–31 ppm NO_2, with analysis of the whole mouse by our method or that of Iqbal et al.[a]

Exp. no.	Method	No. of analyses	NMOR, ng/g tissue	DMNM, ng/g tissue
1	Iqbal	6	140 ± 40	0.3 ± 0.3[b]
2	Iqbal	2	4 ± 0	140 ± 10
3	Our	13	13 ± 8	0.1 ± 0.1

[a] Results are given as mean ± S.E.
[b] cis-DMM was not included here.

In exp. B of Table III, exp. A1 was repeated, but a control group without added morpholine was included. The results demonstrate that morpholine was necessary for a large NMOR peak to be present, but that a small peak with the RT of NMOR was detected in the absence of added morpholine. Only a small amount (0.2 nmol) of the NMOR in exp. B1 is attributed to NMOR in the added morpholine.

Table IV records a study of tissues in which the nitrosating agent occurred. Two mice were exposed to NO_2, killed with CO_2, and dissected to give the skin, liver, lungs, and remainder of the body ("carcass"). Corresponding tissues of the 2 mice were combined and frozen in liquid N_2. The entire tissue, or 5 g of the carcasses (total weight, 41 g) was homogenized, 10 mg

morpholine was added, and the homogenate was worked up by the
Iqbal method. The NMOR yield expressed per g tissue was similar
for the skin and carcass, and much less for the liver and lungs.
This shows that the precursor was not concentrated in the lungs
(the site of NO_2 entry) or the skin (with which NO_2 had direct
contact).

Table III

Effect of adding different amines to a homogenate of mice exposed
to 32 ppm NO_2 for 4 h, with analysis by the Iqbal method

| Exp. no. | Added amine | Nitrosamine yield | |
		Name	nmol/ 5 g tissue
Exp. A			
1	Morpholine	NMOR	8.9
2	cis-DMM	DMNM	32.3
3	Pyrrolidine	NPYR	3.9
		NMOR	7.5
4	All 3 amines	DMNM	19.2
		NPYR	1.4
Exp. B			
1	Morpholine	NMOR	4.36 ± 0.78^a
2	None	"NMOR"	0.71 ± 0.14^a

[a] Mean $\pm$ S.E. for 4 results.

It was possible that the nitrosating agent was NO_2 itself
and that this disappears rapidly from the body after the NO_2
exposure. To check this point, mice were exposed to 29 ppm NO_2
for 4 h, air alone was passed into the chamber at 1.0 L/min for
10 min, and the mice were killed. The homogenate was mixed with
10 mg morpholine and worked up by the Iqbal method. The NMOR
yield was 93 ng/g tissue (66% of the yield in Table II, exp.
1). Hence the nitrosating agent did not disappear rapidly from
the body after the NO_2 exposure.

To check that the NO_2 exposure was necessary for nitrosamine
formation, morpholine was gavaged to mice as usual, but the NO_2
exposure was omitted. Five g mouse homogenate was worked up by
the Iqbal method (after addition of 11 mg cis-DMM) and yielded
neither NMOR nor cis-DMNM. Similarly, when 10 mg morpholine was
added to 5 g homogenate prepared from an untreated mouse and

the mixture was worked up by the Iqbal method, only 9 ng NMOR/g homogenate was obtained.

Table IV

Nitrosating agent in various mouse organs: effect of adding morpholine to homogenates of different organs of mice exposed to 23 ppm NO_2 for 5 h, with analysis for NMOR by the Iqbal method

Organ		NMOR	
Name	Weight (g)	Total ng	ng/g tissue
Skin	10.1	704	70
Liver	2.9	81	28
Lungs	0.67	42	63
Carcass	5.0	335	67

To check that the recovered NMOR was not present in the morpholine sample or derived therefrom in the absence of nitrosating agent, 10 mg morpholine was added to 60 ml 35% methanol in the absence of mouse homogenate, the pH was adjusted to 7.6, and the mixture was worked up by the Iqbal method. We obtained 12 ng NMOR, corresponding to 1.2 ppm NMOR in the morpholine. This is equivalent to 2.4 ng/g mouse homogenate in the standard experiment, which is insignificant relative to the larger yields in Table II. These checks confirmed that NMOR obtained by the Iqbal method was due to the action of a nitrosating agent produced in vivo by NO_2.

Discussion

When mice were gavaged with morpholine and exposed to NO_2, the NMOR yield of 140 ng/g (4.2 μg/30 g mouse) obtained by the Iqbal method (Table II, exp. 1) was similar to that of 2.23 μg/mouse previously reported by Iqbal et al. (7), and hence confirms these results. However, the finding that similar homogenates, when worked up by our method, did not yield a significant amount of NMOR demonstrated that the NMOR formation occurred during the workup, i.e., was an artifact. This confirmed our negative results with rat stomach contents and blood (Table I). A difference from the previous study (7) is that the animals receiving the maximum level of 32 ppm NO_2 were breathing a mixture containing only 14% O_2. We plan to repeat some of our results using NO_2 prepared in air (and not in N_2), but it is unlikely that the different O_2 contents account for the observed differences.

Iqbal et al. (7) reported two checks to demonstrate that NMOR formation was not artifactual. (a) Addition of sodium ascorbate to the blended mouse did not significantly decrease the NMOR yields. However, at the pH of the homogenate (around 7), ascorbate does not reduce nitrite and hence could not have prevented NMOR formation (3). (b) The NMOR yield was reduced > 90% when morpholine was gavaged after exposure of the mice to NO_2, or was added to the homogenate derived from NO_2-exposed mice. We cannot explain these results. In our experiments, morpholine added to the aqueous methanol yielded 1,032 ng NMOR/5 g homogenate (Table III, exp. 1) and morpholine gavaged to mice yielded 700 ng NMOR/5 g homogenate (Table II, exp. 1).

We conclude that exposure of mice to NO_2 produced a nitrosating agent in vivo, that could nitrosate morpholine and other amines when the mouse homogenate was stirred with aqueous methanol, but did not nitrosate morpholine in vivo. The biological significance of this nitrosating agent remains to be established. It is especially important to discover why the agent did not nitrosate morpholine in vivo under our conditions, and whether it could nitrosate other amines or amides in vivo.

The active agent could be nitrite, since N_2O_4 (the dimeric form of NO_2) reacts with water to produce equal amounts of nitrite and nitrate, or even NO_2 itself, which reacts directly with amines in aqueous solution to yield nitrosamines and nitramines (5). Another possibility is nitrite esters, e.g., the nitronitrite esters (nitrosates) produced by adding N_2O_4 to ethylenic groups in unsaturated fatty acids (15,16), analogous to adding NO to yield the nitrosonitrite esters (nitrosites) (17). The nitrosating agent could be important biologically since lung adenomas were induced in mice exposed to 1-2 ppm NO_2 by inhalation and treated with morpholine (1.0 g/L drinking water) (18).

Acknowledgements

We thank Dr. B.C. Challis (Chemistry Dept., Imperial College, London), Dr. D. Nagel (this Institute) for useful discussions, and Dr. T. Lawson (this Institute) for gavaging the mice. This research was supported by grant P01-CA251000 from the National Cancer Institute, National Institutes of Health.

Literature Cited

1.   U.S. Environmental Protection Agency. "Air Quality Criteria for Oxides of Nitrogen", Res. Triangle Park, NC, 1979.
2.   Wynder, E.L.; Hoffman, D. "Tobacco and Tobacco Smoke", Academic Press, New York, 1967.
3.   Mirvish, S.S. Toxicol. Appl. Pharmacol., 1975, 31, 325-351.

4. Iqbal, Z.M.; Krill, I.S.; Mills, K.; Fine, D.H.; Epstein, S.S. In Walker, E.A., et al., Eds.; "Analysis and Formation of N-Nitroso Compounds"; Internat. Agency Res. Cancer: Lyon, 1980, 169–182.
5. Challis, B.D.; Kyrtopoulos, S.A. J. Chem. Soc. Chem. Comm., 1976, 877–878.
6. Mirvish, S.S.; Karlowski, K.; Sams, J.P.; Arnold, S.D. In Walker, E.A., et al., Eds.; "Environmental Aspects of N-Nitroso Compounds"; Internat. Agency Res. Cancer: Lyon, 1978; pp 161–174.
7. Iqbal, Z.M.; Dahl, K.; Epstein, S.S. Science, 1980, 207, 1475–7.
8. Katz, M., Ed.; "Methods of Air Sampling and Analysis"; Am. Public Health Assoc: Washington, DC, 1977; pp 527–534.
9. Mirvish, S.S.; Rose, E.F.; Sutherland, D.M. Cancer Lett., 1979, 6, 159–165.
10. Gingell, R.; Nagel, D.; Kupper, R. Xenobiot., 1978, 8, 439–443.
11. Mirvish, S.S.; Patil, K.; Ghadirian, P.; Kommineni, V.R.C. J. Natl. Cancer Inst., 1975, 54, 869–875.
12. Mirvish, S.S.; Karlowski, K.; Birt, D.F.; Sams, J.P. In Walker, E.A. et al., Eds.; "Analysis, Formation, and Occurrence of N-Nitroso Compounds"; Internat. Agency Res. Cancer: Lyon, 1980; pp 271–279.
13. Mirvish, S.S.; Pelfrene, A.F.; Garcia, H.; Shubik, P. Cancer Lett., 1976, 2, 101–108.
14. Mirvish, S.S.; Issenberg, P.; Sornson, H.C. J. Natl. Cancer Inst., 1976, 56, 1125–1129.
15. Schechter, H.; Gardikes, J.J.; Pagano, A.H. J. Am. Chem. Soc., 1959, 81, 5420–5423.
16. Seifert, W.K. J. Org. Chem., 1963, 28, 125–129.
17. Walters, C.L.; Hart, R.J.; Perse, S. Z. Lebensm. Unters. Forsch., 1979, 168, 177–180.
18. Van Stee, E.W.; Boorman, G.A.; Haseman, J.K. Pharmacologist, 1980, 22, 158.

RECEIVED August 24, 1981.

Blocking Nitrosation Reactions In Vivo

WILLIAM J. MERGENS and HAROLD L. NEWMARK

Hoffmann–La Roche Incorporated, 340 Kingsland Street, Nutley, NJ 07110

$\underline{N}$-Nitroso compounds, once formed and present in vivo, generally do not easily or readily revert back to precursors. Instead, in vivo, they metabolize to, or are converted to alkylating agents as terminal or proximate carcinogens. Control of $\underline{N}$-nitroso compounds, as carcinogenic agents, so far has rested on agents that can block their formation. In order to control carcinogenesis due to $\underline{N}$-nitroso compounds, it is necessary to consider their chemical properties and methods of formation within the body. The in vivo formation of $\underline{N}$-nitroso compounds is directly dependent on the source of nitrosating agent, and recent studies would indicate that they are, indeed, numerous ranging from atmospheric exposure to oxides of nitrogen through bacterial reduction of nitrate in the gastrointestinal tract. The generally accepted mechanism of blocking these reactions is one of competitive kinetics between the susceptible amine and potential blocking agent for the nitrosating agent. The ultimate effectiveness of any given blocking agent then depends on being able to deliver it in sufficient concentration to the organ and phase where the nitrosation takes place. Ascorbic acid and $\underline{dl}$-alpha-tocopherol are water-soluble and lipid-soluble vitamins, respectively, that have been found to be extremely efficient inhibitors of nitrosation reactions in vivo. These mechanism will be presented along with the influence of normal dietary components on nitrosation reactions.

The class of $\underline{N}$-nitroso compounds (i.e., nitrosamines and nitrosamides) is currently considered a unique group that includes members with remarkable carcinogenic properties. Because of their potency and almost ubiquitous presence in certain foods, beverages, and elsewhere in our environment (1, 2, 3), the N-nitroso compounds could represent a significant carcinogenic input in the gastrointestinal system. The carcinogenic and mutagenic properties of the N-nitroso compounds have been reviewed extensively elsewhere (4).

N-nitroso compounds, once formed and present in vivo, generally do not revert easily or readily back to precursors. Instead, in vivo, they are metabolized or otherwise converted to alkylating agents as terminal or proximal carcinogens. Control of N-nitroso compounds as carcinogenic agents so far has rested on agents that can block their formation. It is the intent of this paper to briefly review some of the studies which have been performed on the use of blocking agents to prevent N-nitroso compound formation and describe some recent observations on the mechanism by which these agents function.

Formation of N-Nitroso Compounds

In order to understand how to control carcinogenesis caused by N-nitroso compounds, it is necessary to consider their chemical properties and methods of formation.

The nitrosamines are generally very stable compounds in neutral, alkaline, and weakly acidic solutions. They are uncharged, very soluble, and can readily diffuse through many media and "barriers", including rubber gloves (3, 5, 6). N-nitroso compounds can be decomposed by heating in strong acid or by exposure to ultraviolet light. Their comparatively good stability has permitted development of reliable methods for the ready isolation of nitrosamines in complex analytical schemes (7).

Nitrosamides, however, are generally far less stable, and this has complicated development of reliable analytical methods for measuring their presence in low levels in tissues, foods, etc. However, the reactions caused by their instability do not result in their reversion back to precursors but to re-active intermediates leading to alkylating agents. Thus, in the case of both nitrosamines and nitrosamides, once formed, there is a risk of carcinogenesis by their conversion to an alkylating agent in vivo.

It is, therefore, interesting to determine if (1) forma-tion of these N-nitroso compounds can be prevented or blocked and (2) whether the terminal active alkylating agent can be prevented from attacking the cellular DNA. A large body of reports has appeared in recent years on the successful use of blocking agents, including ascorbic acid (vitamin C) in aqueous media and alpha-tocopherol (vitamin E) in lipid phase, to prevent N-nitroso compound formation (8-29).

In Vitro and In Vivo Studies

Many in vitro and in vivo studies have demonstrated their effectiveness in inhibiting nitrosamine formation in a gastric fluid environment (11, 12, 13, 28). When large single doses of dimethylamine or aminopyrine were gavaged to rats or mice together with a larger $NaNO_2$ dose, acute liver toxicity developed after a few days (30). With both dimethylamine and aminopyrine, the effect is attributed to intragastric formation of dimethylnitrosamine (DMN), which is an acute hepatotoxic agent. When similar experiments were performed with aminopyrine, but with sodium ascorbate added to the amine solution before gavage, hepatotoxicity was completely prevented. In studies where rats received single doses of 1500 mg dimethylamine hydrochloride and 125 mg $NaNO_2$ per kg body weight, sodium ascorbate doses down to 90 mg/kg completely protected the rats from liver necrosis by preventing formation of DMN (19).

Sander and Burkle (31) were the first to induce tumors by feeding nitrite and amines or amides. The tumors were attributed to in vivo formation of the N-nitroso compounds, probably in the stomach. When pregnant rats were gavaged with ethylurea and nitrite, hydrocephalus and nervous system tumors were induced in the offspring. Both these effects were prevented when sodium ascorbate was gavaged together with the ethylurea (21, 32).

Using a piperazine and nitrite system, lung adenoma induction was approximately proportional to piperazine dose and to the square of nitrite dose, when precursor concentrations were varied (33). When sodium ascorbate was added to the food together with the amine or urea and $NaNO_2$ was added to drinking water, the number of lung adenomas was reduced, compared to the group without sodium ascorbate (9).

The incidence of liver tumors due to morpholine and nitrite was reduced from 65% in the absence of to 49% in the presence of sodium ascorbate, and the latent period was nearly doubled, from 54 to 93 weeks, indicating in vivo nitrosomorpholine (NMOR) production was probably about 50% inhibited (10).

Sources of Amines and Nitrosating Agents

In classical organic chemistry, nitrosamines were considered only as the reaction products of secondary amines with an acidified solution of a nitrite salt or ester. Today, it is recognized that nitrosamines can be produced from primary, secondary, and tertiary amines, and nitrosamides from secondary amides. Douglass et al. (34) have published a good review of nitrosamine formation. For the purposes of this presentation, it will suffice to say that amine and amide precursors for nitrosation reactions to form N-nitroso compounds are indeed ubiquitous in our food supply, environment, and particularly in vivo. These precursors of nitrosamines and nitrosamides compose a large part of our living organic world. Therefore, to study means of blocking N-nitroso compound formation, it is necessary to look at the nature of the nitrosating agents and the chemistry of formation of N-nitroso compounds.

There have been some preliminary reports of nitrosamines being found in normal humans, in blood (2, 35), and in urine (36). The precise origin of these substances, if their presence is confirmed, is not fully understood at this time.

An attempt to estimate human daily impact of N-nitroso compounds is shown in Table I. The apparent intake from food of preformed nitrosamines is comparatively low, at least in these surveys of a Western diet in England (3). The intake directly to the respiratory tract from smoking could be somewhat larger. However, if the blood levels reported are confirmed as correct, then inputs of up to 700 mcg per day of at least N-nitrosodimethylamine (NDMA) may be calculated, based on pharmacokinetic considerations of data obtained in animals and extrapolated to man. It should be emphasized that no information is available at present on nitrosamide intake or in vivo formation, largely because of analytical limitations.

Intake of nitrite or nitrosating gases (NO_2, N_2O_3, or generally NO_x given as $NaNO_2$ equivalent) is potentially much higher than preformed nitrosamines. We wish to point particularly to the potential input into the respiratory tract of nitrosating gases, 2-3 mg of NO_2 equivalent from smoking a pack of cigarettes, or up to 100 mg of NO_2 equivalent per day from normal breathing of polluted air with 1 ppm of NO_x content. This level of NO_x is often achieved in cities, and some, like Los Angeles, occasionally reach 2.5 ppm (37).

Tests have demonstrated that inhaled NO_2 is largely absent in expired air, suggesting that it is absorbed rapidly or consumed in reactions in the lung (38). If an individual breathes 1.5 liters per breath, 12 breathes per minute (normal resting breathing for an adult), in an atmosphere of 1 ppm of NO_2, over 1 mmole per day is absorbed.

$$\frac{1500 \text{ ml}}{\text{breath}} \times \frac{12 \text{ breaths}}{\text{minute}} \times \frac{60 \text{ minutes}}{\text{hour}} \times \frac{24 \text{ hours}}{\text{day}} \times$$

$$\frac{1 \text{ } \mu l^*}{10^6 \text{ } \mu l} \times \frac{1 \text{ mM}^{**}}{24.45 \text{ ml}} = \frac{1.06 \text{ mMoles}}{\text{day}}$$

*1 ppm NO_2 in air
**NIOSH Expression for NO_2 at $77°F$ (Ideal Gas Law), 24.45 liters/mole

Depending on the amount of NO absorbed simultaneously, a nitrosation capacity equivalent to 70-140 mg of $NaNO_2$ can be taken directly into the lungs each day.

Table I

N-Nitroso Compounds and Precursors
(Human Daily Impact - Estimate)

	Preformed Nitrosamines (μg/day)	Preformed Nitrosamides	Nitrosating Sources (mg/day) (NaNO$_2$ equiv.)
Food (1, 3)	NDMA (0.5-1.0)	?	3 (2-5) (60)
	NPYR[b] (0.1-0.2)	?	
Saliva	--	?	9 (7-11) (61)
Smoking (57, 58)	NDMA (0.1-1.3)	?	2-3 (1 pack/day)(38)
	NNN[c] (0.8-5.0)	?	
	Others (0.1-0.9)	?	
In Vivo Synthesis[a]	NDMA (up to 700)	?	approx. 10 (62)
Air Pollution	NDMA (1)	?	at 0.1-1.0 ppm NO$_x$ in air = 10-100 mg

[a] Based on blood levels of 0.1-1.5 ppb, $t_{1/2}$ = 30-40 minutes in man and animals.
[b] NPYR = N-nitrosopyrrolidine
[c] NNN = N'-nitrosonornicotine

It is conceivable that nitrosamines can be synthesized in the intestine, since the precursors are present. While the conditions for aqueous nitrosation reactions are not optimum at pH values encountered in the lower gastrointestinal tract, several studies have shown that these reactions can be catalyzed (39, 40, 41). It has been suggested that the intestine might be a site for the formation of nitrosamines by bacterial action (42). Sander (43) has demonstrated the formation of nitrosamines by bacterial action from precursor amines and nitrate at neutral pH and Klubes and coworkers have reported the formation of NDMA upon incubation of ^{14}C-dimethylamine and sodium nitrite with rat fecal contents (44, 45).

Chemistry of Blocking Reactions

Aqueous Systems: In aqueous solution, the optimum condition for nitrosation (46) is usually found to be about pH 3-4 and reflects the mutual optimization of two conditions, (1) the formation of the nitrosating intermediate N_2O_3 and (2) the concentration of the more reactive unprotonated form of the amine which is governed by the following equations:

$$NO_2^- + H^+ \longrightarrow HONO$$

$$2HONO \rightleftharpoons N_2O_3 + H_2O$$

$$R_2NH + H^+ \longrightarrow R_2NH_2^+$$

$$R_2NH + N_2O_3 \longrightarrow R_2NNO + HONO$$

The mechanism, then, by which ascorbic acid functions to block these reactions is one of competitive kinetics with the susceptible amine for the nitrosating agent. Hence, the reactivity of any given amine will be an important parameter.

These above equations suggest that amines with pKa in the range of 4-6 will be more rapidly nitrosated than those with pKa values in the range of 9-11. This has been borne out in practice many times. Amines such as N-methylaniline with a pKa value of 4.84, piperazine (pKa value 5.9, 9.8), and aminopyrine (pKa value 5.04) are much more rapidly nitrosated than piperdine (pKa value 11.2), dimethylamine (pKa value 10.72), and pyrrolidine (pKa value 11.27). Under many of the conditions studied, it has been shown that the reactivity of ascorbic acid is sufficiently rapid that it can successfully compete with most all amines when present in approximately 2 mole ratio excess of the nitrosating agent.

In alkaline aqueous solution, on the other hand, one would expect that nitrosations would not occur at all because of the absence of an active nitrosating intermediate, and no

blocking agent would be necessary. However, Challis et al.
(47) have shown that N-nitrosamines and N-nitramines are
readily formed in aqueous solution between pH 7 and 14 when
the nitrosation agents are the gases N_2O_3 and N_2O_4. Interestingly, amines not only compete effectively with the expected OH⁻ hydrolysis reaction, but the nucleophilic reactivity
of various amines becomes virtually independent of pKa value.
In this case, the alkaline pH of the aqueous solvent exhibits
a "leveling" effect on dissolved amines in that all the amines
are unprotonated and become equivalent in reactivity. Challis
also noted that under their experimental conditions 2 x 10⁻³
M amines competed effectively with 55.5 M H_2O and 0.1 M OH⁻
for the nitrosating agent and suggested that possibly more
reactive isomers of N_2O_3 and N_2O_4 are generated by the gaseous
NO and NO_2 components. Here, N-nitrosamines result from
reaction of the unsymmetrical tautomer (ON-NO₃), whereas the
symmetrical tautomer (O₂N-NO₂) produces an N-nitramine possibly via a four-center transition state. The results for
N_2O_3 may be explained similarly in terms of the corresponding
ON-NO ⇌ ON-ONO tautomers. This conclusion has a prece-
dent (48) in the case of N_2O_4 but not for N_2O_3.
    In moderately acidic aqueous nitrite solution, the nitro-
sating agent is essentially nitrous acid anhydride, N_2O_3,
formed from nitrous acid, HONO (pKa = 3.14 at 25°C), which is
in turn formed from acidification of nitrite ions. Stronger
acidification can yield a very reactive nitrosating agent,
H_2ONO^+. Of even greater interest in nitrosation *in vivo* is
the catalytic effect of certain anions that form more reactive
nitrosating agents (39, 49, 50, 51). Thiocyanate is the most
active such catalyst, followed by halides in the order I⁻, Br⁻,
Cl⁻. This effect is often strong at low pH.

    Nonaqueous Systems: In nonaqueous (nonpolar) solvent
systems, nitrosation also proceeds. In these solvents, alpha-
tocopherol acts as a lipid soluble blocking agent in much the
same fashion as ascorbic acid functions in the aqueous phase.
Alpha-tocopherol reacts with a nitrosating agent and reduces
it to nitric oxide. At the same time, alpha-tocopherol is
oxidized to tocoquinone, which is the first oxidation product
of vitamin E and also a normal metabolite *in vivo*.
    Once the conditions are set for a nitrosating agent and
susceptible amine to enter a lipid nonpolar phase, the reaction is generally extremely rapid. Free amines readily
dissolve in aprotic lipid solvent systems as the unprotonated
base, yielding a high proportion (if not all) of the more
reactive free base form of the molecule for nitrosation. The
active intermediates of the nitrosating agents N_2O_3 or even
NO_2 are gases with appreciable solubility in lipid nonpolar
solvent systems. This is in marked contrast to the poor lipid
solubility of nitrite ion NO_2^-. The existence of a nitrosating

agent in aprotic solvents has been known for a long time and
was demonstrated by Mirvish et al. (52) who recorded the UV
spectra of "dried HNO_2" in methylene chloride.

The formation of nitrosamines in aprotic solvents has
applicability to many practical lipophilic systems including
foods (particularly bacon), cigarette smoke, cosmetics, and
some drugs. The very rapid kinetics of nitrosation reactions
in lipid solution indicates that the lipid phase of emulsions
or analogous multiphase systems can act as "catalyst" to
facilitate nitrosation reactions that may be far slower in
purely aqueous media (41, 53, 54). This is apparently true in
some cosmetic emulsion systems and may have important applicability to nitrosation reactions in vivo, particularly in the
GI tract. In these multiphase systems, the pH of the aqueous
phase may be poor for nitrosation in aqueous media (e.g.,
neutral or alkaline pH) because of the very small concentration of HONO or N_2O_3 that can exist at these pH ranges.
However, the small amount of N_2O_3 readily enters the lipid
where the nitrosation reaction is very rapid. The nitrosation
reaction is almost completely unidirectional towards the
formation of N-nitroso compounds. The net result is that the
presence of the lipid in such an emulsion brings together low
concentrations of free base (which increases in the lipid as
the pH goes up towards alkalinity) and nitrosating agent
(N_2O_3) which react rapidly and probably completely to form N-
nitroso compounds. In physiological lipids, especially in the
GI tract, the free hydroxyl groups of substances such as
monoglycerides, some phosphatides, cholesterol, bile acids,
and salts, etc. may contribute by functioning as transnitrosating agents.

Generally, it can be seen that nitrosamine formation can
take place in both the aqueous and lipophilic phases. Hence,
it becomes extremely important when investigating in vivo
blocking of nitrosamine formation that it is understood where
nitrosamine formation is taking place (e.g., stomach, lung,
intestine) and, additionally, in what phase the reaction
proceeds. Lacking such an understanding of an in vivo system, it would appear prudent to employ blocking agents which
protect both phases of the potential reaction medium.

Indeed, given an improperly designed or understood system, a blocking agent, like ascorbic acid, could be catalytic
toward nitrosamine formation. For example, if the source of
nitrosating agent is nitrite ion and the susceptible amine is
in the lipid phase, conceivably ascorbic acid could cause the
rapid reduction of nitrite ion to nitric oxide which could
migrate to the lipid phase. Subsequent oxidation of NO to NO_2
in the lipid phase could cause nitrosation.

Aside from ascorbic acid and alpha-tocopherol, which have
been shown to be effective blocking agents, there are other
factors which appear important in blocking nitrosamine formation

under in vivo conditions. These are the normal constituents of
food which can be referred to as "reductones". In studies performed using the aminopyrine-nitrite system in the presence of
semisynthetic or conventional animal chow diets, differences were
observed in the degree of nitrosation which occurred, as can be
seen in Table II and III. These results indicate that in the
presence of conventional animal chow (55) the potential for
dimethylnitrosamine formation is considerably reduced over that
which was formed in the semisynethetic diet (56). These are not
totally accountable by the change in pH of the medium brought
about by the buffering effect of conventional chow on the simulated gastric fluid medium. Additionally, we have observed that
these differences are not accountable based on the total vitamin
C and E present in the chow but rather indicate the presence of
other reducing compounds which can divert the nitrosating agent
from attacking the susceptible amine. These results would indicate that the design of any in vivo experiment should include
consideration of not only the amine substrate and its possible
mechanism of nitrosation (i.e., stomach via NO_2^- ion or lung via
NO_2 gas) but, for studies of gastrointestinal nitrosation, the
influence of normal dietary components as well.

Summary

The N-nitroso compounds are a potential source of carcinogenesis in humans. N-nitroso compounds appear to be ubiquitous
in our environment, being present in low levels in foods, cosmetics, drugs, atmosphere, etc., and also appear to be formed
endogenously in vivo.
    Once formed, N-nitroso compounds are converted in vivo to
reactive electrophilic ultimate carcinogens. Therefore, the most
practical method of eliminating carcinogenesis by nitrosamine is
to prevent their formation by diverting potential nitrosating
agents to nonnitrosating substances (e.g., by reduction to NO) by
the use of appropriate blocking agents.
    Ascorbic acid and alpha-tocopherol are effective blocking
agents against N-nitroso compound formation. Ascorbic acid is
effective particularly in aqueous media, and tocopherol effective
particularly in lipid phases. They should be used in conjunction
due to the mutually complementary actions of the two vitamins in
blocking nitrosamine formation in both aqueous and lipid media.
As safe nutrient ingredients in many food systems, as well as
available from commercial synthesis, the combination of vitamins
C and E represents very useful compounds for the nutritional
inhibition of formation of tumorigenic N-nitroso compounds.
    In addition to vitamin C and vitamin E as effective blocking
agents, there are other substances which also are capable of
preventing nitrosamine formation which are present in normal
foods. The influence of this factor on the design of experimental studies should not be overlooked.

Table II

Nitrosation of Aminopyrine

	A	B	C*
Gastric Fluid	50 ml	50 ml	50 ml
Aminopyrine	35 mg	35 mg	35 mg
Chow (control)	--	--	10 g
Synthetic (control)	--	10 g	--
$NaNO_2$	10 mg	10 mg	10 mg
pH	1.1	1.1	1.1

After 1 hour at $37^{\circ}C$

	A	B	C
DMN (mg)	3.89	3.29	1.03
DMN % Theory	35	29	9
DMN % Std	100**	85	26

*Normal pH of this sample was 3.2; pH adjusted from 3.2 to 1.1 for experimental comparison.
**By definition

Table III

Nitrosation of Aminopyrine

	D*	E
Gastric Fluid	50 ml	50 ml
Aminopyrine	35 mg	35 mg
Chow (control)	10 g	--
Synthetic (control)	--	10 g
$NaNO_2$	10 mg	10 mg
pH	3.2	3.2

After 1 hour at $37^{\circ}C$

	D	E
DMN (mg)	1.45	3.44
DMN % Theory	13	31
DMN % Std	37	88

*Normal pH of this sample was 1.1; pH adjusted from 1.1 to 3.2 for experimental comparison.

Literature Cited

1. Preussmann, R.; Eisenbrand, G.; Spiegelhalder, B. "Environmental Carcinogenesis," Elsevier/North-Holland Biomedical Press, Amsterdam, 1979, pp. 51-71.
2. Fine, D. H.; Ross, D.; Rounbehler, D.; Silvergleid, A.; Song, L. Nature 1977, 263, 753-755.
3. Gough, T. A.; Webb, K. S.; Coleman, R. F. Nature 1978, 272, 161-163.
4. Druckrey, H.; Preussmann, R.; Ivankovic, S.; Schmahl, D. Z. Krebsforsch 1967, 69, 103.
5. Sansone, E. B.; Tenari, Y. B. "Environmental Aspects of N-Nitroso Compounds," International Agency for Research on Cancer, Lyon, 1978, pp. 517-529.
6. Walker, E. A.; Castegnaro, M.; Garren, L.; Pignatelli, B. "Environmental Aspects of N-Nitroso Compounds," International Agency for Research on Cancer, Lyon, 1978, pp. 535-543.
7. Fan, S. T.; Krull, I. S.; Ross, R. D.; Wolff, M. H.; Fine, D. H. "Environmental Aspects of N-Nitroso Compounds," International Agency for Research on Cancer, Lyon, 1978, pp. 3-17.
8. Mirvish, S. S.; Wallcave, L.; Eagen, M.; Shubik, P. Science 1972, 177, 65-68.
9. Mirvish, S. S.; Cardesa, A.; Wallcave, I.; Shubik, P. J. Nat. Cancer Inst. 1975, 55, 633-636.
10. Mirvish, S. S.; Pelfrene, A. F.; Garcia, H.; Shubik, P. Cancer Letters 1976, 2, 101-108.
11. Mirvish, S. S. Ann. N.Y. Acad. Sci. 1975, 258, 175-180.
12. Greenblatt, M. J. Nat. Cancer Inst. 1973, 50, 1055-1056.
13. Kamm, J. J.; Dashman, T.; Conney, A. H.; Burns, J. J. Proc. Nat. Acad. Sci. 1973, 70, 747-749.
14. Kamm, J. J.; Dashman, T.; Conney, A. H.; Burns, J. J. "N-Nitroso Compounds in the Environment," International Agency for Research on Cancer, Lyon, 1974, pp. 200-204.
15. Kamm, J. J.; Dashman, T.; Newmark, H. L.; Mergens, W. J. Toxicol.and Appl. Pharmacol. 1977, 41, 575-583.
16. Fong, Y. Y.; Chan, W. C. "Environmental N-Nitroso Compounds: Analysis and Formation," International Agency for Research on Cancer, Lyon, 1976, pp. 461-464.
17. Kinawi, V. A.; Doring, D.; Witte, I. Arzutim.-Forsch. 1977, 27, 747.
18. Preda, N.; Popa, L.; Galea, V.; Simo, G. "Environmental N-Nitroso Compounds: Analysis and Formation," International Agency for Research on Cancer, Lyon, 1976, pp. 301-304.
19. Cardesa, A.; Mirvish, S. S.; Haven, G. T.; Shubik, P. Proc. Soc. Exp. Biol. Med. 1974, 145, 124-128.
20. Kawabata, T.; Shazuki, H.; Ishibashi, T. Nippon Suisan Gakkaishi 1974, 40, 1251.

21. Ivankovic, S.; Preussmann, R.; Schmahl, D.; Zeller, J. W. "N-Nitroso Compounds in the Environment," International Agency for Research on Cancer, Lyon, 1974, pp. 101-102.

22. Archer, M. C.; Tannenbaum, S. R.; Fan, T.-Y.; Weisman, M. J. Nat. Cancer Inst. 1975, 54, 1203-1205.

23. Fan, T. Y.; Tannenbaum, S. R. J. Food Sci. 1973, 38, 1067-1069.

24. Sen, N. P.; Donaldson, B. "N-Nitroso Compounds in the Environment," International Agency for Research on Cancer, Lyon, 1974, pp. 103-106.

25. Sen, N. P.; Donaldson, B.; Charbonneau, C.; Miles, W. F. J. Agric. Food Chem. 1974, 22, 1125-1130.

26. Sen, N. P.; Donaldson, B.; Seaman, S.; Iyengar, J. R.; Miles, W. F. J. Agric. Food Chem. 1976, 24, 397-401.

27. Gray, J. I.; Dugan, L. R. J. Food Sci. 1975, 40, 981-984.

28. Ziebarth, D.; Scheunig, G. "Environmental N-Nitroso Compounds: Analysis and Formation," International Agency for Research on Cancer, Lyon, 1976, pp. 279-290.

29. Mergens, W. J.; Kamm, J. J.; Newmark, H. L.; Fiddler, W.; Pensabene, J. "Environmental Aspects of N-Nitroso Compounds," International Agency for Research on Cancer, Lyon, 1978, pp. 199-212.

30. Lijinsky, W.; Greenblatt, M. Nature 1972, 236, 177.

31. Sander, J.; Burkle, G. Z. Krebsforsch. 1969, 73, 54.

32. Ivankovic, S.; Preussmann, R.; Schmahl, D.; Zeller, J. Z. Krebsforsch. 1973, 79, 145-147.

33. Greenblatt, M.; Mirvish, S. S. J. Nat. Cancer Inst. 1973, 50, 119-124.

34. Douglass, M. L.; Kabacoff, B. L.; Anderson, G. A.; Cheng, M. C. J. Soc. Cosmet. Chem. 1978, 29, 581-606.

35. Lakritz, L.; Simenhoff, M. L.; Dunn, S. R.; Fiddler, W. Food Cosmet. Toxicol. 1980, 18, 77-79.

36. Kakizoe, T.; Wang, T. T.; Eng, V. W. S.; Furrer, R.; Dion, P.; Bruce, W. R. Cancer Res. 1979, 39, 829-832.

37. Wayne, L. G. "Technical Progress Report Vol. III," Los Angeles County Air Pollution Control District, Los Angeles, 1962.

38. Bokhoven, C.; Niessen, H. J. Nature 1961, 192, 458-459.

39. Boyland, E.; Nice, E.; Williams, K. Food Cosmet. Toxicol. 1971, 9, 639-643.

40. Keffer, L. K.; Roller, P. P. Science 1973, 18, 1245-1247.

41. Okun, J. D.; Archer, M. C. J. Nat. Cancer Inst. 1977, 58, 409-411.

42. Magee, P. N.; Montesano, R.; Preussmann, R. "Chemical Carcinogens," American Chemical Society, 1976, pp. 491-625.

43. Sander, J. Hoppe-Seyler's Z. Physiol. Chem. 1968, 349, 429-432.

44. Klubes, P.; Jondurf, W. R. Chem. Pathol. Pharmacol. 1971, 2, 24-34.

45. Klubes, P.; Cerna, I.; Rabinowitz, A. D.; Jondurf, W. R. Food Cosmet. Toxicol. 1972, 10, 757-767.
46. Mirvish, S. S. Toxicol. and Appl. Pharmacol. 1975, 31, 325-351.
47. Challis, B. C.; Edwards, A.; Hunma, R. R.; Kryptopolous, S. A.; Outram, J. R. "Environmental Aspects of N-Nitroso Compounds," International Agency for Research on Cancer, Lyon, 1978, pp. 127-142.
48. Steel, F.; Nogradi, J.; Breit, H. Z. Anorg. Chem. 1952, 269, 102.
49. Boyland, E.; Walker, S. A. "N-Nitroso Compounds in the Environment," International Agency for Research on Cancer, Lyon, 1974, pp. 132-136.
50. Fan, T.-Y.; Tannenbaum, S. R. J. Agric. Food Chem. 1973, 21, 237-240.
51. Schweinsberg, F. "N-Nitroso Compounds in the Environment," International Agency for Research on Cancer, Lyon, 1974, pp. 80-85.
52. Mirvish, S. S.; Karlowski, K.; Sams, J. P.; Arnold, S. D. "Environmental Aspects of N-Nitroso Compounds, International Agency for Research on Cancer, Lyon, 1978, pp. 161-174.
53. Kim, Y.-K.; Tannenbaum, S. R.; Wishnok, J. S. "N-Nitroso Compounds: Analysis, Formation and Occurrence," International Agency for Research on Cancer, Lyon, 1980, pp. 207-214.
54. Mergens, W. J.; Chau, J.; Newmark, H. L. "N-Nitroso Compounds: Analysis, Formation and Occurrence," International Agency for Research on Cancer, Lyon, 1980, pp. 259-267.
55. Ralston-Purina Rat Chow
56. Wogan, G. N.; Newberne, P. M. Cancer Res. 1967, 27, 2370-2376.
57. Brunnemann, K. D.; Yu, L.; Hoffmann, D. Cancer Res. 1977, 37, 3218-3222.
58. Brunnemann, K. D.; Hoffmann, D. "Environmental Aspects of N-Nitroso Compounds," International Agency for Research on Cancer, Lyon, 1978, pp. 343-356.
59. Fine, D. H.; Rounbehler, D. P.; Belcher, N. M.; Epstein, S. S. Science 1976, 192, 1328-1330.
60. White, J. W., Jr. J. Agric. Food Chem. 1975, 23, 886-891.
61. Tannenbaum, S. R.; Weisman, M.; Fett, D. Food Cosmet. Toxicol. 1976, 14, 549-552.
62. Archer, M. C.; Saul, R. L.; Lee, L.-J.; Bruce, W. R. "Gastrointestinal Cancer: Endogenous Factors," Cold Spring Harbor Laboratory, Long Island, 1981, pp. 321-330.

RECEIVED August 10, 1981.

Occurrence of *N*-Nitrosamines in the Workplace

Some Recent Developments

D. H. FINE and D. P. ROUNBEHLER

New England Institute for Life Sciences, 125 Second Avenue, Waltham, MA 02254

The largest known human exposures to exogenous
nitrosamines have been shown to occur in the
work place, particularly in the rubber and
leather tanning industries. Recent data for
amines, nitrosating potential and nitrosamines
will be presented in the framework of assessing
the extent of the various exposures.

Excluding tobacco and tobacco smoke (1,2,3), the largest
known human exposure to N-nitrosamines is in the industrial
sector (4). The area or process air samples containing the
largest amount of nitrosamines are in a tire factory, NMOR at the
250 $\mu g/m^3$ level (5), a leather tannery, NDMA at the 47 $\mu g/m^3$ level
(6), and a rocket fuel factory, NDMA at the 36 $\mu g/m^3$ level (7).
The highest actually measured personal (breathing zone) exposure
was 25 $\mu g/m^3$ of NMOR (time-weighted average) for a worker in the
feed mill and calender operation of a tire factory (5).
Recently, much of our work has concentrated on minimizing,
or possibly eliminating, these extraordinarily high human exposures to nitrosamines. This has been achieved by first - improv-
ing the analytical procedures so that reliable data can be obtained, second - using the new data base to understand the broad scope
of the problem and three - applying the new knowledge and understanding in the work place so as to minimize the human exposure.
Measurement of airborne nitrosamines is complicated by the fact
that nitrosamines are usually found to be present in a vast sea
of precursor amines and oxides of nitrogen. A proper understanding of the problem, therefore, requires the simultaneous artifact-
free measurement of nitrosamines, amines and oxides of nitrogen.
The capability to carry out all three determinations on the same
sample of air has only recently been developed.

Nitrosating Potential. The nitrosating potential of a sample
is defined as its capacity to nitrosate amines to produce N-

nitrosamines. It may be determined quantitatively by measuring
the amount of an added marker amine which is converted to the
corresponding marker nitrosamine. In air, we determine nitrosating potential by passing a known volume of air over a ThermoSorb/A
cartridge (8) to which had been added a known weight (approximately 10 mg) of thiomorpholine (or 2,6 dimethlmorpholine). The
amount of the corresponding N-nitroso derivative, N-nitrosothiomorpholine (or N-nitroso 2,6, dimethylmorpholine) found on the
cartridge is a measure of the nitrosating potential of the air
sample. Nitrosating potential of the air sample may be calibrated to the nitrogen dioxide content of the sample by passing
air containing a known amount of nitrogen dioxide through a
cartridge containing the marker amine. A typical calibration,
showing the amount of N-nitrosothiomorpholine produced per liter
of air, versus the square of the nitrogen dioxide concentration,
is shown in Figure 1.

Amines. A modified ThermoSorb/N cartridge (8), called
ThermoSorb/A (Thermo Electron Corporation), is used to collect
airborne amines. Instead of eluting with an organic solvent such
as methanol/dichloromethane, the ThermoSorb/A cartridge is eluted
with 0.1N KOH. The eluate contains both the amine and the nitrosamine, and is injected directly into a GC interfaced to an absolute organic nitrogen detector, the TEA 610 (Thermo Electron
Corporation). This new GC detector operates by completely com-
busting all organics with oxygen, to give carbon dioxide, water
vapor, and nitric oxide, NO (9). The NO content (and hence the
amine content of the sample) is determined by the intensity of the
chemiluminescent reaction of the NO with ozone. Instead of an
absolute GC-nitrogen detector, a GC detector with enhanced selectivity to nitrogen, such as the Hall or AFID can be used instead,
provided that non-nitrogen containing compounds do not co-elute
with the compound of interest.

Tire Factories. Airborne amine and nitrosamine data from a
typical tire factory are shown in Figures 2 and 3. Figure 2, a
GC-TEA (N-mode) shows the presence of large concentrations of
dimethylamine and morpholine, together with traces of trimethylamine, diethylamine and triethylamine, together with small
amounts of several unidentified N-containing compounds. Figure 3,
the equivalent GC-TEA chromatogram of the same sample of air,
shows that the only two nitrosamines which were detected were
the N-nitroso derivatives of the two amines which were present
are included in Table I. In most cases, the airborne nitrosamine concentration was about 1% of the airborne amine concentration. It is perhaps surprising to note that although diethyl-
and triethylamine were found, N-nitrosodiethylamine was not
detected.
        Due to a combination of company and union concern for worker
safety, and parallel involvement of the National Institute of

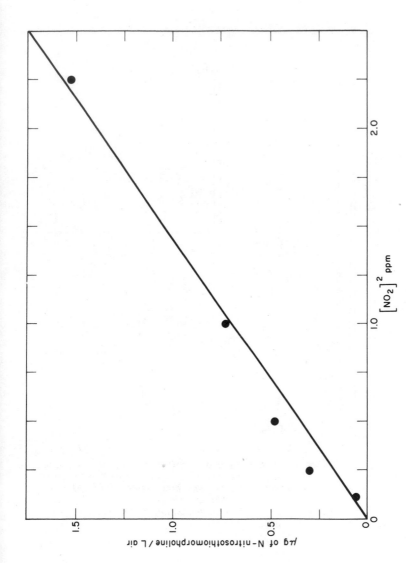

Figure 1. Calibration plot for nitrosation potential showing amount of N-nitroso-thiomorpholine formed vs. the square of the nitrogen dioxide concentration. Standard conditions of 50% relative humidity, 25°C, and 1 L/min flow rate for 30 min were used.

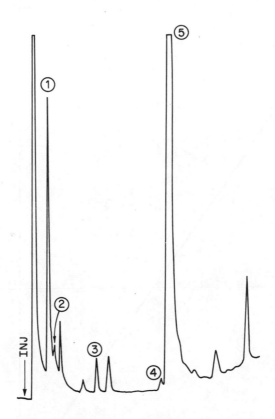

Figure 2. GC-TEA (N mode) chromatogram of a tire factory air sample. The column was a 5.5 m glass tube, 2mm. i.d., packed with Carbopak B(4% Carbowax 20 M, with 0.8% KOH on charcoal). Carrier gas flow rate was 15 mL/min. Column temperature was held at 40°C for 2 min, and then increased by 8°/min to 180°C. Peak identity 1-dimethylamine, 2-trimethylamine, 3-diethylamine, 4-triethylamine, and 5-morpholine.

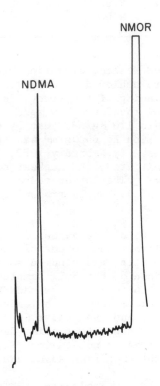

Figure 3.   Conventional GC-TEA (nitrosamine mode) chromatogram of the tire factory air sample shown in Figure 2 above. Two nitrosamines are shown to be present, N-nitrosodimethylamine and N-nitrosomorpholine.

Table I
Airborne Amine and Nitrosamine Concentrations in a Tire Factory
Concentrations in $\mu g/m^3$

Methyl-			Propyl-		Ethyl			Morpholine	
DMA	TMA	NDMA	NPA	NNPA	DEA	TEA	NDEA	MOR	NMOR
2	5	0.12	10	N.D.	0	330	N.D.	20	0.7
130	27	1.5	9	N.D.	50	16	N.D.	2600	66

Occupational Safety and Health, dramatic reductions in airborne
nitrosamine levels were achieved in only 7 months (5). Figure 4
shows that at the beginning of the study, in August, 1979, the
highest area NMOR level found was 250 $\mu g/m^3$. Within two months,
this had been reduced to 120 $\mu g/m^3$, just by venting. Two months
later, a further reduction to 63 $\mu g/m^3$ was achieved, mainly by the
use of exhaust canopies. By February 1980, just seven months
after the study was initiated, the highest detectable area NMOR
level was only 14 $\mu g/m^3$. The latter reduction was achieved by
discontinuing the use of diphenylnitrosamine, and using a phthalimide derivative instead (5).

Leather Industry. Using the capability of simultaneous measurement of airborne nitrosation potential, amines and nitrosamines,
a complete tanning operation was studied so as to ascertain the
source of the NDMA. None of the bulk samples which were collected
at the same time (see Table II), contained NDMA (sensitivity limit
0.5 $\mu g/ml$), including the fresh dimethylamine sulfate (DMAS), the
hide unhairing solution, and even water on the floor near the unhairing operation. This negative finding demonstrates conclusively that the source of NDMA is not an impurity in the DMAS; nor is
NDMA being formed in the unhairing solutions.

Table II
Analyses of Bulk Samples from a Leather Tannery

Sample Description	NDMA ($\mu g/ml$)	DMA ($\mu g/ml$)
Floor Water near unhairing	N.D.*	120
Fresh hide unhairing solution containing DMAS	N.D.	790
Hide unhairing solution – no DMAS added	N.D.	8
4-day old hide unhairing solution containing DMAS	N.D.	630
40% solution of DMAS	N.D.	120,000

* N.D. – None detected – detection limit 0.5 $\mu g/ml$

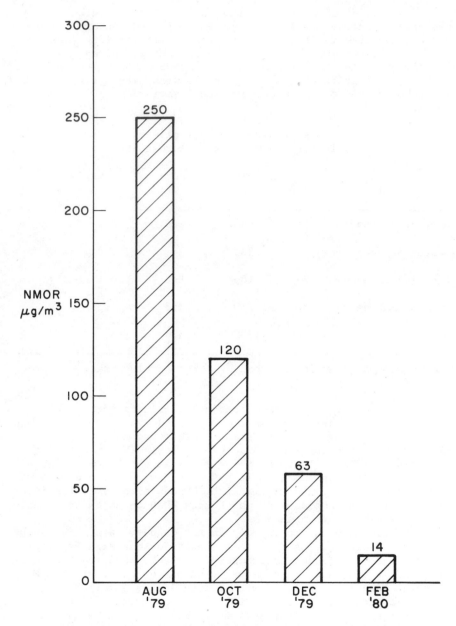

Figure 4.   Highest area sample NMOR levels in a tire factory, showing the reduction achieved.   The October 1979 reduction was attributed to venting, the
December 1979 reduction to the installation of exhaust canopies, and the February
1980 reduction to use of a phthalimide derivative instead of diphenylnitrosamine
(data from Ref. 5)

However, at the time the bulk samples were collected, NDMA was present in the air at all test sites in the tannery. The airborne data (see Table III), shows that dimethylamine (DMA) was always present. The amount of airborne NDMA was always approximately equal to 1% of the amount of airborne DMA. The apparent lack of correlation of the NDMA levels with nitrosating capacity is probably due to the relatively small variation in the measured NO_x levels.

Table III
Analyses of Air Samples* Collected
at a Tannery in Milwaukee, August 1980

Sample Location	NO_x ppb	DMA_3 $\mu g/m^3$	$NDMA_3$ $\mu g/m^3$	NDMA/DMA %
Unhairing vats	58	488	4.6	0.9
Center of unhairing	37	280	5	1.8
Fleshing machine	67	183	1.2	0.6
Head splitting	88	260	3.3	1.3
Bating area	80	280	3.6	1.3

*Each data point is the mean of three samples taken on three different days

A total of eight separate leather facilities were surveyed for the presence of N-nitroso compounds and two of these were resurveyed. Four of the eight plants were found to have airborne NDMA at levels greater than 0.5 $\mu g/m$. Table IV summarizes the operations of each plant and the highest level of NDMA found at that plant. The use of DMAS is associated with the presence of airborne NDMA. Even a facility which had recently discontinued the use of DMAS, and another which used DMAS on an experimental basis, contained airborne NDMA.

There is an apparent anomoly in the tannery data. As can be seen from Table IV, DMAS, used to gently loosen hairs from the hides is clearly needed to produce detectable NDMA levels. Yet the DMAS itself, and its aqueous solutions do not contain NDMA. Airborne NDMA is associated with a much higher level of DMA, as well as an adequate airborne nitrosation capacity. Thus, it must be concluded that the NDMA is formed from DMA outside the solution - either in the gas phase, or heterogeneously on surfaces. The major source of the DMA is presumably the DMAS.

The source of the nitrosating agent is probably oxides of nitrogen, formed by the combustion of fossil fuels in gas powered fork-lift trucks, or in open gas heaters. This could not be demonstrated unambiguously, since the tanneries which were

visited and which used DMAS, all had a plausible source of oxides
of nitrogen (Table IV).

Table IV
Summary of NDMA Findings at Eight Leather Manufacturers

Description of Tannery	DMAS Used	NO_x Source	Highest NDMA observation $\mu g/m^3$
All operations	Yes	fork-lift trucks	47
All operations	Yes	fork-lift trucks	11
All operations	No	fork-lift trucks	0
All operations	No	fork-lift trucks	0
Partial-wet	recently discontinued	fork-lift trucks	8
Partial-wet	used experimentally	open gas heaters	3
Partial-dry	No	fork-lift trucks	0.05
Partial-dry	No	None	0

Acknowledgements

We thank John Fajen for many valuable and illuminating discussions. The work described here was supported, in part, by
the National Institute for Occupational Safety and Health (NIOSH)
under Contract 210-77-0100. Any opinions, findings, conclusions
and recommendations experessed are those of the authors and do
not necessarily reflect the views of NIOSH.

Literature Cited

1. Hoffman, D., Adams, J.D. Brunnemann, K.D., Hecht, S.S.,
   Cancer Res., 1979, 39, 2505-2509
2. Hoffmann, D., Adams, J.D., Piade, J.J., Hecht, S.S., IARC
   Sci. Publ. 1981, 31, 507-616.
3. Brunnemann, K.D., Yu, L., Hoffmann, D., Cancer Res., 1977,
   37, 3218-3222.
4. Fine, D.H., Oncology, 1980, 37, 199-202
5. McGlothlin, J.D., Wilcox, T.C. Fajen, J.M. and Edwards, G.S.,
   "Chemical Hazards in the Workplace", American Chemical
   Society Symposium Series #149, Washington, D.C.,1981; p.283

6. Rounbehler, D.P., Krull, I.S., Goff, U.E., Mills, K.M.,
    Morrison, J., Edwards, G.S., Fine, D.H., Fajen, J.M., Carson
    G.A. and Reinhold, V., Fd. Cosmet. Toxicol., 1979, 17, 487–
    491.
7. Fine, D.H., Rounbehler, D.P., Pellizzari, E.D., Bunch, J.E.,
    Bereley, R.W., McCrae, J., Bursey, J.T., Sawicki, E., Krost,
    K. and DeMarrais, G.A., Bull. Environ. Contam. Toxicol.,
    1976, 15, 739–746.
8. Rounbehler, D.P., Reisch, J.W., Coombs, J.R. and Fine, D.H.,
    Anal. Chem., 1980, 52, 273–276.
9. Fine, D.H., British Patent Number 1513007, issued 1978.   Filed
    1974.

RECEIVED September 16, 1981.

Reduction of Human Exposure to Environmental *N*-Nitroso Compounds

R. PREUSSMANN, B. SPIEGELHALDER, and G. EISENBRAND

Institute of Toxicology and Chemotherapy, German Cancer Research Center, Im Neuenheimer Feld 280, 6900 Heidelberg, Federal Republic of Germany

The complex pattern of human exposure to environmental N-nitroso compounds is summarized. Recent results are given in three areas, where a significant reduction of human exposure has been achieved after elucidation of its causes: 1. N-Nitrosodimethylamine in beer. 2. Volatile N-nitrosamines in baby nipples and pacifiers and 3. occupational exposure in the rubber industry.

Determination of environmental carcinogens is not an end in itself. Although often misunderstood or misinterpreted, the aim of scientific work in this area is to collect data on human exposure, to pinpoint emission and immission sources and to determine the causes of potential health risks to humans. The ultimate aim is always elimination or reduction of human exposure and thereby prevention of human cancer. In view of the slow progress in the treatment of human cancer this approach is an important strategy in the fight against this disease.

Among environmental carcinogens N-nitroso compounds are considered to be important contributors to the total burden of carcinogens, since they are potent carcinogens causing tumors in almost every organ of experimental animals and since they are present in the human environment. Human exposure to carcinogenic N-nitroso compounds may occur by preformed nitrosamines from various environmental sources and by endogenous formation of such compounds from precursors in the human body. The in-vivo formation, unique for N-nitroso compounds, is treated by other authors during this symposium. We will therefore confine ourselfs to exogenous exposure and present some recent data from our

0097-6156/81/0174-0217$05.00/0

own work. We will show that the identification of exposure sources inevitably leads to investigations into the causes of such exposure and finally to suggestions for preventive measures. In the examples given the suggested technological changes seem to be neither complicated nor unjustifiably expensive; they have been put into practice in many cases.

The most important sources of preformed environmental nitrosamines according to present knowledge, are given in Table I.

TABLE I:

Human exposure to preformed N-nitroso compounds and potential preventive measures

Life-style	Occupational
Food	Rubber industry
Tobacco smoke	Leather tanneries
Cosmetics	Chemical industry
Household commodities	Cutting fluids
(esp. rubber products)	Hydraulic fluids
Drugs	Pesticide production
Other sources	and application

Total exposure

Reduction

Technological improvement	Avoidance of nitrosating
(Malt, Rubber)	chemicals
Reduction in nitrite use	Reduction of NO_x pollution
Selective filtering	Use of "safe" amines
(Tobacco smoke)	Ventilation at working
Legal regulations	places

First example: N-Nitrosodimethylamine in Beer

We have previously shown in a survey on the occurence
of volatile nitrosamines in commercial food of the
German market that beer is the most important source
of N-nitrosodimethylamine (NDMA) (Spiegelhalder et al.,
1979; Preussmann et al., 1980; Spiegelhalder et al.,
1980). The investigation of about 3000 food samples
allowed the calculation of an average daily intake
of 1.1 μg NDMA and of 0.1-0.15 μg N-nitrosopyrrolidine
(NPYR) for a male adult in the year 1978. For NDMA
64% (= 0.7 μg/day) of the total daily intake resulted
from the consumption of beer.

    A subsequent investigations on the origin of
NDMA in beer was done by analyzing every stage of the
brewing process together with all used ingredients.
Invariably we found that malt was practically the
only source of NDMA-contamination. Therefore a wide
variety of different malt samples was analyzed and it
was found that NDMA is formed during kilning (drying)
of green malt. The extent of nitrosamine formation
clearly depended on the type of heating technique used
as shown in Table II.

TABLE II

NDMA concentration (μg/kg) in malt depending on
different drying techniques

Type of malt	Heating technique		
	indirect	direct	
		oil burners	gas burners
Pale malt	1	0.5 - 10	15 - 80
Dark malt	2 - 8	6 - 10	80 - 320

It is obvious that the highest formation of NDMA
occured in malts dried in direct-fired kilns. Unfortunately, gas-fired kilns are predominant in most
countries at present. The logical next step was to try
and prevent this NDMA formation during kilning. In cooperation with brewing experts it was then found out
that sulfur burning ("sulfuring"), i.e. SO_2-treatment
during drying significantly reduced NDMA formation. As
shown in Figure 1, this method is quite effective, but

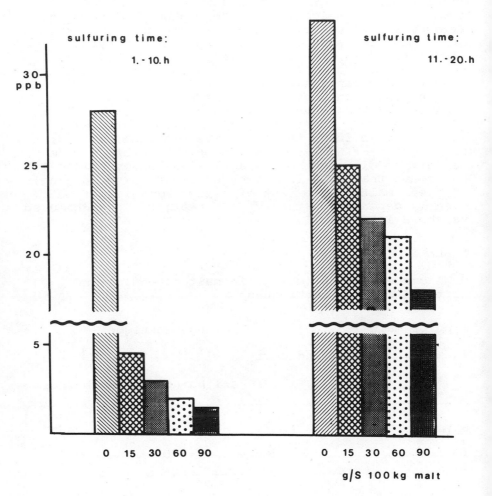

Figure 1.   Reduction of NDMA in malt by sulfuring.

only if applied during the first ten hours, when the malt is still wet. In the later stage sulfuring is much less effective.

Since NO_x is a precursor in NDMA formation and high combustion temperatures (usually from 1500 to 1800°C) yield high reaction rates between oxygen and nitrogen, a decrease in NDMA formation can also be achieved by lowering the flame temperature. Excess air seems to be the most economic way to reduce flame temperature and NO_x synthesis. In a new type of burner developed on this principle the resulting air had only 0.05 - 0.1 mg NO_x/m^3 as compared with 14 mg/m^3 in conventional burners. Accordingly, malt dried with such burners contains only 1 to 3 mg/kg NDMA, a 15-30 fold reduction of the NDMA concentration.

Using these improved methods for malt kilning, it is now possible to produce malt with low NDMA contamination even with existing direct-fired kilns. A comparison of NDMA concentration in commercially available beer in 1978/79 and 1981, after intro- duction of the new techniques is given in Table III.

Second example: <u>Volatile nitrosamines in baby nipples and pacifiers</u>.

Although a contamination of rubber stoppers with volatile nitrosamines had been mentioned, though not thoroughly examined in the literature (Ireland et al., 1980), rubber commodities used in households have not been examined in detail. During the last year we analysed a large number of baby nipples and pacifiers for a contamination with volatile nitrosamines. About 70 samples of commercially available products of the German market, including samples from other European countries, the USA and Japan were analyzed; all samples were contaminated and concentrations ranged from 1 to 230 μg/kg rubber material. Most products contained more than 1 volatile nitrosamine. Some representative values are summarized in Table IV.

A migration test was developed to simulate human exposure patterns. 5 g of rubber material was cut into 1-2 mm stripes and then immersed in 20 ml of standard test solution of artificial saliva (4.2 g $NaHCO_3$, 0.2 g K_2CO_3, 0.5 g NaCl, 0.03 g $NaNO_2$ ad 1000 ml with aqua dest.). After 24 h incubation at 40°C volatile nitrosamines were determined in an aliquot and determined after distillation with a standard technique (GC-TEA-method).

The nitrosamine contamination of rubber products originates from amine-containing accelerators and

TABLE III

NDMA in beer. Comparison of determination in 1978 before and in 1981 after introduction of new malt drying techniques

Type of beer	1978/79				1981			
	N	% positive (>0.5 ppb)	mean ppb	max ppb	N	% positive (>0.5 ppb)	mean ppb	max ppb
Pilsen lager	54	65	1.2	7	169	24	0.43	6.5
Pale lager & export lager	42	47	1.2	7	179	26	0.39	2.0
Pale strong lager	25	76	1.9	8	38	26	0.42	1.6
Top fermented pale ales	22	23	1.0	6.2	19	5	0.32	0.7
Top fermented dark ales	25	76	2.7	11	21	24	0.69	7.0
Dark lager & dark strong lager	22	68	6.0	47	25	32	0.51	4.0
"Rauchbier" (from smoked malt)	9	100	18.0	68	3	100	1.50	2.0*
All types	199	66	2.5	68	454	24	0.44	7.0

TABLE IV

Migration of N-nitrosamines from nipples and pacifiers after incubation in a standard test with artificial saliva for 24 h at 40°C. Concentrations in μg/kg rubber product (- = <1 ppb)

Article		NDMA	NDEA	NDBA	NPIP	NMPhA
A		1	-	230	-	-
B		150	-	-	45	-
C		70	-	-	65	-
D	natural rubber	-	-	10	-	-
E	products,	2	-	15	-	25
F	dipped and pressed	12	-	-	14	-
G		5	-	-	60	120
H		1	10	30	-	-
I		5-10	-	-	-	-
K	silicone rubber	50	6	-	-	-

NDMA = N-nitrosodimethylamine; NDEA = N-nitrosodiethylamine; NDBA = N-nitroso-dibutylamine; NPIP = N-nitrosopiperidine; NMPhA = N-nitrosomethylphenylamine

stabilizers used in the vulcanisation process. The
more important ones are summarized in Table V.

TABLE V

Amine-containing accelerators and stabilizers used in
the polymerisation of rubber latex

Thiuram-mono- and -disulfides

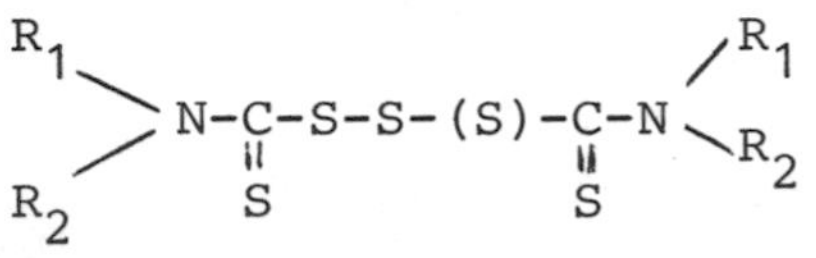

$$R_1 \diagdown \diagup N-\overset{S}{\underset{\|}{C}}-S-S-(S)-\overset{S}{\underset{\|}{C}}-N \diagup \diagdown R_2 \qquad R_1, R_2 = methyl$$

R_1, R_2 = methyl

R_1 = methyl

R_2 = phenyl

Dithiocarbamates

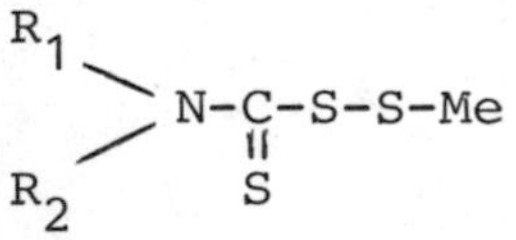

$$R_1 \diagdown \diagup N-\overset{S}{\underset{\|}{C}}-S-S-Me \diagup R_2$$

Me = Zn, Na, K, NH_4

R_1, R_2 = methyl, ethyl,
N-butyl, penta-
methylene

R_1 = methyl; R_2 = phenyl

Such amine derivatives are always contamined with the
corresponding nitrosamine in varying concentrations.
However, the greatest part of nitrosamine is formed
by reaction with nitrosating agents present in the
vulcanisation mixture or by reaction with nitrogen
oxides from polluted indoor atmospheres. The crucial
role of accelerants can be seen from data of Table VI:
amounts of dithiocarbamates in a polymerisation mixture determine nitrosamine concentration in the rubber
product. A decrease in the accelerator concentration
results in a decreased nitrosamine formation. Exchange of the amine-containing accelerators by other
types, which do not generate amines, eliminates nitrosamine contamination as expected. Table VI gives also
data on nitrosatable precursors migrated into the
artificial saliva medium and nitrosated after acidification (pH 1; 30 min at ambient temperatures).

TABLE VI

Effect of decreasing amounts of accelerators on nitrosamine and nitrosatable amine content in nipple rubber

Amount of accelerator %			Nitrosamine in extract ppb		Nitrosatable amine in extract ppb after nitrosation	
LDA	LDB		NDEA	NDBA	NDEA	NDBA
0.4	0.5		25	100	3800	8000
0.15	0.35		10	30	1600	5100
0.15	0.25		10	–	240	180
0.15	0.0 + 0.2 merc.		8	–	130	–

LDA = Zn-diethyldithiocarbamate; LDB = Zn-di-n-butyldithiocarbamate; merc. = Mercaptobenzothiazole

The amount of such nitrosatable precursors migrating from rubber nipples has to be taken into consideration, because such compounds can migrate during the normal use of nipples and, after being swallowed with nitrite containing saliva, can be nitrosated in the stomach. Elimination of amine-containing accelerators would therefore eliminate this problem. Technological experience until now indicated that conventional accelerators, such as thiuramsulfides or dithiocarbamates, though they cannot be replaced totally yet, can be added at strongly reduced concentrations, For such cases the concept of "safe" amines might be helpful. It proposes the use such secondary amines or derivatives which after nitrosation give rise to non-carcinogenic, or at best to weakly carcinogenic nitrosamines. Some of these amines are summarized in Table VII.

In West Germany a regulation is being discussed allowing no more than 10 ppb of preformed nitrosamines and 200 ppb of nitrosatable amines in migrates of nipples and pacifiers.

TABLE VII

"Safe" secondary amines resulting in non-carcinogenic N-nitroso compounds after nitrosation (German Patent Application No. P 3029 318.6)

Third Example: Rubber Industry

Fajen et al. (1979) discovered the presence of air-borne nitrosamines in rubber factories. Nitrosomorpholine (NMOR) in low concentrations ($<$ 1 $\mu g/m^3$) was found. Our own studies in 19 rubber and tire plants confirmed the significance of these findings. NDMA and NMOR were the nitrosamines which could be found regularly. In most cases the levels ranged from 1-20 $\mu g/m^3$. In one working area (tire tube curing) an extremely high NDMA concentration of 140 $\mu g/m^3$ could be detected. At another place (conveyor belt curing) 40-90 $\mu g/m^3$ NDMA and 120-380 $\mu g/m^3$ NMOR could be detected. The measurement was made by personal monitoring.

The origin of nitrosamine formation is the reaction of nitrosating agents with amine based accelerators. Until now the following nitrosating agents could be detected:

1. Nitrosodiphenylamine (a retarder)
2. Nitrogen oxides from combustion processes (forklift trucks)
3. Nitrite-nitrate salt bath curing

The result of preventive measures is shown in Table VIII. In a tire and tube factory nitrosodiphenylamine (NDPhA) was substituted by cyclohexylthiophthalimid (PVI) to remove the major nitrosating agent.

TABLE VIII

Change in airborne nitrosamine levels ($\mu g/m^3$) after introduction of a new retarder in one factory at three different times

Area	1. NDPhA in use	2. PVI in use	3.	
Tube curing area	50 - 130	1	1 - 2	NDMA
Tube warehouse	6 - 20	1-2.5	3	NDMA
Tire warehouse	4 - 10	2-4.5	4	NMOR

In other working areas the reduction of airborne nitrosamines is possible, too. In a special curing process using a nitrite-nitrate salt bath for tube

and profile curing the nitrosamine problem can be
solved by using peroxides as accelerators (Table IX).

TABLE IX

Nitrosamines in the air of a salt bath curing area
($\mu g/m^3$)

	NDMA	NDEA	NMOR
Curing of profiles	1 - 10	trace	trace
Curing of tubes	9 - 18	3 - 5	3
Curing of profiles (compounded with peroxides)	0.1	-	-

Trace $\simeq 0.05$ $\mu g/m^3$
- $= <0.05$ $\mu g/m^3$

Literature Cited

1. Spieglehalder, B.; Eisenbrand, G.; Preussmann, R.
   Fd. Cosmet. Toxicol. 1979, 17, 29-31.
2. Preussmann, R.; Spiegelhalder, B.; Eisenbrand, G.
   in: B. Pullman; P.O.P. Ts'o; H. Gelboin (Eds.)
   "Carcinogenesis: Fundamental Mechanisms and Environmental Effects": Reidel Publ. Co., Boston
   1980 p. 273-285.
3. Spiegelhalder, B.; Eisenbrand, G.; Preussmann, R.
   Oncology, 1980, 37, 211-216.
4. Ireland, C.B.; Hytrek, F.P.; Lasoski, B.A. Amer.
   Ind. Hyg. Ass. H. 1980, 41, 895-900.
5. Fajen, J.M.; Carson, G.A.; Rounbehler, D.P.; Fan,
   T.Y.; Vita, R.; Goff, U.E.; Wold, M.H.; Edwards,
   G.S.; Fine, D.H.; Reinhold, V.; Biemann, K.
   Science 1979, 205, 1262-1264.

RECEIVED August 26, 1981.

N-Nitrosamines in Beer

MARIO M. MANGINO and RICHARD A. SCANLAN
Department of Food Science and Technology, Oregon State University,
Corvallis, OR 97331

TERENCE J. O'BRIEN
Great Western Malting Company, P.O. Box 1529, Vancouver, WA 98668

N-Nitrosodimethylamine (NDMA) levels in malt
and beer have been significantly reduced, and methods
for the inhibition of NDMA formation are discussed.
NDMA in beer originates from malt dried by direct-
fired kilning. Nitrosating agent(s) appear to be
nitrogen oxides incorporated into the drying air from
the thermal oxidation of ambient nitrogen. Possible
amine precursors of NDMA include dimethylamine,
trimethylamine and the alkaloidal tertiary amines
hordenine and gramine. At pH 4.4 or pH 6.4, 65°C,
hordenine and gramine were both nitrosated to yield
NDMA. Gramine was highly susceptible to nitrosation
at pH 4.4, giving a yield of NDMA equal to that obtained from dimethylamine.

N-methyltyramine, the immediate biosynthetic
precursor of hordenine, was synthesized and nitrosated
in aqueous acid. The major product was 4-hydroxy-3-
nitro-N-nitroso-N-methylphenethylamine.

In the last fifteen years there has been considerable interest in the analysis of volatile N-nitrosamines in foods. The
primary focus has been on meat cured with nitrite (1) although
nitrosamines have been shown to occur occasionally in other foods
such as fish and cheese (2, 3). Recently, attention has been
directed to volatile nitrosamines in beer and other alcoholic
beverages. The purpose of this paper is to review current information on the presence of nitrosamines in beer, and to discuss
work done in our laboratory and elsewhere on the mode of formation
of nitrosamines in beer.

Since the first report of N-nitrosodimethylamine (NDMA) contamination in beer (4), there have been a number of surveys on
volatile nitrosamine occurrence in beers of different types and
origins. The data in Table I includes NDMA analyses on light
beers, dark beers, ales, and malt liquors. Table I shows that the

0097-6156/81/0174-0229$05.00/0
© 1981 American Chemical Society

Table I. Surveys on NDMA in Beer (ppb)

Investigators	Source	No. Positive/No. Samples[a]	Range	Mean[b]
Spiegelhalder et al. (4)	Europe + Import	111/158[c]	ND–68	2.7
Walker et al. (5)	Europe + Import	63/75[c]	ND–10.8	1.9
Goff and Fine (6)	USA + Import	18/18[c]	0.4–7	2.8
Scanlan et al. (7)	USA	23/25[c]	ND–14	5.9
Sen et al. (8)	Canada + Import	21/22[c]	ND–4.9	1.5
Fazio et al. (9)	USA + Import	62/64[c]	ND–7.7	2.8
Kawabata et al. (10)	Japan	29/29	tr–13.8	5.1
Kann et al. (11)	USSR	78/146	ND–56	–

[a] Does not represent the number of brands.

[b] Mean for all samples analyzed in each survey.

[c] NDMA was confirmed by GC–MS for some samples.

occurrence of NDMA in beer was widespread, not limited to a particular point of origin or type of beer. NDMA was by far the most predominant volatile nitrosamine found, but the presence of nitrosodiethylamine (NDEA) was reported in a very small number of samples in two studies (4, 5), and nitrosopyrrolidine (NPYR) was reported in two samples of Japanese beer (10).

The data in Table I are also significant in terms of the type of analysis to determine the presence of NDMA. In all cases analysis was done using gas chromatography coupled with a Thermal Energy Analyzer, a sensitive, relatively specific nitrosamine detector (12). Further, in six of the studies, the presence of NDMA in several samples was confirmed by gas chromatography-mass spectrometry (GC-MS). The mass spectral data firmly established the presence of NDMA in the beer samples.

Because all of the information summarized in Table I was obtained several years ago, the data do not necessarily reflect the current levels of NDMA in beer. For example, in our original survey, conducted two years ago, we found a mean value of 5.9 ppb for 25 beer samples representing 18 different labels (Table I). We recently completed a new survey of 25 beers representing the same labels from similar markets in Oregon and found a mean value of 0.2 ppb, which is more than an order of magnitude lower than the original mean value. The reduction is a direct result of measures taken by the industry to reduce nitrosamine levels in beer. Reduction of nitrosamine levels in beer will be further discussed in the final section of this paper.

Sources Of NDMA In Beer

The discovery that NDMA was present in a high percentage of beer samples prompted studies to determine the origin of NDMA in beer. Analyzes of the raw materials of brewing as well as analyzes of the NDMA content at different stages of beer production have been reported by several groups (11, 13, 14). The results of all of the investigations pointed to malt as the source of NDMA. With the exception of a very few samples of hops, the other raw materials of brewing and the brewing additives (filter aids, adjuncts, salts, and enzymes) were found to be essentially free from NDMA contamination. When attention was turned to the stages of the brewing process, it was shown that in all steps following the addition of malt there was no increase in the NDMA content of beer (13).

Invariably, the malting process was shown to be the source of NDMA in beer (11, 13, 14). Since the malting process is itself a multistep operation, the possibility of NDMA formation at each step was considered. A diagram of the malting process is shown in Figure 1. Work carried out in our laboratory in cooperation with members of the malting industry showed that raw barley and the green malt obtained after germination contained negligible

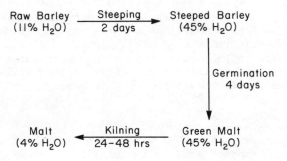

Figure 1.   Unit operations in the malting process.

amounts of NDMA. Finished malt did contain NDMA (Table II), therefore NDMA formation occurred during the kilning (drying) operation.

Table II. NDMA Formation During Malting

Sample	ppb
Barley	0.3
Green Malt	0.4
Malt	54

The type of heating system used during kilning was found to drastically effect the level of NDMA in the finished malt. The direct-fired process in which the products of combustion come in direct contact with the green malt being dried produced the highest levels of NDMA (13, 14). Indirect-firing in which hot air is generated from steam or electric coils produced much lower levels of NDMA. Typically, direct-firing resulted in 40 to 70 times greater NDMA formation in finished malt than did the indirect-firing process depending on the type of malt (13).

During the direct-fired process, the products of combustion are directly incorporated into the drying air. Nitrogen oxides (NO_x) are produced when ambient air containing N_2 is passed through flames at a temperature of at least 1800-2000°C. These nitrogen oxides come into contact with green malt during the kilning cycle. The relationship between NO_x generation and NDMA formation in malt has been substantiated by use of a pilot kiln in which heat is produced indirectly from electric coils and nitrogen oxides (NO and NO_2) are added to the drying air from an external source at a concentration of a few ppm. With this system, malts can be produced which contain NDMA at levels comparable to those found in production malts kilned by direct-firing (15).

Because of the complicated chemistry involved in formation of oxides of nitrogen and the complicated dynamics of air flow during kilning, it is not possible to state with certainty the identity of the nitrosating species. However, a series of reactions which could lead to the formation of nitrosating agents in direct-fired drying air is as follows:

$$N_2 + O_2 \xrightarrow{\Delta} 2NO \quad [1]$$

$$2NO + O_2 \longrightarrow 2NO_2 \quad [2]$$

$$NO_2 + NO \rightleftharpoons N_2O_3 \quad [3]$$

$$NO_2 + NO_2 \rightleftharpoons N_2O_4 \quad [4]$$

Nitric oxide (NO) has been shown to be a very poor nitrosating agent (16). In the gas phase at low partial pressures
N_2O_3 and N_2O_4 are both largely dissociated into their component
gases (17), but in the condensed phase the equilibrium constants
for recombination are very favorable, being 1.4×10^4 for reaction [3] at $20°C$ (18) and 6.5×10^4 for reaction [4] at $20°C$
(19). Dinitrogen trioxide (N_2O_3) is well known as a nitrosating
agent in acidic solution, and both N_2O_3 and dinitrogen tetraoxide
(N_2O_4) have been shown to be effective nitrosating agents in
neutral or alkaline solution (17). In fact, it is quite possible
that nitrosamine formation from secondary alkyl amines is very
much faster in neutral solution than the conventional reaction in
acidic solutions when preformed N_2O_3 or N_2O_4 are the nitrosating
agents (16). Whether there is enough moisture on the surface of
malt during the latter stages of kilning to effect a condensed
phase nitrosation reaction is not known. Even if the moisture
content is too low, the possibility exists that NO and NO_2 form
more N_2O_3 and N_2O_4 than would be predicted from their gas phase
equilibrum constants due to "third body" collisions with the
surface of the malt. More research will be required to identify
the nitrosating agent(s) responsible for nitrosamine formation
during the kilning of malt.

The relationship between NDMA in direct-fired malt and the
moisture content and kilning temperature of malt have been
followed in a tracking experiment (20). Figure 2 shows the amount
of NDMA in malt during a 42 hr drying cycle on a two floor commercial kiln. It was found that over half of the NDMA accumulation took place in the final six hours of drying, which corresponds to the lowest moisture content of malt. This may indicate the importance of the dehydration process for accelerating
the formation of NDMA during kilning. Dehydration could enhance
nitrosamine formation by at least two modes of action: dehydration
allows for a dramatic increase in the temperature at the malt
surface and dehydration promotes the physical migration of amine
precursors to the malt surface where these precursors come in
contact with nitrosating agents in the drying air.

Amine Precursors In Green Malt

Spiegelhalder et al. suggested that nitrosation of either
dimethylamine or a germination product such as hordenine (Figure
3) by NO_x gases in the drying air could lead to NDMA formation
in malt (13). Kann and co-workers held that nitrosation of
hordenine was the more likely source of NDMA formation in malt
(11).

A group of amines which have the potential to produce NDMA
upon nitrosation and an estimate of the relative amount of each
amine in green malt is listed in Table III.

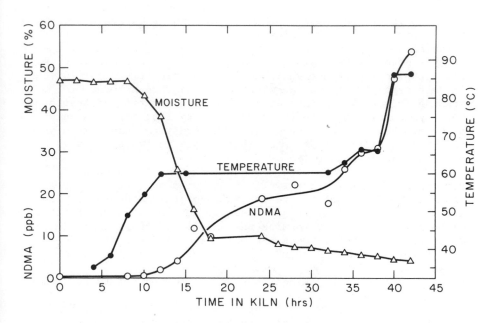

Figure 2.   Content of NDMA as a function of malt temperature and moisture during a 42-h kilning cycle (20).

Table III. Potential Precursors of NDMA in Green Malt

Amine	Estimate of Amount in Green Malt (ppm)	Ref.
Dimethylamine	1-5 (estimated from beer)	25, 26
Trimethylamine	"presence probable"	25
Hordenine	67	23
Gramine	15-20 (estimated from acrospires)	24

Hordenine, formed biosynthetically from tyrosine (Figure 3), is the principle alkaloid formed in malt roots during germination. During biosynthesis, tyrosine is enzymatically decarboxylated to tyramine which is methylated in two successive steps using the methyl- donor S-adenosylmethionine (SAM) (21). Gramine, formed biosynthetically from tryptophan, is the principle alkaloid found in malt acrospires after germination. The biosynthesis of gramine (Figure 4) is initiated by the condensation of tryptophan with pyridoxal phosphate (PLP) followed by loss of the α-carbon of tryptophan as a glycine unit. The resulting indole moiety is aminated to a primary amine and methylated in two successive steps to yield gramine (22).

Quantitative estimates are available for the amount of hordenine in malt roots as well as in dried malt (23) For gramine, only a quantitative estimate for the amount found in acrospires is available (24). Because non-specific colorimetric methods were used to determine the amounts of these alkaloids in malt, the quantitative data can only be considered approximate.

Dimethylamine and trimethylamine are both reported to be present in beer (25, 26). Malt is the most likely source of these volatile amines, since they were reported not to be formed during fermentation (25).

From the mechanistic standpoint, the formation of NDMA from dimethylamine follows the classic reaction for nitrosamine formation from secondary amines (27). For the tertiary amines hordenine, gramine, and trimethylamine the formation of NDMA could result via nitrosative dealkylation followed by nitrosation of an intermediate secondary amine according to the mechanism described by Smith and Loeppky (28), and summarized schematically in Figure 5. According to this mechanism, a tertiary amine reacts with a nitrosating agent to give a nitrosammonium ion which decomposes to give an intermediate N,N-disubstituted immonium ion which is hydrolyzed to a carbonyl compound and a secondary amine. The secondary amine is nitrosated to give the corresponding N-nitrosamine. Keefer has offered a variation of this mechanism in which the substituted immonium ion undergoes nucleophilic attack by nitrite ion following by rearrangement and collapse to

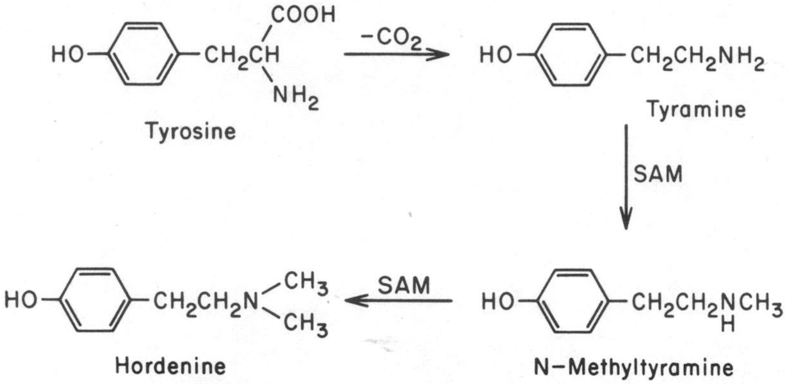

Figure 3. Biosynthesis of hordenine.

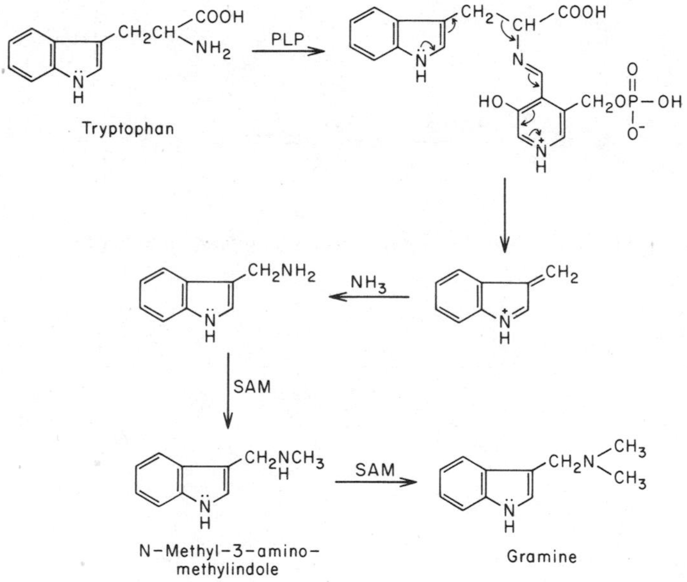

Figure 4. Biosynthesis of gramine (22).

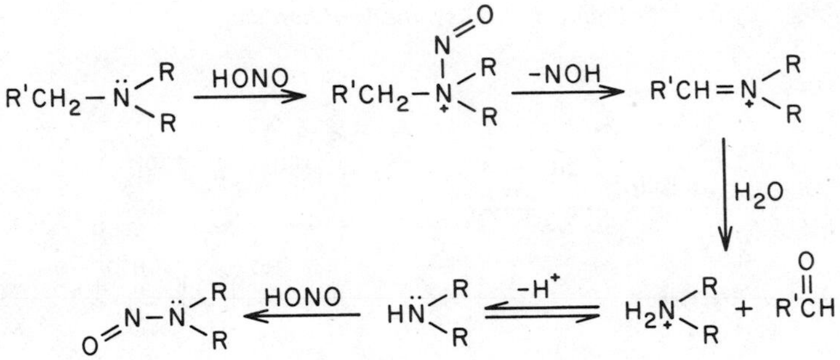

Figure 5. Nitrosative dealkylation of tertiary amines (28).

yield a nitrosamine directly without the necessity for secondary amine formation (29).

Since the amines hordenine and gramine are formed in malt during germination, it was necessary to determine whether they yield NDMA during kilning. Widmaier (30) indicated that the NDMA content of malt increased when hordenine or gramine was applied to the surface of green malt and then dried in a pilot direct-fired kiln. Nitrosation of hordenine and gramine was carried out in our laboratory in aqueous buffer at pH 4.4 and pH 6.4, 65°C for 16 hrs. These conditions represent the extremes of pH and inter-mediates of temperature and time which are encountered during kilning. The yields of NDMA from both compounds using 0.1M amine and 0.5M nitrite are shown in Table IV. The yields of NDMA from dimethylamine and trimethylamine under the same conditions are also included.

Table IV. Yields (%) of NDMA from Nitrosation of Malt Precursors

Amine	pH 4.4[a]	pH 6.4[b]
Dimethyl-	78	65
Trimethyl-	8	0.8
Hordenine	11	2
Gramine	76	5

Conditions: 0.1M Amine
0.5M NaNO$_2$

[a]Acetate buffer

[b]Citrate-phosphate buffer

The results show that hordenine and gramine are both readily nitrosated to give NDMA, especially at pH 4.4. Gramine was shown to be extremely susceptible to nitrosation at pH 4.4, giving a yield of NDMA an order of magnitude larger than the yield of NDMA obtained from trimethylamine, the most commonly reported tertiary amine in foods. In fact, gramine was nitrosated to yield NDMA as readily as dimethylamine at pH 4.4. In a separate experiment conducted under the same conditions described here, except using 0.1M amine and 0.3M NaNO$_2$, we found that nitrosation of gramine at pH 4.4 gave a larger yield of NDMA than did the nitrosation of dimethylamine. This large preference for NDMA formation from gramine is not predicted on the basis of the usual steric pre-ferences seen for the nitrosative dealkylation of tertiary amines (28, 31). For example, Smith and Loeppky found that nitrosation of N,N-diethylbenzylamine resulted mainly in loss of an ethyl group so that the most predominant nitrosamine formed (in a ratio of 4:1) was N-nitrosoethylbenzylamine. By analogy, we expected that loss of a methyl group rather than loss of the indole carbinyl group would be favored during nitrosation of gramine if the usual nitrosative dealkylation mechanism were operating.

Consequently, we postulate that the nitrosation of gramine proceeds through an N,N-dimethylnitrosammonium ion which decomposes
via direct displacement of NDMA from the indole-3-carbinyl system
(Figure 6). This mechanism is analogous in concept with the
elimination of trimethylamine from the quaternary salts of gramine
(32). The dimethylamino- group of gramine was shown to be labile,
since attempts to isolate the methiodide failed, and trimethylamine and tetramethylammonium iodide were obtained as by-products
(33).
        The relationship between the results obtained here for nitro-
sation of hordenine and gramine and the mechanism of NDMA formation
from malt precursors during kilning is now under active investigation in our laboratory.

Nonvolatile Nitrosamines From Kilned Malt

        To this point we have discussed only volatile nitrosamine
formation from the kilning of green malt. A review of the biosynthesis of hordenine and gramine (Figures 3 and 4) shows that
the intermediate secondary amines N-methyltyramine and N-methyl-
3-aminomethylindole are encountered on the pathway to hordenine
and gramine, respectively. Both of these secondary amines are
present in significant concentration in malt after four days of
germination (34), and one would predict that both would be
susceptible to nitrosation. We have synthesized a sample of N-
methyltyramine and carried out its nitrosation in acetic acid
at room temperature. The major product in 24% yield is 4-hydroxy-
3-nitro-N-nitroso-N-methylphenethylamine (Figure 7). We postulate
that the nitro- group arises by oxidation in situ of a preformed
o-nitrosophenol which we have not been able to isolate Challis
and co-workers studied the nitrosation of p-substituted phenols
and were able to isolate only o-nitrophenols (35). They attributed
this to rapid oxidation of initially formed o-nitrosophenols by
dilute acid.
        The nitrosation of N-methyltyramine represents an interesting
case of N-nitrosation and C-nitrosation in the same molecule. In
relatively strong acid (below pH 5), C-nitrosation is predicted to
be more rapid (36); but in dilute acid where there is a considerable
amount of unprotonated amine available, it is not yet clear which
type of nitrosation is favored. N-methyltyramine would serve as
a useful model to study the competitive kinetics of N-nitrosation
and C-nitrosation in dilute acid. Details of the synthesis and
nitrosation of N-methyltyramine will be published elsewhere.

Reduction Of NDMA Formation In Malt And Beer

        It was stated earlier that the level of NDMA in beer which we
analyzed recently was considerably lower than the NDMA level in
beers analyzed in our laboratory two years ago. Our observation

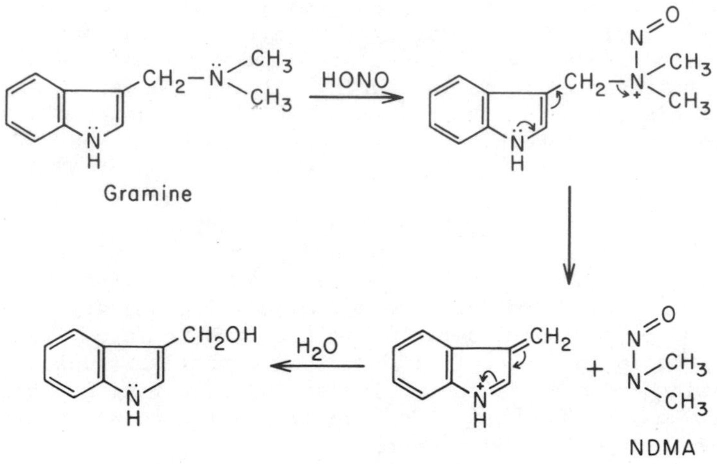

Figure 6.   Proposed mechanism for nitrosation of gramine.

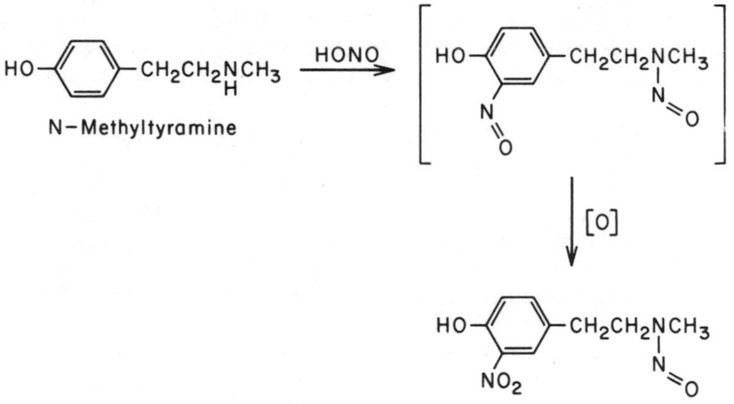

Figure 7.   Nitrosation of N-*methyltyramine.*

was supported by the results of Havery, Hotchkiss, and Fazio, who recently conducted a nationwide survey of domestic (U.S.) and imported beers (37). They found a mean value of less than 1 ppb for 180 samples of domestic beer and 1 ppb for 80 samples of imported beer. Both levels represented a reduction in NDMA levels compared to their original survey (Table I). It was suggested that the lower levels were due to steps taken by maltsters to minimize nitrosamine formation during kilning. An accompanying paper in this symposium by Preussmann et al. indicates that NDMA levels in European beers have also been sharply reduced.

Among the methods suggested for the reduction of NDMA formation in beer, the most effective are changes in the technology of the malting process at the kilning stage. The conversion to indirect-fired kilning causes a dramatic reduction in the level of NDMA in malt (13, 14). In the indirect-firing process, combustion products are not mixed with the drying air. The conversion to indirect-firing is costly both in capital investment and in increased fuel costs; nevertheless, much of the U.S. malting industry is converting to indirect-firing because it ultimately may be the most effective method for reducing nitrosamine levels in malt and in beer.

The elimination of amine precursors by making changes in the malting process or by altering the genetic characteristics of the barley seed are possible solutions. The use of germination inhibitors, such as bromate salts, to retard the growth of malt roots has been suggested as a method to retard hordenine synthesis during malting (38). Because of differences among barley varieties, it might also be possible to develop barley varieties with a deficiency in the capacity to produce hordenine and gramine and thereby lower the level of nitrosatable amines synthesized during germination.

The use of sulfur dioxide application or the burning of elemental sulfur during direct-fired kilning is currently being employed with considerable success by the malting industry to retard NDMA formation during kilning. Historically, sulfuring of green malt at the start of kilning was done to increase the solubility of the proteins in malt (39). Sulfur application to retard NDMA formation during kilning is done either by burning elemental sulfur to produce SO_2 or by the direct injection of SO_2 into the drying air. Injection of SO_2 has been shown to be the more effective method for application of sulfur during kilning, since the injection of SO_2 can be accurately controlled by mechanical metering (20). Detailed experiments conducted by O'Brien et al. showed that the most effective inhibition of NDMA formation occurred when SO_2 was applied during the first 8 hours of the kilning cycle. Any time delay between the start of kilning and the application of SO_2 resulted in much less effective inhibition of NDMA formation (20).

The use of SO_2 during malt kilning can retard nitrosamine formation by two modes of action. SO_2 dissolves in the aqueous phase on the surface of green malt thereby forming acid which lowers the surface pH of the malt (20). The lower pH leads to an increase in the level of protonated amines, whereas only amines in the unprotonated form can be nitrosated. Secondly, SO_2 in solution is in equilibrium with bisulfite ion. The bisulfite ion is a reducing agent in food systems (40), and may chemically reduce nitrosating agents on the surface of malts. In this respect, the action of bisulfite would be analogous to the in-hibition of nitrosamine formation by ascorbate (41, 42).

ACKNOWLEDGEMENTS

This investigation was supported in part by Grant Number CA 25002, awarded by the National Cancer Institute, DHHS and in part by a grant from the United States Brewers Association, Inc. We also thank the National Cancer Institute, DHHS for loan of the Thermal Energy Analyzer under Contract No. NO1-CP-85610. Oregon Agricultural Experiment Station Technical Paper No. 5993.

LITERATURE CITED

1. Scanlan, R.A. CRC Crit. Rev. Fd. Technol. 1975, 5, 357.
2. Crosby, N.T.; Foreman, J.K.; Palframan, J.F.; Sawyer, R. Nature, London 1972, 238, 342.
3. Gough, T.A.; Webb, K.S.; Coleman, R.F. Nature, London 1978, 272, 161.
4. Spiegelhalder, B.; Eisenbrand, G.; Preussmann, R. Food Cosmet. Toxicol. 1979, 17, 29.
5. Walker, E.A.; Castegnaro, M.; Garren, L.; Toussaint, G.; Kowalski, B. J. Natl. Canc. Inst. 1979, 63, 4, 947.
6. Goff, E.U.; Fine, D.H. Food Cosmet. Toxicol. 1979, 17, 6, 569.
7. Scanlan, R.A.; Barbour, J.F.; Hotchkiss, J.H.; Libbey, L.M. Food Cosmet. Toxicol. 1980, 18, 27.
8. Sen, N.P.; Seaman, S.; McPherson, M. J. Food Safety 1980, 2, 1, 13.
9. Fazio, T.; Havery, D.C.; Howard, J.W. Proc. VI Inter. Sym. N-Nitroso Compds. 1979, 419-431. Eds: Walker, E.A.; Griciute, L.; Castegnaro, M.; Borzsonyi, M.; Davis, W. International Agency for Research on Cancer Public. No. 31; Lyon, France, 1980.
10. Kawabata, T.; Uibu, J.; Ohshima, H.; Matsui, M.; Hamano, M.; Tokiwa, H. Proc. VI Inter. Sym. N-Nitroso Compds. 1979, 481-490. Eds: Walker, E.A.; Griciute, L.; Castegnaro, M.; Borzsonyi, M.; Davis, W. International Agency for Research on Cancer Public. No. 31; Lyon, France, 1980.

11. Kann, J.; Tauts, O.; Kalve, R.; Bogovski, P. Proc. VI Inter. Sym. N-Nitroso Compds. 1979, 319-326. Eds: Walker, E.A.; Griciute, L.; Castegnaro, M; Borzsonyi, M.; Davis, W. International Agency for Research on Cancer Public. No. 31; Lyon, France, 1980.
12. Fine, D.H.; Rounbehler, D.P. J. Chromat. 1975, 109, 271.
13. Spiegelhalder, B.; Eisenbrand, G.; Preussmann, R. Proc. VI Inter. Sym. N-Nitroso Compds. 1979, 467-477. Eds: Walker, E.A.; Gricuite, L.; Castegnaro, M.; Borzsonyi, M; Davis, W. International Agency for Research on Cancer Public. No. 31; Lyon, France, 1980.
14. Hickman, D.H.; Hardwick, W.A.; Ladish, W.J.; Meilgaard, M.C.; Jangaard, N.O. "Technical Report of the United States Brewers Association Ad Hoc Committee on Nitrosamines" Master Brewers Association of America, Technical Quart. In press (1981).
15. Scanlan, R.A.; O'Brien, T.J. Unpublished data.
16. Challis, B.C.; Kyrtopoulos, S.A. J. Chem. Soc. Chem. Comm. 1976, 877.
17. Challis, B.C.; Kyrtopoulos, S.A. J.C.S. Perkin II 1979, 1296.
18. Gratzel, M; Taniguchi, S.; Henglein, A. Ber. Bunsengesellschaft Phys. Chem. 1970, 74, 488.
19. Gratzel, M.; Henglein, A.; Lilie, J.; Beck, G. Ber. Bunsengesellschaft Phys. Chem. 1969, 73, 646.
20. O'Brien, T.J.; Lukes, B.K.; Scanlan, R.A. Mast. Brew. Assoc. Amer., Tech. Quart 1980, 17, 4, 196.
21. Leete, E.; Bowman, R.M.; Manuel, M.F. Phytochem. 1971, 10, 3029.
22. Scott, A.I. "Biosynthesis of Indole Alkaloids", in MTP Inter. Rev. Sci., Org. Chem. Series I: Alkaloids 1973, 105-141. Eds: Wiesner, K. University Park Press.
23. McFarlane, W.D. Proc. Europ. Brewery Conv. Stockholm, 1965, 387-395.
24. Schneider, E.A.; Wightman, F. Can. J. Biochem. 1974, 52, 698.
25. Drews, B.; Just, F.; Drews, H. Proc. Europ. Brewery Conv. Copenhagen, 1957, 167.
26. Singer, G.M.; Lijinsky, W. J. Agric. Food Chem. 1976, 24, 3, 550.
27. Mirvish, S.S. J. Natl. Cancer Inst. 1970, 44, 633.
28. Smith, P.A.S.; Loeppky, R.N. J. Amer. Chem. Soc. 1967, 89, 5, 1147.
29. Keefer, L.K. Proc. Amer. Chem. Soc. Sym. on N-Nitroso Compds. 1978, 91-108. Eds: Anselme, J.P. ACS Symposium Series No. 101, Washington, D.C. 1979.
30. Widmaier, R. Personal Communication.
31. Lijinsky, W.; Keefer, L.; Conrad, E.; Van de Bogart, R. J. Natl. Cancer Inst. 1972, 49, 1239.
32. Marion, L. "The Indole Alkaloids", in The Alkaloids: Chemistry and Physiology Vol. II, 371-398. Eds: Manske, R.H.F.; Holms, H.L. Academic Press, New York 1952.

33. Madinaveitia, J. J. Chem. Soc. 1973, 1927.
34. Mann, J.D.; Steinhart, C.E.; Mudd, S.H. J. Biol. Chem. 1963, 238, 2, 676.
35. Challis, B.C.; Higgins, R.J. J.C.S. Perkin II 1973, 2365.
36. Challis, B.C. Nature 1973, 244, 466.
37. Havery, D.C.; Hotchkiss, J.H.; Fazio, T. J. Food Sci. 1981, 46, 501.
38. Hough, J.S.; Briggs, D.E.; Stevens, R. "Malting and Brewing Science"; Chapman and Hall, Ltd., London, 1971, 156-157.
39. Pomeranz, Y. CRC Crit. Rev. Food Tech. 1974, 4, 3, 377.
40. Roberts, A.C.; McWeeny, D.J. J. Food Tech. 1972, 7, 221.
41. Gray, J.; Dugan, L.R. J. Food Sci. 1975, 40, 981.
42. Mirvish, S.S. Ann. N.Y. Acad. Sci. 1975, 258, 175.

RECEIVED July 20, 1981.

Formation, Occurrence, and Carcinogenicity of N-Nitrosamines in Tobacco Products

DIETRICH HOFFMANN, JOHN D. ADAMS, KLAUS D. BRUNNEMANN, and STEPHEN S. HECHT

Naylor Dana Institute for Disease Prevention, American Health Foundation, Valhalla, NY 10595

Tobacco and tobacco smoke contain three types of N-nitrosamines. These are the volatile nitrosamines (VNA), nitrosamines deriving from residues of agricultural chemicals on tobacco, and the tobacco specific nitrosamines (TSNA). All three types of nitrosamines are formed during tobacco processing and during smoking. The VNA can be selectively reduced in cigarette smoke by cellulose acetate filter tips. Of the nitrosamines deriving from agricultural chemicals, we have so far only detected the animal carcinogen nitrosodiethanolamine in concentrations of 0.1-6.8 ppm in tobacco and 10-40 ng per cigarette in the smoke. TSNA are formed during tobacco processing and smoking from nicotine, nornicotine and anatabine. Their concentrations range from 1-100 ppm in tobacco and from 1-10 µg/cig. in cigarette smoke. Selective removal of TSNA from mainstream smoke by cellulose acetate filters was demonstrated and can be enhanced by perforation of the filter tip. In mice, rats and hamsters, N'-nitrosonornicotine is a moderately active carcinogen while 4-(methylnitrosamino)-1-(3-pyridyl)-1-butanone is a strong carcinogen. N-nitrosoanatabine is being bioassayed at this time.

N-Nitrosamines are formed during processing and smoking of tobacco products. Proteins, agricultural chemicals and alkaloids in tobacco products serve as major precursors for volatile, non-volatile, and tobacco-specific nitrosamines (Figure 1). In this review we will summarize the progress achieved in respect to tobacco nitrosamines since the last ACS symposium in Boston in June of 1978 (1). Additional papers will review the metabolism of cyclic N-nitrosamines, including that of N'-nitrosonornicotine (2) and the correlation between tobacco and alcohol consumption and cancer of the upper alimentary tract (3).

N-NITROSAMINES IN TOBACCO PRODUCTS

I. VOLATILE NITROSAMINES

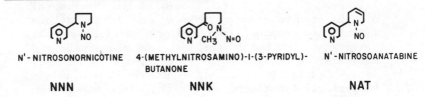

$R = R' =$ CH_3, C_2H_5, C_3H_7 or C_4H_9

2. NITROSAMINES FORMED FROM AGRICULTURAL CHEMICALS

$$HO-CH_2-CH_2$$
$$N-NO$$
$$HO-CH_2-CH_2$$

N-NITROSODIETHANOLAMINE (NDELA)

3. TOBACCO SPECIFIC NITROSAMINES

N'-NITROSONORNICOTINE   4-(METHYLNITROSAMINO)-I-(3-PYRIDYL)-   N'-NITROSOANATABINE
                                    BUTANONE

NNN   **NNK**   **NAT**

Figure 1.   N-Nitrosamines found in tobacco products.

Volatile N-Nitrosamines

In Tobacco. At the time of harvesting, fresh tobacco leaves do not contain measurable amounts of nitrosamines (<5 ppb). However, these compounds are formed during curing, aging and fermentation. Their concentrations depend primarily on the content of proteins, alkaloids, agricultural chemicals and nitrate in the tobacco, as well as on the processing conditions which lead to the reduction of the nitrates.

The first step in the analysis is extraction of the tobacco with buffer solution (pH 4.5) containing 20 mM ascorbic acid. The nitrosamines are then concentrated by partition with dichloromethane, and a chromatographic clean-up on alumina. In the final step, the concentrate is analyzed by GC-TEA and confirmation of the nitrosamines is obtained by GC-MS (4). If isolated amounts of the nitrosamines are below levels needed for GC-MS confirmation, we employ confirmatory techniques proposed by Krull et al. (5).

TABLE I.   **VOLATILE NITROSAMINES IN TOBACCO** (ppb) (4)

	NDMA	NDEA
Cigar (Pennsylvania)	6.9	n.d.[a]
Robinson, high NO_3	9.5	14.9
Catterton, high NO_3	15.6	12.1
French cigarette	188	12
Fine-cut chewing tobacco	56	8.6

[a]n.d.=not determined   Cancer Research - Table 5

Table I lists the concentrations of nitrosodimethylamine (NDMA) and nitrosodiethylamine (NDEA) in some of the tobacco products which had greater than 0.5% nitrate content. Tobacco with less than 0.5% nitrate content, such as Bright tobaccos, yielded NDMA, NDEA and nitrosopyrrolidine (NPYR) generally below 5 ppb. The relatively high concentrations of NDMA in fine cut tobaccos and in snuff are possibly of significance in the increased risk for oral cancer among snuff dippers who use these tobacco types repeatedly each day by placing a pinch of the product directly into the gingival buccal fold (6,7).

In Smoke. We compared the gas chromatograms of nitrosamines in matching aliquots of mainstream smoke derived from Burley type cigarettes that were identical except for the degree of nitrate fertilization during cultivation (Figure 2).   This comparison supports the concept that the nitrate concentration in tobacco is a determining factor for the nitrosamine yields in the smoke. The data in Table II confirm this concept. These studies have

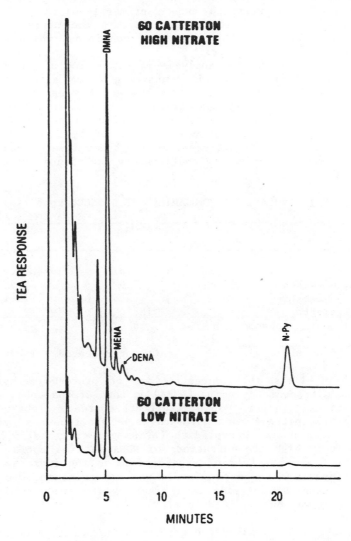

Figure 2.   Effect of nitrate in tobacco on nitrosamine yield in mainstream smoke
(4).

Table II.

VOLATILE N-NITROSAMINES IN CIGARETTE SMOKE

Cigarette	Volatile Nitrosamines (ng/cig)			
	NDMA[e]	NEMA[f]	NDEA[g]	NPYR[h]
Mainstream smoke				
Burley, NF[a]	76	9.1	2.5	52
Bright, NF	13	<0.1	1.8	6.2
Commercial, NF	13	1.8	1.5	11
Commercial, FA[b]	5.7	0.4	1.3	5.1
Kentucky 1R1, NF	9.0	1.5	2.0	6.6
Catterton, high NO$_3$, NF	97	8.0	4.8	42
Catterton, low NO$_3$, NF	20	1.2	2.3	4.1
French, NF	29	2.7	0.6	25
French, FA	4.3	0.48	0.1	11
French, FP[c]	14	2.1	0.4	11
Commercial I, FA	6.8	0.5	0.8	8.5
Commercial I, minus FA	27	2.2	1.2	33
Commercial II, FC[d]	14	0.6	7.6	7.6
Commercial II, minus FA	19	1.2	8.3	14
Sidestream smoke				
Commercial, NF	680	9.4	53	300
Commercial, NF	820	30	8.2	205
Commercial, FA	730	10	73	390
Commercial, NF	1040	10	63	210

[a]Nonfilter.
[b]Cellulose acetate filter.
[c]Paper filter.
[d]Charcoal-cellulose acetate filter.
[e]N-nitrosodimethylamine.
[f]N-nitrosoethylmethylamine.
[g]N-nitrosodiethylamine.
[h]N-nitrosopyrrolidine.

further demonstrated that practically all of the volatile nitrosamines (VNA) are formed during smoking and that cellulose acetate filter tips can selectively reduce VNA from mainstream smoke up to 90% (4,8,10,11). In this context it is of interest that about 85% of the cigarettes on the U.S. market have cellulose acetate filter tips.

The generation of VNA, and especially of NDMA in sidestream smoke (SS) is much higher than that in mainstream smoke (MS). Consequently, VNA levels in SS exceed those in MS up to 50 times (4,9,10; Table II). The possible significance of this observation in respect to indoor air pollution has been discussed by us earlier (12).

Nitrosamines from Agricultural Chemicals.

Among the agricultural chemicals used for the cultivation of tobacco crops we find several amines, amides and carbamates. These include dimethyldodecylamine acetate (Penar), maleic hydrazide-diethanolamine (MH-30), and carbaryl (Sevin) as a representative of the methyl urethanes (Figure 3; 13,14). It is known that small quantities of these agents are found as residues in harvested tobacco (15). To date, only diethanolamine (DELA), the water-solubilizer for maleic hydrazide in MH-30, has been studied as a possible precursor for nitrosamines in tobacco and in tobacco smoke. In 1976, more than 1,400 metric tons of maleic hydrazide had been used on U.S. tobacco (16), most of which had been applied as the MH-30 formulation with diethanolamine (14,16).

NDELA Analysis. Using a conventional analytical method which we developed in 1977, we determined that MH-30 treated tobaccos contain 100-170 ppb of N-nitrosodiethanolamine (NDELA; 17). The availability of the TEA detector made it possible to develop an analytical method which permitted routine monitoring of NDELA in tobacco and its smoke.

This method requires about 40 g of tobacco which are extracted with ethyl acetate in the presence of ascorbic acid. A trace amount of ^{14}C-NDELA is added as an internal standard for quantitative analytical work. The filtered extract is concentrated and NDELA is enriched by column chromatography of the concentrate on silica gel. The residues of fractions with β-activity are pooled and redissolved in acetonitrile. Initially, we attempted to separate NDELA on a 3% OV-225 Chromosorb W HP column at 210°C using a GC-TEA system with direct interface similar to the technique developed by Edwards et al. for the analysis of NDELA in urine (18). We found this method satisfactory for reference compounds; however, it was not useful for an optimal separation of NDELA from the crude concentrate of the tobacco extract (Figure 4). Therefore, we silylated the crude concentrate with BSTFA and an aliquot was analyzed by GC-TEA with direct interface. The chromatographic conditions were 6 ft glass column filled with 3% OV-

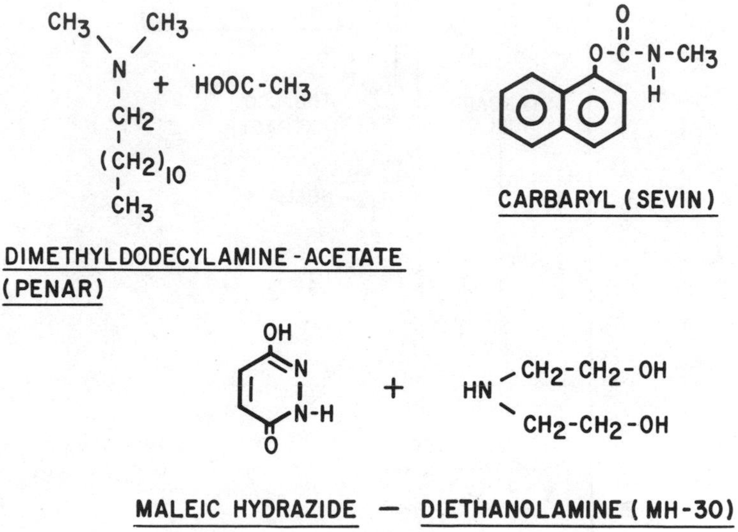

Figure 3. *Agricultural chemicals for tobacco cultivation.*

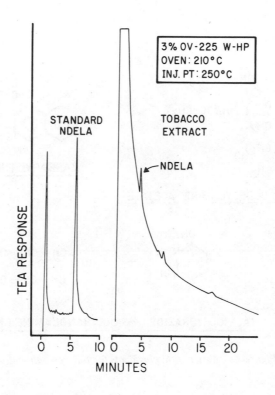

Figure 4.   NDELA analysis via direct GC-TEA.

225 on Chromosorb W HP 80-100, operated at 130°C (Figure 5). NDELA in tobacco smoke was determined by directing the mainstream smoke of 100 cigarettes or little cigars through 2 gas wash bottles in sequence which were filled with ethyl acetate and ascorbic acid. Cigars were smoked under appropriate standard conditions developed for these products (20). Enrichment steps and GC-TEA procedures were similar to those described for the tobacco extracts. Table III lists some of our results for NDELA in commercial as well as in experimental tobacco products.

In tobacco, we found the highest NDELA values for fresh and aged snuff with 6.8 and 3.2 ppm, respectively. The fermentation process appears to increase NDELA, as was also observed for VNA concentrations. There was clear evidence that tobaccos which had not been treated with MH-30, and cigarette smoke obtained from these tobaccos, were free of NDELA, whereas all MH-30 treated tobaccos and cigarette smoke derived from them showed measurable quantities of this nitrosamine (19).

In order to ascertain that the NDELA formation does not occur as a result of trapping of the smoke or during the analysis, we added diethanolamine to tobacco prior to extraction with ethyl acetate in the presence of ascorbic acid. The control value for NDELA was 121 ppb and the experiment with 5.5 mg diethanolamine addition yielded 113 ppb NDELA. For control of the smoke analysis we added 5.5 mg of DELA in the solvent trap and smoked cigarettes known to be free of DELA. Analysis of the trapped material showed no significant quantities of NDELA, so that artifactual formation of this nitrosamine during smoke collection and analysis can be ruled out.

Thus, we conclude that the diethanolamine in MH-30 is the major precursor for NDELA in processed U.S. tobaccos and tobacco smoke. In fact, NDELA concentrations of 600-1,900 ppb were already present in five agricultural spray formulations of MH-30 which we analyzed (19).

Currently, we are studying the transfer rate of NDELA from tobacco into smoke as well as the potential of DELA and NDELA to serve as precursors for other nitrosamines in smoke.

Carcinogenicity of NDELA. Our special interest in NDELA as a constituent of tobacco products and as an environmental agent relates to the observation that this nitrosamine induces carcinoma of the liver as well as of the kidney in rats (21,22) and carcinoma of the nasal cavity and papillomas of the trachea in hamsters (23). Recently, Lijinsky et al reported that NDELA administration in drinking water for 34 weeks in concentrations of 3,900 ppm induces hepatocarcinomas in all rats in the test. Thus, the authors consider NDELA to be a potent carcinogen in rats (22). In an ongoing bioassay in Syrian golden hamsters, we observe that NDELA induces carcinomas of the nasal cavity independent of route of administration, i.e. subcutaneous injection, skin painting or oral swabbing in minimum doses of 58 mg/kg (24).

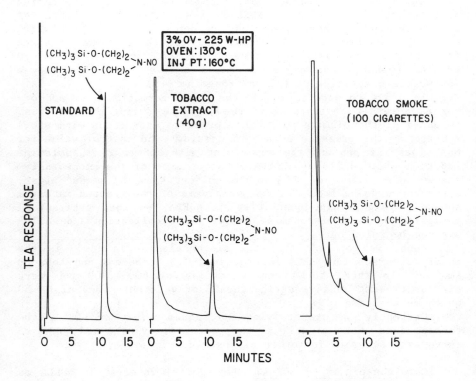

Figure 5.   NDELA analysis via silylation GC-TEA.

Table III.

NITROSODIETHANOLAMINE (NDELA) IN TOBACCO AND TOBACCO SMOKE*

Tobacco Product	Tobacco** (ppb)	Mainstream (ng/cig.)
a. Experimental Cigarettes		
Kentucky 1R1, NF, 85 mm	86	30
Kentucky 2R1, NF, 85 mm	84	51
Handsuckered 1970 Burley, NF, 85 mm	n.d.	n.d.
MH-30 treated 1970 Burley, NF, 85 mm	227	290
USDA M-6, Bright, pesticide treated, NF, 85 mm	19	20
USDA L-8, Bright, pesticide free***, NF, 85 mm	n.d.	n.d.
b. Commercial Products		
Cigarette A, NF, 85 mm	115	36
Cigarette B, F, 85 mm	194	24
Little Cigar, F, 85 mm	419	68
Large Cigar (7.7 g), NF, 125 mm	108	10
Snuff (Fine Cut), aged	3,180	–
Snuff (Fine Cut), fresh	6,840	–
Chewing Tobacco A	285	–
Chewing Tobacco B	224	–

*   Data corrected for recovery using isotope dilution method.
** Tobacco data are reported per dry weight.
*** Grown in Prince Edward Island, Canada.
n.d. = not detected (below detection limit), NF = non-filter,
F = filter.

This then confirms the fact that NDELA penetrates the skin to act as an organ specific carcinogen as was also evident from the studies by Lijinsky et al on rats (25), from the reports by Edwards et al on cosmetics in man (18) and from studies with Syrian golden hamsters in our laboratory (24).

Tobacco-Specific N-Nitrosamines

Precursors and Formation. Tobaccos used for commercial products in the U.S.A. contain between 0.5 and 2.7% alkaloids. Nicotine constitutes 85-95% of the total alkaloids (14,26,27). Important minor alkaloids are nornicotine, anatabine, anabasine, cotinine and N'-formylnornicotine (Figure 6). Several of these alkaloids are secondary and tertiary amines and, as such, amenable to N-nitrosation. The N-nitrosated alkaloids identified to date in tobacco and tobacco smoke include N'-nitrosonornicotine (NNN), 4-(methylnitrosamino)-1-(3-pyridyl)-1-butanone (NNK) and N'-nitrosoanatabine (NAT; Figure 7). In model experiments, nitrosation of nicotine also yielded 4-(methylnitrosamino)-4-(3-pyridyl)butanal (NNA; 28).

In a study for precursor determination, we stem-fed individual Burley leaves with nicotine-2'-^{14}C or nornicotine-2'-^{14}C (29). Subsequently, the leaves were air cured, dried and analyzed for NNN and NNN-^{14}C. Recovery of the β-activity in the form of NNN-^{14}C amounted to 0.009% and 0.007%, respectively of the stem-fed label. This demonstrates that both alkaloids give rise to NNN. More importantly, it points to the fact that the tertiary amine, nicotine, which constitutes 0.5-2.6% of commercial tobaccos (26,27), is the major precursor for the carcinogenic tobacco-specific NNN, while the secondary amine, nornicotine is of lesser importance because it amounts to only 0.005-0.06% in tobacco (Figure 8).

We thus assume, that the biochemical processes during tobacco curing may be different from the in vitro N-nitrosation of secondary and tertiary amines (30).

Nonvolatile Nitrosamines In Tobacco. A method which we developed several years ago for the analysis of tobacco-specific nitrosamines (TSNA; 31) involves extraction of tobacco with buffered ascorbic acid (pH 4.5) followed by partition with ethyl acetate, chromatographic clean-up on silica gel, and analysis by HPLC-TEA (Figure 9). Results obtained with this method for a large spectrum of tobacco products (Table IV), strongly support the concept that the levels of nitrate and alkaloids, and especially the methods for curing and fermentation, determine the yields of TSNA in tobacco products. Recent and as yet preliminary data from snuff analyses indicate that aerobic bacteria play a role in the formation of TSNA during air curing and fermentation.

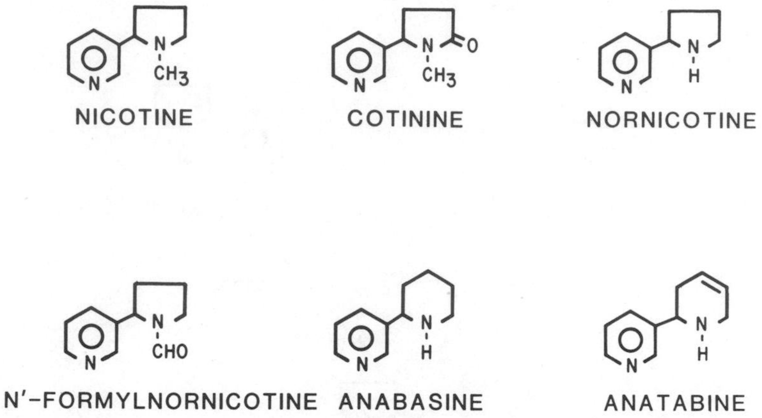

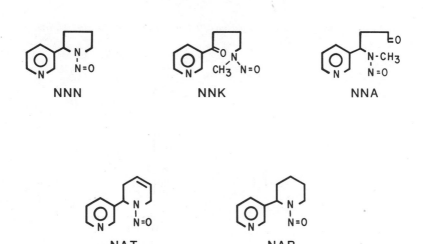

NICOTINE COTININE NORNICOTINE

N'-FORMYLNORNICOTINE ANABASINE ANATABINE

Figure 6. Common tobacco alkaloids in tobacco and/or smoke.

NNN NNK NNA

NAT NAB

Figure 7. Some nitrosamines which can be derived from the tobacco alkaloids.

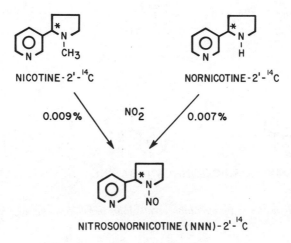

Figure 8.   *Conversion of labeled nicotine and nornicotine during curing.*

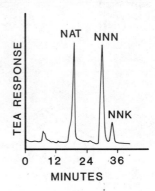

Figure 9.   *HPLC-TEA trace of NAT, NNN, and NNK in snuff.*

TABLE IV.

TOBACCO SPECIFIC N̲-NITROSAMINES IN TOBACCO PRODUCTS (31̲)

Tobacco product[a],[b]	Tobacco (ppm)			Mainstream (μg/cig)		
	NAT	NNN	NNK	NAT	NNN	NNK
Burley cigarette NF	3.2	7.0	nd[c]	4.6	3.7	0.32
Bright cigarette NF	0.44	0.22	0.37	0.41	0.62	0.42
Commercial cigarette NF	1.6	1.7	0.74	0.33	0.24	0.11
Commercial cigarette F-A	1.3	1.4	0.70	0.37	0.31	0.15
Kentucky IRI NF	0.62	0.63	0.13	0.53	0.39	0.16
Commercial French cigarette NF, 70 mm	1.8	2.9	0.52	0.18	0.48	0.44
Commercial French cigarette F-A, 70 mm	1.5	2.7	0.37	0.18	0.49	0.36
French cigarette NF	2.0	11.9	1.1	0.64	3.2	0.43
French cigarette F-A	2.0	11.9	1.1	0.19	1.0	0.19
French cigarette F-P	2.0	11.9	1.1	0.16	0.73	0.12
Little cigar F-A	13.	45.	35.	1.7	5.5	4.2
Columbia cigar (5,7 g)	3.3	10.7	1.1	1.9	3.2	1.9

[a]All cigarettes and the little cigars were 85 mm long, except when otherwise stated.
[b]F-A = Cellulose acetate filter tip; F-P = Paper filter tip; NF = Non-filter cigarette.
[c]nd = not detected;

Abbreviations see footnote 2 of Table V.

Cancer Research

The alkaloid derived nitrosamines in cigarettes, cigars and pipe tobacco contribute to the quantities of these carcinogens in the smoke because of their partial transfer (32).

As constituents of chewing tobacco and snuff, carcinogenic TSNA come into direct contact with tissues of the oral cavity, when the user places the tobacco product in the gingival buccal fold. Snuff use is associated with a significantly increased risk for cancer at this site (7,38). Snuff dipping has significantly gained in popularity among young people in the U.S.A. and in Sweden (33,34) and has been associated with an increased risk for cancer of the oral cavity in the southern United States (34). The relatively high concentrations of the carcinogenic TSNA in snuff motivated us to examine this type of tobacco product in more detail. Our first analysis pointed out that products purchased at a given time in several different regions had significant variations in TSNA content. This is the case for 4 Swedish snuff products listed in Table V (35). TSNA values in American snuff products varied from 6-85 ppm. The fact that a recently marketed brand of snuff contains only 6.6 ppm TSNA indicates that a method for significant TSNA reduction in snuff is available. However, it is important that the snuff product is wrapped in airtight packages in order to avoid exposure to air and airborne microorganisms since this exposure has been shown to increase TSNA formation in snuff (35).

Although we concur with other scientists (36,37) that snuff use may be a feasible alternative to cigarette smoking, we feel that no efforts should be spared to reduce the concentration of alkaloid derived N-nitrosamines in snuff, since the use of this type of tobacco product has been associated with an increased risk for cancer of the oral cavity. According to a recent NCI study, the relative risk for oral and pharyngeal cancer among snuff dipping women in the Southern United States was 4.2 and among chronic users the risk approached fifty fold that of a non-smoker for cancer of the gum and of the buccal mucosa (38).

Nonvolatile Nitrosamines In Saliva.   In vitro experiments had indicated that the tobacco-specific nitrosamines are formed also during snuff dipping (26). Therefore, we analyzed the saliva of snuff dippers and tobacco chewers. A comparison of the results demonstrated the presence of TSNA in saliva at a wide range of concentrations (Table VI), which could be ascribed to differences in the product, but also to differences in the manner of chewing, and, lastly, to individual factors in each person's saliva.

In order to verify some of these associations, we gave 4 women, who had been habitual snuff dippers for more than 10 years, identical samples of a snuff product. The women used this product under controlled conditions and we collected saliva twice on subsequent days. Table VII shows the differences observed

Table V.

**TOBACCO SPECIFIC N—NITROSAMINES
IN COMMERCIAL U.S., BAVARIAN AND SWEDISH SNUFF[1]**

Snuff Origin			Tobacco Specific N-Nitrosamines[2](TSNA)			
			ppm			ppm
			NNN	NNK	NAT	Total TSNA
U.S.	I		39	2.4	44	85.4
	II		26.5	4.65	22.7	43.8
	III		3.5	1.3	1.8	6.6
Bavaria	I		6.72	1.54	4.37	12.6
	II		6.08	1.50	3.92	11.5
Sweden	I	Umeå	8.59	2.19	5.56	16.3
	I	Uppsala	6.05	1.44	4.29	11.8
	I	Lund	8.86	2.25	5.55	16.7
	II	Umeå	8.35	1.82	3.20	13.4
	II	Uppsala	5.19	0.59	3.14	8.9
	II	Lund	10.3	2.12	6.96	19.4
	III	Umeå	7.84	2.53	3.09	13.5
	III	Uppsala	3.53	0.85	1.21	5.6
	III	Lund	3.79	0.90	0.78	5.5
	IV	Umeå	9.71	2.14	6.03	17.9
	IV	Uppsala	77.1	3.82	25.1	106
	IV	Lund	8.17	1.55	4.96	14.7
	V	Umeå	4.73	1.28	1.02	7.0

1. Values are given for dry snuff (moisture content ≈50%).
2. NNN = N'-nitrosonornicotine; NNK = 4-(methylnitrosamino)-1-(3-pyridyl)-1-butanone; NAT = N'-nitrosoanatabine

Cancer Research

Table VI.

DETERMINATION OF TOBACCO SPECIFIC N-NITROSAMINES
IN SALIVA OF SNUFF DIPPING WOMEN

Snuff Dipper		Saliva Collected	Tobacco Specific N-Nitrosamines[2] (ng/g)		
No.		(mg)	NNN	NNK	NAT
	Age				
1	45	840	5.0	7.5	6.6
2	43	900	27.1	20.1	14.2
3	40	1,500	21.7	2.1	28.8
4	53	1,600	125	201	147

1. Abbreviations see footnote 2 of Table V.

Table VII.

ANALYSIS OF SNUFF AND SALIVA OF SNUFF DIPPERS (35)[1,2]

Snuff Dipper		Alkaloids mg/g			Nitrosamine µg/g			Ratio	Day of	SNUFF Nicotine	SALIVA Nitrosamines ppb			Ratio
No.	Age	Nicotine	Nornico-tine	Anata-bine	NNN	NNK	NAT	NNN/Nico-tine	Sampling	ppm	NNN	NNK	NAT	NNN/Nico-tine
1.	41	23.4	0.05	0.54	26.5	4.65	22.7	1:890	1	189	140	26.2	210	1:1,350
									2	448	106	21.0	470	1:4,200
2.	37	23.4	0.05	0.54	26.5	4.65	22.7	1:890	1	73	30	13	17.3	1:2,400
									2	430	132	<10	510	1:3,250
3.	44	23.4	0.05	0.54	23.1	4.33	20.4	1:1010	1	1150	420	96	370	1:2,600
									2	1560	323	62	320	1:4,900
4.	52	23.6	0.06	0.60	24.8	5.17	22.6	1:1040	1	211	25.9	10.6	12.5	1:8,150
									2	430	56.8	22.5	45.9	1:7,570

1. Saliva of 3 non-snuff-dipping women (controls) was free of nicotine and tobacco-specific N-nitrosamines.
2. The women were long-term snuff dippers (>10 years). Analysis was completed for the same snuff which the specific volunteer consumed. Abbreviations see footnote 2 of Table V.

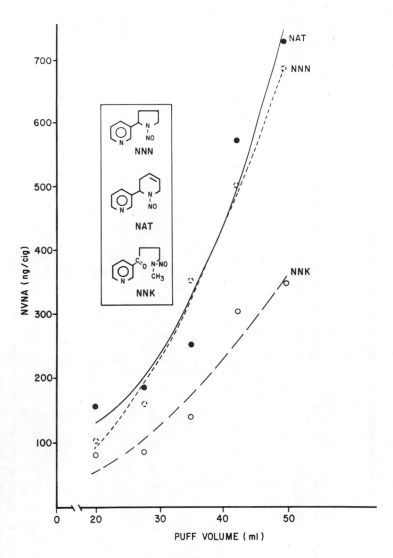

Figure 10. Tobacco specific nitrosamines as a function of puff volume (other smoking conditions were kept constant: puff duration, 2 s; puff frequency, once a minute; and butt length, 23 mm).

TABLE VIII

SELECTIVE FILTRATION OF TOBACCO SPECIFIC N-NITROSAMINES
EXPERIMENTAL CIGARETTES WITH BLACK TOBACCO (85mm)

Parameter	Cigarette 1 Nonfilter	Cigarette 2 20mm – Filter (Cellulose Acetate)		Cigarette 3 20mm – Filter (Cellulose)	
	Value	Value	% Reduction	Value	% Reduction
"Tar"–FTC (mg)	18.2	10.9	40	8.1	50
Nicotine (mg)	2.1	1.2	42	1.0	52
NNN, ng/cig	3,200	1,000	69	730	77
NNK, ng/cig	430	190	56	120	72
NAT, ng/cig	640	120	72	160	75
Total TSNA, ng/cig	4,270	1,310	69	1,010	76

[a]Number in parentheses represents the number of the last puff taken.
Abbreviations: see footnote 2 of Table V.

from one individual to the next, as well as the different TSNA
values in the saliva of a given donor. A comparison of the ratio
of nicotine to NNN in the snuff product and in the saliva revealed that in saliva the ratio increases in favor of nicotine.
Model studies have shown that this was not due to slow extraction
of NNN from tobacco, but that it could be due to faster degradation of the TSNA within the oral cavity, or to more rapid absorption of the unprotonated TSNA by the oral tissues (35).

Nonvolatile Nitrosamines In Tobacco Smoke. Although there
are more than 10 million exsmokers in the U.S.A., 53 million
adults continue to smoke cigarettes and an additional 10 million
still smoke cigars or pipes (39). The cigarette smokers are exposed to about 10 ng of volatile nitrosamines, 20-40 ng of NDELA
and, most importantly, to 1-10 μg of tobacco specific N-nitrosamines with each cigarette smoked (Table IV). Similar quantities
of the TSNA are found in sidestream smoke. The quantities of
TSNA in the smoke are dependent on nitrate, nitrite, tobacco
alkaloids and on NNN, NNK and NAT in the tobacco itself (31).

Since cigarette tobacco already contains several micrograms
of the TSNA, we determined the transfer rate of NNN into the
smoke by spiking the tobacco column with NNN-2'-^{14}C. The smoke
from such radiolabeled cigarettes is then analyzed by HPLC and
the amount of unchanged NNN-2'-^{14}C is determined by liquid scin-
tillation counting. Independent of the smoke pH, about 11% of
the radioactive NNN is found in the mainstream smoke; thus 41-46%
of mainstream smoke NNN stems from the tobacco NNN and 54-59% are
pyrosynthesized (11).

Reduction in Tobacco Smoke. One of our research goals is
the development of concepts for the reduction of TSNA in cigarette smoke. Although the nicotine and alkaloid levels in tobacco
smoke, especially in the smoke of U.S. cigarettes, have decreased
by more than 40% during the last 2 decades (39,40), it is unlikely that further reduction of the habituating nicotine will be accepted by smokers. In fact, observations of smokers have demonstrated that further reduction of smoke nicotine below 0.6 and
1.0 mg/cigarette will lead to more intense smoke inhalation by
the consumer in an effort to reach his/her "satisfaction level"
for nicotine (41, 42). An increase in smoking intensity (increase of puff volume per second) and puff frequency leads to
elevated TSNA levels in the smoke beyond those observed in the
smoke of cigarettes yielding higher amounts of "tar"-nicotine
(Figure 10).

The smoke analysis of cigarettes made from the same tobacco
blend, but with and without filter tips revealed that cellulose
acetate retains TSNA selectively (Table VIII). This phenomenon is
clearly established for a large number of filter cigarettes.

TABLE IX.

CARCINOGENIC ACTIVITY OF TOBACCO-SPECIFIC NITROSAMINES

Compounds	Species	Application	Principal Organ Affected	References
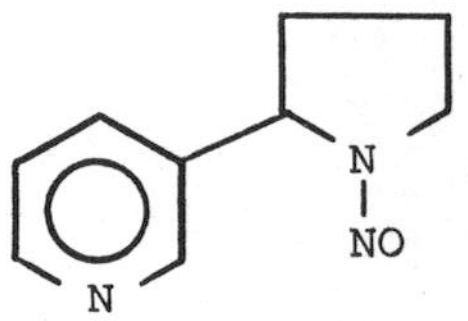	Mouse	i.p.	Lung (Adenoma, Adeno-carcinoma)	29,47,48
			Salivary glands (?)	29
	Rat	s.c.	Nasal Cavity (Carcinoma)	44
	Rat	p.o. (water)	Esophagus (Papilloma, carcinoma)	49
			Pharynx (Papilloma)	
			Nasal Cavity (Carcinoma)	50
	Hamster	s.c.	Trachea (Papilloma)	23,45
			Nasal Cavity (Carcinoma)	
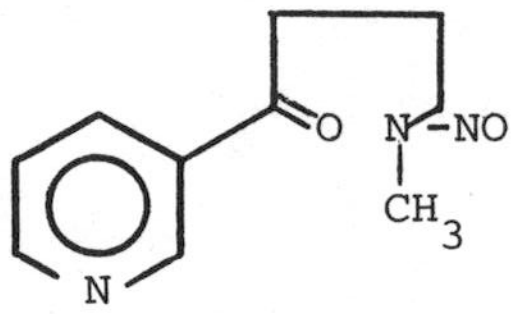	Mouse	i.p.	Lung (Adenoma, Adenocarcinoma)	29
	Rat	s.c.	Nasal Cavity (Carcinoma)	44
			Liver	
			Lung (Adenoma, Carcinoma)	
	Hamster	s.c.	Lung (Adenoma, Adenocarcinoma)	45
			Trachea (Papilloma)	
			Nasal Cavity (Carcinoma)	

NNN

NNK

Even more promising appears to be the selective reduction of TSNA by perforated cellulose acetate filter tips with high air dilution. Although the smoker will compensate for the reduction in nicotine by smoking more intensely, it appears that the TSNA yield in smoke does, in this case, not increase to the level observed in smoke of an unperforated filter cigarette.

Carcinogenicity. Our major interest in the tobacco specific N-nitrosamines has been induced by the carcinogenic activity of NNN and NNK in mice, rats and hamsters (43-45; Table IX). Whereas NNN is a moderately active animal carcinogen, NNK is a strong carcinogen and is even more active than N-nitrosopyrrolidine (46). NNN induces lung adenomas in mice, benign and malignant tumors of the nasal cavity in rats and tracheal tumors in the Syrian golden hamster. NNK induces lung adenomas in mice, and carcinomas in the nasal cavity, liver and lung of rats. In Syrian golden hamsters, NNK induces tracheal papillomas and carcinomas of the nasal cavity and lung. When NNN is given to rats in the drinking water, it induces primarily tumors of the esophagus.

ACKNOWLEDGEMENTS

We thank Dr. T.C. Tso of the U.S.D.A. Tobacco Laboratory, Beltsville, MD for valuable advice and suggestions and for supplying experimental cigarettes. We would also like to thank Mrs. Ilse Hoffmann and Mrs. Constance Hickey for their editorial assistance.

This research program was supported by Public Health Service Contract NO1-CP-55666 and National Cancer Institute Grant 1P01 CA 29580.

LITERATURE CITED

1. Hecht, S.S.; Chen, C.B.; McCoy, G.D.; Hoffmann, D. "Tobacco Specific N-Nitrosamines: Occurrence, Carcinogenicity and Metabolism", ACS Symp. Ser. 1979, 101, 125-152.
2. Hecht, S.S.; McCoy, G.D.; Chen, C.B.; Hoffmann, D. "Metabolism of Heterocyclic Nitrosamines." ACS Symp. Ser. 1981
3. Weisburger, J.H. "N-Nitroso Compounds: Diet and Cancer Trends. An Approach to the Prevention of Gastric Cancer. ACS Symp. Ser. 1981.
4. Brunnemann, K.D.; Yu, L.; Hoffmann, D. Cancer Res. 1977, 37, 3218-3222.
5. Krull, I.S.; Goff, E.U.; Hoffman, G.G.; Fine, D.H. Anal. Chem. 1979, 51, 1706-1709.
6. Axell, T.; Moernstad, H.; Sundstroem, B. Laekartidningen 1978, 75, 2224-2226.
7. Pindborg, J.J. "Oral Cancer and Precancer." John Wright and Sons, Ltd., Bristol, England, 1980, p. 177.
8. Morie, G.P.; Sloan, C.H. Beitr. Tabakforsch. 1973, 7, 61-66.
9. Brunnemann, K.D.; Fink, W.; Moser, F. Oncology 1980, 37, 217-222.
10. Ruehl, C.; Adams, J.D.; Hoffmann, D. Anal. Toxicol. 1980, 4,255-259.
11. Hoffmann, D.; Adams, J.D.; Piade, J.J.; Hecht, S.S. IARC Sci. Publ. 1980, 31, 507-516.
12. Brunnemann, K.D.; Hoffmann, D. IARC Sci. Publ. 1978, 19, 343-356.
13. Wynder, E.L.; Hoffmann, D. "Tobacco and Tobacco Smoke - Studies in Experimental Carcinogenesis." Academic Press, New York, 1967, p. 730.
14. Tso, T.C. "Physiology and Biochemistry of Tobacco Plants." Dowden, Hutchinson and Ross, Inc., Stroudsburg, PA., 1972, p. 393.
15. Sheets, T.J.; Leidy, R.B. Recent Advan. Tobacco Sci. 1979, 5, 83-131.
16. U.S. Department of Agriculture "The Biologic and Economic Assessment of Maleic Hydrazide." U.S. Dept. Agric. Tech. Bull. 1980, 1634, p. 106.
17. Schmeltz, I.; Abidi, S.; Hoffmann, D. Cancer Lett. 1977, 2, 125-132.
18. Edwards, G.S.; Peng, M.; Fine, D.H.; Spiegelhalder, B.; Kann, J. Toxicol. Lett. 1979, 4, 217-222.
19. Brunnemann, K.D.; Hoffmann, D. Assessment of the Carcinogenic N-Nitrosodiethanolamine in Tobacco Products and Tobacco Smoke. Submitted.
20. International Committee for Cigar Smoke Study "Machine Smoking of Cigars." Coresta Inf. Bull. 1974, 1, 31-34.
21. Druckrey, H.; Preussmann, R.; Ivankovic, S.; Schmaehl, D. Z. Krebsforsch. 1967, 69, 103-201.

22. Lijinsky, W.; Reuber, M.D.; Manning, W.B. Nature 1980, 288, 589-590.
23. Hilfrich, J.; Schmeltz, I.; Hoffmann, D. Cancer Lett. 1977, 2, 169-176.
24. Hoffmann, D.; Rivenson, A.; Hecht. S.S. Unpublished Data.
25. Lijinsky, W.; Losikoff, A.M.; Sansone, E.B. J. Natl. Cancer Inst. 1981, 66, 125-127.
26. Hecht, S.S.; Ornaf, R.M.; Hoffmann, D. J. Natl. Cancer Inst. 1974, 54, 1237-1244.
27. Piade, J.J.; Hoffmann, D. J. Liquid Chromatog. 1980, 3, 1505-1515.
28. Hecht, S.S.; Chen, C.B.; Ornaf, R.M.; Jacobs, E.; Adams, J.D.; Hoffmann, D. J. Org. Chem. 1978, 43, 72-76.
29. Hecht, S.S.; Chen, C.B.; Hirota, N.; Ornaf, R.M.; Tso, T.C.; Hoffmann, D. J. Natl. Cancer Inst. 1978, 60, 819-824.
30. Mirvish, S.S.; Sams, J.; Hecht, S.S. J. Natl. Cancer Inst. 1977, 59, 1211-1213.
31. Hoffmann, D.; Adams, J.D.; Brunnemann, K.D.; Hecht, S.S. Cancer Res. 1979, 39, 2505-2509.
32. Hoffmann, D.; Dong, M.; Hecht, S.S. J. Natl. Cancer Inst. 1977, 58, 1841-1844.
33. Schmaehl, D. Arzneimitt. Forsch. 1965, 15, 704-705.
34. Christen, A.G.; McDaniel, R.K.; Doran, J.E. Texas Dent. J. 1979, 97, 6-10.
35. Hoffmann, D.; Adams, J.D. Carcinogenic Tobacco Specific N-Nitrosamines in Snuff and in the Saliva of Snuff Dippers. Submitted.
36. Pindborg, J.J.; Axelsen, N.H. The Lancet 1980, 1, 775.
37. Schievelbein, H. Deut. Med. Wochschrift; Praxis-Forum 1980, 105, 183.
38. Winn, D.M.; Blot, W.J.; Shy, C.M.; Pickle, L.W.; Toledo, A.; Fraumeni, J.F.,Jr. New Engl. J. Med. 1981, 304, 745-749.
39. U.S. Department of Health, Education and Welfare, "Smoking and Health". A Report of the Surgeon General, DHEW Publ. No. (PHS) 79-50066, 1979.
40. Hoffmann, D.; Tso, T.C.; Gori, G.B. Prev. Med. 1980, 9, 287-296.
41. Russel, M.A.H. "A Safe Cigarette?", Banbury Report 3, Cold Spring Harbor Laboratory, Cold Spring Harbor, N.Y., 1980, 297-325.
42. Hill, P. ; Marquardt, H. Clin. Pharmacol. Therap. 1980, 27, 652-658.
43. Hoffmann, D.; Chen, C.B.; Hecht, S.S. "A Safe Cigarette?", Banbury Report 3, Cold Spring Harbor Laboratory, Cold Spring Harbor, N.Y., 1980, 113-127.
44. Hecht, S.S.; Chen, C.B.; Ohmori, T; Hoffmann, D. Cancer Res. 1980, 40, 298-302.
45. Hoffmann, D.; Castonguay, A.; Rivenson, A.; Hecht, S.S. Cancer Res. 1981, 41, 2386-2393.

46. McCoy, G.D.; Hecht, S.S. Unpublished Data.
47. Boyland, E.; Roe, F.J.C.; Gorrod, J.W. Nature 1964, 202, 1126.
48. Hoffmann, D.; Hecht, S.S.; Ornaf, R.M.; Wynder, E.L. IARC Sci. Publ. 1976, 14, 307-320.
49. Hoffmann, D.; Raineri, R.; Hecht, S.S.; Maronpot, R.; Wynder, E.L. J. Natl. Cancer Inst. 1975, 54, 977-981.
50. Singer, G.M.; Taylor, H.W. J. Natl. Cancer Inst. 1976, 57 1275-1276.

RECEIVED July 20, 1981.

N-Nitrosamine Formation in Soil from the Herbicide Glyphosate and its Uptake by Plants

SHAHAMAT U. KHAN

Chemistry and Biology Research Institute, Research Branch, Agriculture Canada, Ottawa, Ontario, Canada K1A OC6

The herbicide glyphosate was nitrosated in aqueous solution by third order kinetics to *N*-nitrosoglyphosate. The nitrosation at 25°C was maximum at the reaction pH of 2.5 and had a pH-dependent rate constant of $2.4 M^{-2} sec^{-1}$. An activation energy of 9.5 Kcal mole^{-1} indicated that glyphosate in water is nitrosated very readily. Formation of *N*-nitrosoglyphosate was also observed when different soils were treated with sodium nitrite and the herbicide glyphosate at elevated levels. However, at low levels of glyphosate (5 ppm) and nitrite nitrogen (2 ppm) the formation of *N*-nitrosoglyphosate was not observed. The residues of *N*-nitrosoglyphosate can be assimilated by the roots of oat plants and translocated to the shoots. Formation of *N*-nitrosoglyphosate in soil and its uptake by plants under normal field conditions is not expected.

In recent years greater attention has been given to nitrogen containing pesticides and the possibility of their nitrosation in soil. The *N*-nitrosamines that form may arise from the parent pesticide or from a pesticide metabolite. The reaction calls for favourable pH conditions (pH 3-4) and excess nitrite. Under field conditions, the nitrosable residues are usually present in traces and only small quantities of these will actually be nitrosated in soils. However, the possibility exists that the small amounts of *N*-nitrosamines could be assimilated by plants.

Formation of *N*-nitroso compounds in soil from nitrite and nitrosable pesticides, such as s-triazine (1), dithiocarbamate and carbamate (2) and dinitroaniline (3) has been demonstrated under in vitro conditions. The herbicide glyphosate [*N*-(phosphonomethyl) glycine] (1) contains a secondary amine nitrogen, which can undergo *N*-nitrosation to form *N*-nitrosoglyphosate (2). Glyphosate has been widely used to provide effective control of most herbaceous perennial weeds through postemergence application (4). For this purpose the herbicide is applied to weed foliage at rates up to 4.2 kg/ha, often as a spot treatment.

0097-6156/81/0174-0275$05.00/0

$$HO-\underset{\underset{OH}{|}}{\overset{\overset{O}{\|}}{P}}-CH_2-\underset{\underset{H}{|}}{N}-CH_2-COOH \quad \xrightarrow[H_3O^+]{NaNO_2} \quad HO-\underset{\underset{OH}{|}}{\overset{\overset{O}{\|}}{P}}-CH_2-\underset{\underset{NO}{|}}{N}-CH_2-COOH$$

GLYPHOSATE   N-NITROSOGLYPHOSATE

<u>1</u>   <u>2</u>

We initiated an investigation to examine the possibility of
N-nitrosamine formation from the herbicide glyphosate in different
soils. It was also of interest to determine whether uptake of
N-nitroso compound by plants would occur from soil.

Formation and Persistence of *N*-Nitrosoglyphosate in Soil

Formation of *N*-nitrosoglyphosate was demonstrated by thoroughly mixing 10 g air-dried soil with 1 mg of sodium nitrite
(20 ppm nitrite nitrogen) and 10 mg glyphosate (740 ppm acid
equivalent). Some of the properties of the soils used in this
study are shown in Table I. Sufficient distilled water was added
to bring the soils to field capacity and the samples were incubated in the dark at 25°C. At regular intervals the soil samples
were analyzed for *N*-nitrosoglyphosate. The method involves
aqueous extraction of *N*-nitrosoglyphosate from the soil sample,
column clean-up of the extract, thin-layer chromatographic

TABLE I. Selected Properties of Soils[a]

Location	Soil series	pH[b]	Organic matter %	Field capacity % water
Ontario	Fox	5.8	1.1	10.1
	Grenville	6.1	7.5	19.0
	Brookston	4.3	4.4	25.4
Quebec	De l'Anse	3.8	18.0	55.5
Alberta	Grandin	4.4	1.1	18.9

[a][adopted from Khan and Young (6)]

[b]Soil-water 1:1 (w/v)

separations on silica gel, degradation of *N*-nitrosoglyphosate to
aminomethylphosphonic acid with ultraviolet light, and fluorophore
formation from this primary amine with fluorescamine (5).

Under the experimental conditions used the method gave a limit
of detection of 5 ng. Recoveries of N-nitrosoglyphosate from
the fortified soil samples (10 g) at the 1 and 5 ppm levels were
nearly quantitative (6).
    To ascertain that aminomethylphosphonic acid (3) but not
glycine (4) is a photolysis product, the primary amino product
was identified by comparison of TLC Rf values. Possible products
and their formation pathways from photolytic cleavage of N-
nitrosoglyphosate are shown in Figure 1 (5). The TLC Rf values
on silica gel for different solvent systems are shown in Table II
(5). The data clearly indicate that aminomethylphosphonic acid

TABLE II. Thin-Layer Chromatographic Rf values on Silica Gel[a]

Compound	Solvent system[b]			
	A	B	C	D
N-Nitrosoglyphosate (2)	0.30	0.28	0.55	0.50
UV irradiated 2	0.05	0.00	0.18	0.38
Aminomethylphosphonic acid (3)	0.05	0.00	0.18	0.38
Glycine (4)	0.15			0.38

[a][adopted from Young et al. (5)].

[b]A 95% Ethanol - benzene-water (4:1:1); B Methanol - chloroform
(4:1);   C Acetonitrile - 95% ethanol-glacial acetic acid-water
(12:4:3:1); D 1-Propanol - water (1:1).

and not glycine is a photolysis product. Glyoxylic acid (5)
standards could be detected at the 50 ng level using O-aminodi-
phenyl spray technique (7) and at about 0.5 μg using the aniline-
imine technique (8). However, glyoxylic acid was not detected
as a photolysis product of N-nitrosoglyphosate by either of
these two methods.
    Since the herbicide glyphosate is relatively persistent when
applied to irrigation water (9) and under certain conditions
nitrite can accumulate in soil (10) or be a constituent in run-
off water (11), a possibility for N-nitrosoglyphosate formation
may exist. The extent of N-nitrosoglyphosate formation will
depend largely upon the kinetics. In our initial studies we
examined the kinetics of nitrosation of the herbicide glyphosate
in aqueous solution (12). Figure 2 shows the results from three
kinetic runs at constant and varying initial concentration of
glyphosate and nitrite, respectively. As the initial nitrite
concentration was lowered, the initial rate was no longer a
maximum at the outset. The effect of pH on the rate of nitrosation was also determined. Figure 3 shows that for glyphosate,
the maximum rate of nitrosation of glyphosate showed a maximum
at pH 2.5 and decreased rapidly at pH ≥ 3.5 (Figure 3). At

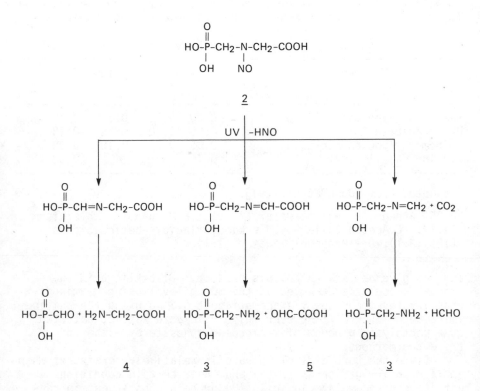

Figure 1. Possible photolytic degradation pathways for N-nitrosoglyphosate (5).

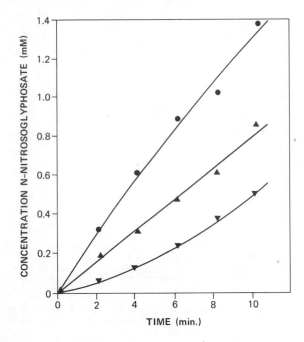

Figure 2. Formation of N-*nitrosoglyphosate after treatment with sodium nitrite at pH 3.0 and 25°C (12). Initial concentrations: glyphosate, 10.5mM; sodium nitrite, 10.1mM (●), 7.5mM (▲), and 5.0mM (▼).*

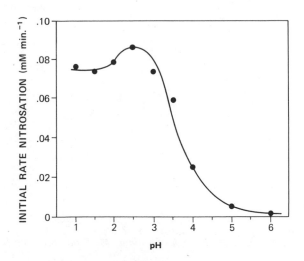

Figure 3. Effect of pH on initial rate of nitrosation of glyphosate at 25°C (12). Initial concentrations of glyphosate and sodium nitrite, 10.5 and 7.5mM respectively.

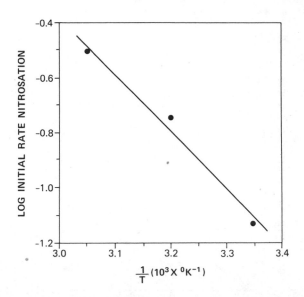

Figure 4. Effect of temperature on the initial rate of nitrosation of glyphosate at pH 3.0 (12). Initial concentrations of glyphosate and sodium nitrite, 10.5 and 7.5mM, respectively.

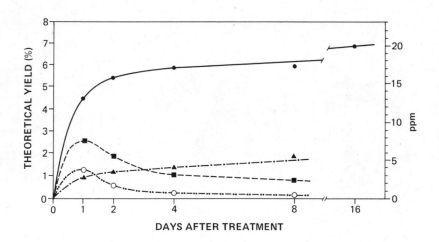

Figure 5. Formation of N-nitrosoglyphosate in soils incubated at 25°C with 20 ppm of nitrite nitrogen as sodium nitrite and 740 ppm glyphosate (free acid equivalent) (6). Soil symbols: ●, Fox; ▲, Grandin; ■, Brookston; and ○, Granville.

higher reactant concentrations, the overall yields of N-nitrosoglyphosate remained steady from pH 1-3 and decreased only slightly
(10-15%) at pH 4 and 5. Figure 4 shows a normal Arrhenius plot
for the effect of temperature on the reaction rate. The activation energy of nitrosation was about 9.5 Kcal mole^{-1}. A comparison of the activation energy for glyphosate with those of other
amino acids (Table III) shows that glyphosate is nitrosated very
readily.

Figure 5 shows the formation of N-nitrosoglyphosate in
soils incubated at 25°C with 20 ppm of nitrite nitrogen as sodium
nitrite and 740 ppm glyphosate (free acid equivalent). The
formation of N-nitrosoglyphosate was very rapid during the first
day of incubation. The levels of N-nitrosoglyphosate in Fox and
Grandin soils rose slowly with increase in incubation time,
whereas a decline was observed after the first day for Brookston
and Grenville soils. The formation of N-nitrosoglyphosate was
generated in soil samples only when both the herbicide and sodium
nitrite were present. In other studies, the herbicide atrazine
(1) and butralin (3) were also found to form N-nitrosamines in
soil when high levels of sodium nitrite were added. No N-
nitrosations of these herbicides were observed when ammonium
nitrate was substituted for sodium nitrite.

In our earlier studies, an optimum pH of 2.8 - 3.0 was found
for the formation of N-nitrosoglyphosate in solution (5). However, pH dependence of the nitrosation of glyphosate in soils of

TABLE III.   Kinetic Data for Nitrosation of Secondary Amino
             Acids at pH 2.5

Amino Acid	Activation-Energy (Kcal.mole^{-1})
Proline[a]	13
4-Hydroxyproline[a]	11
Sarcosine[a]	11
Glyphosate[b]	9.5

[a][adopted from Mirvish et al. (24)].

[b][adopted from Young and Khan (12)].

pH range 3.8 - 6.1 (Table I) was not observed. Mills and
Alexander (13) also reported that the amount of dimethylnitrosamine formation in soil was not affected by pH. However, Kearney
et al. (1) observed that the chemical conversion of atrazine to
N-nitrosoatrazine increased as the pH of the soil decreased.

It has been shown that N-nitrosation was enhanced by the
increase in organic matter content of the soil (13). It was
postulated that a simple constituent of soil organic matter may
be involved as a catalyst. For example, thiocyanate has been

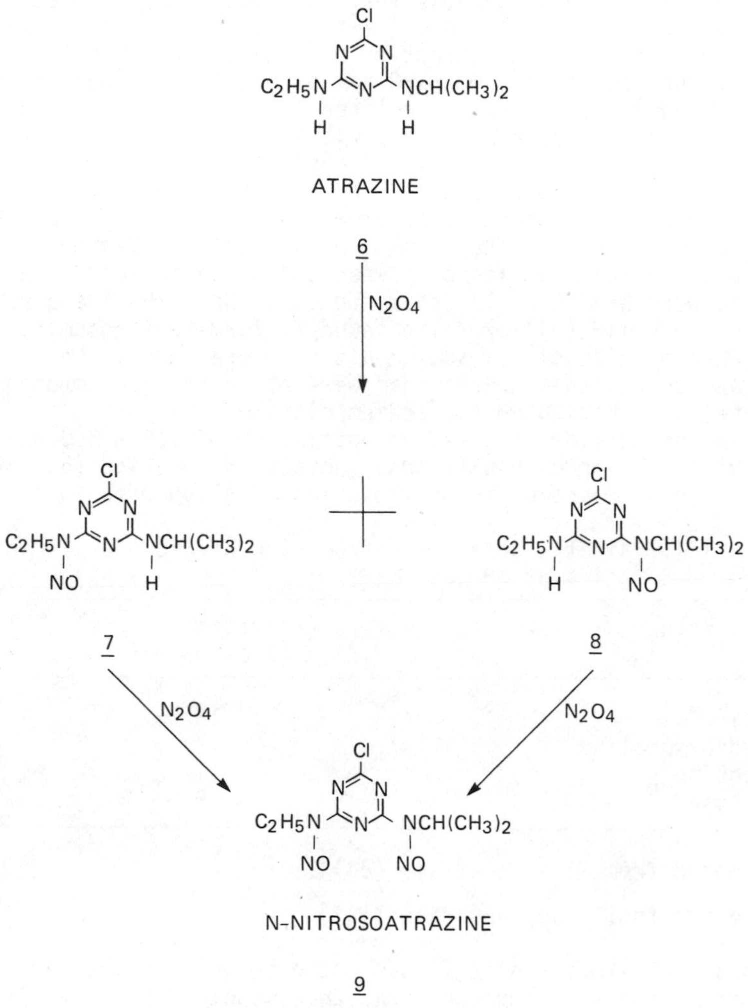

Figure 6. *Formation of* N-*nitrosoatrazine from atrazine and dinitrogen tetroxide (1).*

shown to promote nitrosation (14). However, in our study greater
N-nitrosation was observed in soils with low organic matter content. Thus in Fox soil containing 1.1% organic matter about 17 ppm
of N-nitrosoglyphosate was detected at 8 days. In contrast,
formation of N-nitrosoglyphosate in De 1,Anse soil containing 18%
organic matter was not observed (Figure 5).

Whether soil microorganisms are capable to promote, either
directly or indirectly, N-nitrosations is still a matter of con-
jecture. It is known that microorganisms are responsible for the
dealkylation of trimethylamine to yield dimethylamine and for the
reduction of nitrate to nitrite. However, little evidence has
been presented to demonstrate that microorganisms are responsible
for an enzymatic combination of secondary amines and nitrite to
form N-nitrosamines in a soil environment. Verstraete and
Alexander (15) observed that autoclaved soil did not accumulate
dimethylnitrosamine when treated with dimethylamine and nitrite,
although nonsterile soil generated detectable levels of dimethylnitrosamine. Ayanaba and Alexander (16) were the first to provide evidence for enzymatic N-nitrosation by using the soluble
fraction of an extract made from Cryptococcus sp. that catalyzed
N-nitrosamine synthesis at pH 7.5, a pH value considered to be
unfavourable for spontaneous N-nitrosation. Subsequently, Mills
and Alexander (13) observed that dimethylnitrosamine was formed
in similar quantities in sterilized and nonsterilized soil
samples. They concluded that although microorganisms may carry
out an enzymatic N-nitrosation in some soils, the dimethylnitrosamine could be formed nonenzymatically even at neutral pH's.

The persistence of the N-nitrosamine that may be formed in
soil will depend on a host of conditions, such as soil type,
organic matter content, clay content, pH, the microflora present
in the soil, moisture content and temperature, etc. Superimposed
on all these factors will be the chemical nature of the pesticide.
The N-nitrosoatrazine (9) formed in soil from the herbicide
atrazine (6) was shown to be rapidly disappeared (1). Thus, in
soil N-nitrosoatrazine was observed after one week, but was
absent 4 and 10 weeks later (Table IV). In contrast, N-nitrosobutralin (11) persisted much longer than N-nitrosoatrazine (9)
under the same conditions (Table V) and was still detectable
after 6 months (3). Our studies demonstrated that N-nitrosoglyphosate is persistent in the soil. Fox soil treated with
20 ppm of nitrite nitrogen and 740 ppm glyphosate contained about
7 ppm of N-nitrosoglyphosate even after 140 days (6).

It should be recognized that the high levels of the pesticides and nitrite nitrogen employed in several studies to demonstrate the formation of N-nitroso compounds in soil are not
likely to be encountered in practical agriculture. For example,
the average recommended rate of application of the herbicide
glyphosate is about 2.24 kg/ha. At this level of application we
cannot envisage the formation of N-nitrosoglyphosate in soil
under normal field conditions. In our laboratory experiments we

were unable to detect *N*-nitrosoglyphosate when Fox soil was
incubated over an 8-day period with 5 ppm of glyphosate and 2 ppm
of nitrite nitrogen.   However, in view of the fact that *N*-
nitroso compounds are even of concern in the parts-per-billion
range, the possibility exists that trace amounts could be assimilated by plants or be magnified by food chain.

Plant Uptake of *N*-nitrosoglyphosate

Very little information is available on the uptake of *N*-
nitrosoamine from soils by plants.   Uptake of *N*-nitrosodimethyl-
amine and *N*-nitrosodiethylamine from soil by wheat, barley and

TABLE IV.   Nitrosation of Atrazine in Matapeake Loam Soil[a]
          with Sodium Nitrite[b]

Soil incubation time (weeks)	Yield of nitrosoatrazine (%)
1	0.75
4	0
10	0

[a]Soil pH 4.5
[b][adopted from Kearney et al. (1)].

BUTRALIN   N-NITROSOBUTRALIN

10   11

N-nitrosodiethylamine from soil by wheat, barley and several vegetable crops has been demonstrated (17,18).   Rapid disappearance
of the *N*-nitrosoamine absorbed by plants was observed.   Sander
et al. (19) observed that several *N*-nitrosoamines could be
removed from water by cress, but the residues rapidly decreased
when *N*-nitrosamine containing water was replaced with non-
contaminated water.   Dean-Raymond and Alexander (20) reported

translocation of N-nitrosodimethylamine from soil into vegetable crops. Field studies by Kearney et al. (21) showed no ^{14}C residues uptake in plants grown in ^{14}C labelled N-nitrosodiprophylamine treated soil. However, low levels of ^{14}C were found in plants grown in N-nitrosopendimethalin amended soil.

TABLE V. Recoveries of N-Nitrosobutralin from Incubation of ^{14}C-Butralin in Matapeake Loam Soil Amended with Sodium Nitrite[a]

Soil incubation time (weeks)	% of total ^{14}C recovered as N-Nitrosobutralin
1	4.4
4	1.6
10	1.7
26	0.3

[a][adopted from Oliver and Kontson (3)].

Formation of the N-nitroso derivative of glyphosate in various nitrite-treated soils under in vitro conditions was demonstrated in our laboratory (12). It became of interest to determine whether uptake of N-nitrosoglyphosate by plants would occur from soil. Previously steam sterilized soil was spread evenly in plastic trays and 25 ml solution containing N-nitrosoglyphosate or the herbicide glyphosate from 5 to 300 ppm levels were applied over the surface. The samples were dried, thoroughly mixed and used for greenhouse experiments. Oats were grown in this soil; after 2 weeks, plants were harvested and roots and shoots were analyzed. Table VI showed the residues of glyphosate and N-nitrosoglyphosate in roots and shoots of oat plants grown in the treated soil. It was observed that the herbicide glyphosate can be absorbed from soil (Table VI). However, it was also observed that N-nitrosoglyphosate is not strongly retained by the soil but moves more readily into the root and shoot of oat plants than glyphosate. N-nitrosoglyphosate was detected in the shoot until the soil concentration reached 25 ppm. Thus, our study demonstrated that N-nitrosoglyphosate can be assimilated by the roots of oat plants and translocated to the shoots.
    Whether the N-nitrosation of glyphosate in soil to form N-nitrosoglyphosate will occur under natural conditions is still a matter of conjecture. However, based on our studies (12), the formation of detectable amounts of N-nitrosoglyphosate in soil under normal field conditions is not expected. It was observed that high concentrations of the herbicide glyphosate and nitrite were essential to get measureable amounts of N-nitrosoglyphosate in soil. Thus, to produce 5 ppm N-nitrosoglyphosate in the soil used in our study, one requires about 185 ppm of the herbicide

TABLE VI.   Residues (ppm on Fresh Weight Basis) of Glyphosate and
            N-Nitrosoglyphosate in Roots and Shoots of Oat Plants
            Grown in the Treated Soil[a]

Treatment Level ppm	Glyphosate[b]		*N*-Nitrosoglyphosate[b]	
	Root	Shoot	Root	Shoot
0	ND[c]	ND	ND	ND
5	ND	ND	4.7	ND
10	ND	ND	9.1	ND
25	4.8	ND	21.3	4.4
50	8.6	1.4	40.3	7.9
100	17.0	3.9	72.7	15.4
200	39.1	10.1	135	25.1
300	59.8	16.8	213	45.6

[a][adopted from Khan and Marriage (22)].

[b]Mean values for triplicate samples

[c]Not detected

glyphosate. This is about 90-100 times higher than would be required under normal use of the herbicide. Thus the possibility
of *N*-nitrosoglyphosate formation in agricultural soils and their
subsequent uptake by crops seems extremely remote under normal
conditions.

Acknowledgements

     I am grateful to Drs. J.C. Young and P.B. Marriage for their
cooperation in this study. This paper is a contribution No. 1253
of the Chemistry and Biology Research Institute.

Literature Cited

1.   Kearney, P.C.; Oliver, J.E.; Helling, C.S.; Isensee, A.R.;
     Kontson, A. J. Agric. Food Chem. 1977, 25, 1177-1180.
2.   Tate, R.L.; Alexander, M. Soil Sci. 1974, 118, 317-321.
3.   Oliver, J.E.; Kontson, A. Bull. Environ. Contam. Toxicol.
     1978, 20, 170-173.
4.   Spurrier, E.C., Pans 1973, 19, 607.
5.   Young, J.C.; Khan, S.U.; Marriage, P.B. J. Agric. Food Chem.
     1977, 25, 918-922.
6.   Khan, S.U.; Young, J.C. J. Agric. Food Chem. 1977, 25,
     1430-1432.
7.   Nakai, T.; Ohta, T.; Takayama, M. Agric. Biol. Chem. 1974,
     38, 1209-1213.
8.   Young, J.C. J. Chromatogr. 1977, 130, 392-396.

9. Comes, R.D.; Bruns, V.F.; Kelley, A.D. Weed Sci. 1976, 24, 47-51.

10. Chapman, H.D.; Liebig, G.F. Soil Sci. Soc. Am. Proc. 1952, 16, 276-278.

11. Tabatabai, M.A. Commun. Soil Sci. Plant Anal. 1974, 5, 569-575.

12. Young, J.C.; Khan, S.U. J. Environ. Sci. Health 1978, B13, 59-72.

13. Mills, A.L.; Alexander, M. J. Environ. Qual. 1976, 5, 437-440.

14. Fan, T.Y.; Tannenbaum, S.R. J. Agric. Food Chem. 1973, 21, 237-240.

15. Verstraete, W.; Alexander, M. J. Appl. Bacteriol. 1971, 34, IV.

16. Ayanaba, A.; Alexander, M. Appl. Microbiol. 1973, 25, 862-868.

17. Dressel, J. Landwirtsch. Forsch. Sonderk, 1976, 33, 326-334.

18. Dressel, J. J. Qual. Plant Foods Hum. Nutr. 1976, 25, 81-390.

19. Sander, J.; Ladenstein, M.; La Bar, J.; Schweinsberg, F.J. IARC Sci. Publ. 1975, 9, 205-210.

20. Dean-Raymond, D.; Alexander, M. Nature 1976, 262, 294-296.

21. Kearney, P.C.; Oliver, J.E.; Kontson, A.; Fiddler, W.; Pensabene, J.W. J. Agric. Food Chem. 1980, 28: 633-635.

22. Khan, S.U.; Marriage, P.B. J. Agr. Food Chem. 1979, 27, 1398-1400.

23. Sprankle, P.; Meggitt, W.F. Pemmer, D. Weed Sci. 1975, 23, 224-228.

24. Mirvish, S.S.; Sames, J.; Fan, T.Y.; Tannenbaum, S.R. J. Natt. Cander Inst. 1973, 51, 1833-1839.

RECEIVED July 20, 1981.

Reaction of Meat Constituents with Nitrite

R. G. CASSENS, K. IZUMI, M. LEE, M. L. GREASER, and J. LOZANO

Department of Meat and Animal Science, University of Wisconsin, Madison, WI 53706

Meat is a complex biological system, and it is well-established that when nitrite is added to it, chemical reactions occur with various of the constituents of the meat. Numerous factors such as composition and previous history of the meat, morphology and processing techniques affect the reactions. In general, fifty percent or more of the nitrite added to meat in a usual curing procedure disappears from analytical detection and is assumed to be converted to other forms or be reacted with or bound to constituents of the meat. A great deal of evidence exists regarding the reaction of nitrite with the heme protein myoglobin and considerable information is now in the literature regarding reaction of nitrite with non-heme proteins -- possible sites of reaction including sulfhydryl groups, amide linkage and imidazole nitrogen of tryptophan have been established. Nitrite also reacts with the lipid components of meat. On a quantitative basis, the reaction with protein is the most important.

More than a decade has passed since the use of nitrite as a meat curing agent was questioned because of the possibility that N-nitrosamine could be formed by reaction of nitrite with constituents of the meat. Indeed, some detections of extremely low levels of N-nitrosamines were made; much of the initial concern has diminished, however, because of several lines of research which were pursued in the interim. Alteration of ingoing levels of nitrite, modification of processing procedures and addition of suitable blocking agents have resulted in cured products having traditional visual, organoleptic and keeping properties and which have non-detectable levels of N-nitrosamines (or in the case of cooked bacon, comply with Governmental limits).

Another area of research has been to elucidate mechanisms of action of nitrite in meat products, and while progress has been

made, it has not been so rapid. Undoubtedly, real answers to the
questions about formation of N-nitroso compounds in foods, the
body and the environment, and the importance of their occurrence,
will only be obtained after the reaction mechanisms are understood.
Therefore, it seems appropriate that a Symposium on N-nitroso compounds has been organized at this time to present a forum for exchange of scientific information by those actively engaged in research and to inform those more peripherally involved of the state
of the situation.

Our objective in this manuscript is twofold -- first to describe the biological setting (meat) in which the chemistry of nitrite occurs, and second to provide an analysis of research
accomplishments relating directly to what is known about reaction
of nitrite with the constituents of meat. In the relatively short
time span of the ten years mentioned above, a substantial amount
of literature has appeared. It would be duplicative effort to repeat all of the pertinent details here. Rather, the reader is
referred to a recent review on reactions of nitrite in meat (1)
which provides extensive documentation and enumerates several
other pertinent major reviews.

Composition and Structure of Meat

Meat is a complex biological system, a fact which in and of
itself has greatly hindered progress in learning about reaction of
nitrite with the constituents normally found in meat. The approximate composition of lean meat is 70% water, 20% protein, 9% fat
and 1% analyzable ash. These constituents may vary considerably
depending on species, nutritional status of the animal, stage of
maturity, etc. Within the protein category, numerous specific examples exist ranging from the well-known proteins of muscle such
as myosin, collagen and myoglobin to the complete array of enzymes
associated with energy metabolism and including such examples as
lactic dehydrogenase and phosphorylase. The proteins are classi-
fied as water-soluble, salt-soluble or insoluble. Free amino
acids, peptides, vitamins, carnosine and nucleotides are some examples of non-protein nitrogen constituents present in muscle.

The lipid in muscle is composed primarily of triglycerides
(depot fats) and of phospholipids (membrane components), and is a
constituent which varies enormously not only in amount present,
but also in properties such as degree of saturation (species
dependent). The ash of lean meat is comprised of various minerals
such as phosphorus, potassium, sodium, magnesium, calcium, iron
and zinc. Carbohydrate was not noted in the proximate composition
because while some may be present, it is normally there in low
concentration compared to the other constituents. Glycogen is the
carbohydrate occurring in greatest concentration in muscle but is
normally degraded soon after the animal is sacrificed.

It is emphasized that the array of chemical constituents is
not static. As living muscle is converted to meat, substantial

changes occur, not the least of which is decline in pH from about
7.0 to the range 5.5 to 6.0. Subsequent to this, further changes
due to inherent proteolysis, as well as degradation from microbiological contamination takes place. These processes undoubtedly
further complicate the situation by resulting in production of
intermediates or breakdown products thereby providing more constituents in the meat which have the potential for reaction with
nitrite.

Another point requires emphasis and that is that the chemical
constituents are not uniformly distributed but arranged in a
highly organized structure. An extensive connective tissue framework holds the cells in place. The muscle cells (myofibers) are
elongated, multinucleate and highly specialized containing the
contractile units which give muscle its function. The result is
that when chemistry of nitrite is placed within the morphology of
muscle, some complications may arise because of the highly organized structure. For example, there are built-in barriers, as well
as channels, which may hinder or help the movement of a substance
such as nitrite into the interior of a piece of meat. The circu-
latory system provides open pipelines as do the clefts of inter-
cellular space. Barriers to passage of substances are set up from
the extensive connective tissue layers ranging from that found
surrounding entire muscle to the delicate investments encasing
each cell. To pass substances through the cell membrane proper,
generally requires an active transport, but, of course, in
postmortem muscle, that feature is lost and is replaced by simple
diffusion.

Of probably greater importance is the effect of local concentration gradients. For example, analysis for a given constituent
in the entire meat mass does not reflect the real concentration at
a given point. For example, DNA is localized in the nuclei and
lipid is localized predominantly in the adipose cells. Another
factor of potential influence in reaction schemes for nitrite is
the fact that polar-nonpolar interfaces are present as a result of
structural compartmentalization. In an adipose cell, the lipid is
contained as the body of the cell, but it is surrounded by a thin
layer of sarcoplasmic protein. Therefore, large surface areas are
involved.

The myofibers of muscle are not homogenous but can be categor-
ized into red and white on the basis of ATPase activity (2). Not
only the functional properties but also the composition of the two
fiber types varies widely. Red fibers, for example, have considerably more myoglobin and lipid than do white type fibers.

Several of these morphological factors are illustrated in
Figure 1. Figure 1A is of the fat portion of bacon and has been
stained for connective tissue. It is noted that fat tissue is not
all lipid but has an extensive connective tissue component ranging
from fairly thick layers to delicate layers defining each adipose
cell. Figure 1B is from a finely chopped emulsion. Connective
tissue pieces are stained dark, the protein matrix is gray and the

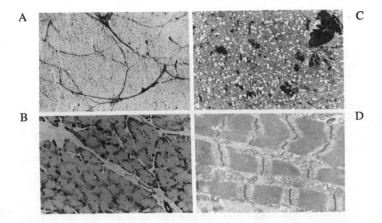

Figure 1. A composite photomicrograph illustrating several morphological factors that may play a role in reaction of nitrite with constituents of meat. A. Adipose tissue from bacon (57). Heavy dark bands of connective tissue are clearly visible, and, in addition, a layer of connective tissue is found surrounding each individual cell.   B. Commercial wiener in which connective tissue is stained darkly, the protein matrix is gray and the open white areas represent the location of lipid globules although not specifically stained.   C. Distribution of fiber types in skeletal muscle. White fibers are darkly stained and red fibers lightly stained.   D. Electron micrograph of skeletal muscle showing myofibrils with the typical banding pattern and sarcoplasmic components such as mitochondria. The sample in A was fixed in 10% formalin, frozen sectioned, and stained with Van Giesons method. The sample in B was frozen sectioned, formal-calcium fixed, and stained with picro-ponceau. C was stained for myosin ATPase after acid preincubation. A, B, and C are about 12× and D is about 6500×.

white open spots are considered to be lipid (lipid was not stained). In this case, an interphase between lipid and protein has been formed mechanically instead of the natural occurrence shown in Figure 1A. Figure 1C is a low power photomicrograph of skeletal muscle in which the red fibers are stained lightly and the white fibers are stained darkly. Figure 1D is an electron micrograph illustrating that the myofibrils of muscle are composed of a highly ordered array of proteins.

Properties of Nitrite and the Curing Procedure

The functions of nitrite added to meat for the purpose of curing are now well known by many people because of the extensive publicity given to the process via the popular press. Nitrite added to meat results in a typical color and a characteristic flavor, provides microbiological protection especially against outgrowth of C. botulinum spores and may play a role in textural characteristics.

There has been a long standing discussion whether nitrite produces a cured flavor or retards development of an off-flavor. A review by Bailey and Swain (3) cited several references which indicated nitrite was responsible for cured meat flavor. These same authors presented chromatograms of volatiles from cured and uncured hams and while the chromatograms were similar, some quantitative differences led to the conclusion that the major difference due to nitrite was its reactivity to retard lipid oxidation. Greene and Price (4) suggested, however, that sodium chloride was the major factor responsible for cured meat flavor rather than sodium nitrite or an absence of lipid oxidation. It has been concluded from other recent work (5) that nitrite was necessary to produce a typical ham aroma and flavor as well as to retard the development of off-odors and flavors during storage of cooked cured meat.

Nitrite is an extremely reactive chemical and is soluble in the aqueous phase of meat. It is usually used for curing in the form of the sodium salt. The nitrite ion is the conjugate base of nitrous acid (a weak acid) and has a PK of 3.36. The usually mild acid conditions found in meat give formation of only a small quantity of nitrous acid when nitrite is added to the meat (6). Nitrite is a weak reducing agent and is oxidized to nitrate by strong chemical oxidants or by nitrifying bacteria. It also oxidizes reduced substances.

Several reviews of the basic chemistry of nitrite have been published in the chemical literature (7, 8, 9), and the reaction schemes for nitrosation, diazotization and deamination have been studied in detail in pure systems; in regard to meat, the mechanism of nitrosation has been of greatest interest and that primarily for N-nitrosation even though C- and S- nitrosation may occur. N-nitrosation of amines has been studied most extensively since, with secondary amines, the reaction results in N-nitrosamines (also the case with tertiary amines) while with primary aliphatic amines, the reactions result in a wide variety of deamination products.

It is probably safe to conclude from the literature that, in meat, the nitrosating agent is either NOX or N_2O_3 even though it is common to refer to nitrite or nitric oxide as the active nitrosating agent. A second conclusion is that the nitrosation process consists of several steps, may take a number of different courses and may occur simultaneously via more than one reagent. The obvious point to be drawn from this discussion is that nitrosation chemistry in a pure system is complex; when the chemistry is placed in a complex biological system, the complexity is multiplied many times, and, in fact, explains the relatively slow progress made in arriving at definitive conclusion about mechanisms of reaction in cured meat.

A great variety of cured meat products results from selection of differing starting materials and then subjecting them to widely different processing procedures. A knowledge of at least the generalities of the curing procedure is also prerequisite to an appreciation of the problem. Even though nitrite is the key ingredient, sodium chloride is invariably added at concentrations up to 2.5 to 3.0% of the product. The end product is obviously influenced by the seasoning, and, in addition, so called curing adjuncts such as ascorbate or erythrobate are often added. The curing ingredients may be added as a dry powder and rubbed onto the surfaces of the meat. This is so-called dry curing and requires long time for diffusion of the chemicals into the piece of meat. More rapid procedures involve grinding or mixing the curing ingredients into the meat or dissolving them in water and injecting the solution into the meat. The product is almost always heat processed (often together with smoking) to a temperature of 155-160°F (results in a perishable product) or more severely to a commercially sterile canned type product.

Reaction with Heme Proteins

Since curing results in a typical color and since the pigment myoglobin, which is primarily responsible for color of meat, is a well studied protein, it is not surprising that a considerable amount is known about reaction of nitrite with the meat constituent myoglobin. The knowledge base is explained well in a review by Giddings (10), and the more practical aspects of the function of nitrite as a stabilizer of color in red meats have also been reviewed recently (11). Even though myoglobin is the major pigment responsible for meat color, hemoglobin may be present in trapped red blood cells and also contributes. Other pigments such as cytochrome c occur in much lower concentration and an analysis is given later in this manuscript regarding their possible role in color of cured meat.

A description of the heme group which is a nonpolypeptide responsible for the color of myoglobin and hemoglobin has been provided by Stryer (12). The organic portion of the heme is called protoporphyrin and is made up of four pyrrole groups linked by methene bridges to form a tetrapyrrole ring. Four methyl, two vinyl, and two propionate side chains are attached to the tetrapyrrole ring. They can be arranged in 15 different ways, but the one known as protoporphyrin IX is the only one present in biological systems. The iron binds to the four nitrogens in the center of the protoporphyrin ring and can form two additional bonds, one on either side of the heme plane. These bonding sites are termed the 5th and 6th coordination positions and the iron atom can be in the ferrous (+2) or ferric (+3) state. The 5th coordination position is normally occupied by the histidine side chain of the globin protein of the myoglobin molecule. The 6th position is usually occupied by a water molecule, or oxygen or, in the case of cured meat, by NO. From the literature (10), there seems little doubt that in cured meat, NO binds to the heme iron (6th position) via the nitrogen atom to form an extremely stable paramagnetic adduct. However, the series of reactions leading to the product are not so well established -- transient intermediates are involved and enzymatic systems must be considered as well as the effect of heat. There is information (13, 14) to suggest that the cooked cured pigment is a dinitrosylhemochrome. This means that the globin portion is detached by heating and the two coordination positions of iron could both be occupied by NO.

Practical experience in meat curing has demonstrated that when nitrite is added to meat, the purplish-red color of myoglobin (reduced) is converted to the brown metmyoglobin (oxidized) form. With time and reducing conditions, the color is converted to the reddish hue of nitrosylmyoglobin. Heat denaturation converts the pigment to the more stable nitrosylhemochrome form which is pinkish red. These observations can be explained chemically (15, 16, 17). Nitrite is a strong heme oxidant and, therefore, when it is added to meat, the heme is oxidized changing the color from a red to brown. With time, endogenous reducing agents in the meat reduce the oxidized heme, and the red color is regained. This reduction process obviously proceeds much more rapidly if a reducing agent, such as ascorbate, is added to the curing mixture.

Cytochrome c is a heme containing protein which occurs in muscle at lower concentrations than does myoglobin. It was demonstrated some time ago (18) that oxidized cytochrome c reacts with gaseous nitrite oxide to produce a nitrosyl compound. Recent work (19, 20, 21) has examined the reactions of cytochrome c with nitrite and the contribution of the product formed to cured meat color in considerably more detail. The general conclusion is that even at the pH normally encountered in meat, the reaction can take place in the presence of ascorbic acid but probably does not affect meat color because of the unstable nature of the reaction product and the low concentration.

Reaction with Non-Heme Proteins

There is now sufficient information in the literature to
conclude that nitrite reacts with the non-heme proteins of meat.
In fact, as will be indicated later in this manuscript, the proteins of meat are the major constituent with which nitrite reacts
and explain the largest proportion of the nitrite lost from analytical detection during curing. While considerable discussion has
occurred about this so called protein bound nitrite, little has
been substantiated about identification and quantitation of the
reaction products. Protein bound nitrite has been of concern in
analysis for free nitrite because depending on conditions of
analysis, some portion of it may be released and measured.
Further, the protein bound nitrite represents a substantial pool
of nitrosating ability that may function in transnitrosation
(22, 23, 24).
    One obvious reaction that proteins may undergo when exposed
to nitrite is with the α-amino group (Van Slyke reaction).
Nitrite may also react with the ε-amino group of lysine, the
indolic nitrogen of tryptophan and with tyrosine (C-nitrosation)
(25). A great amount of interest has been expressed regarding
reaction of nitrite with sulfhydryl groups of protein to form
nitrosothiols. Mirna and Hoffman (26) worked with a meat system
in which the molar ratio of $NaNO_2$/SH was about 1/9 and concluded
that reaction did occur; they demonstrated that some nitrite could
be released by heavy metal ions such as Hg^{++}. Olsman and Krol
(27), however, blocked sulfhydryl groups before treatment with
nitrite and concluded that the major loss of nitrite was not
caused by reaction with sulfdryls. Kubberód et al. (28) studied
the reaction using the isolated protein myosin and also concluded
that formation of nitrosothiols could account for only a small
proportion of the total loss of nitrite observed in cured meat.
Olsman (29) later hypothesized that the formation of RSNO is
reversible and that it might act as an intermediate nitrosating
agent to form other stable nitrosated compounds as free nitrite
declined in the system. Dennis et al. (30) have shown that N-
Methylanaline could be transnitrosated with nitrite and S-
nitrosocysteine and also by a simulated protein bound nitrite. In
the latter case, an important factor was the local concentration
of nitrosothiol groups on the matrix. The effects of S-
nitrosocysteine as an inhibitor of lipid oxidation, as a color
developer, and as an anticlostridial, have been reported recently
in a turkey product (31). The Molar concentration of RSNO
equating to 25 ppm nitrite gave similar results for color and inhibition of lipid oxidation but had less anti-clostridial activity.
Transnitrosation between RSNO and heme protein was demonstrated.
    Under appropriate condition, the peptide bond may react with
aqueous nitrous acid to form a nitrosamide (32). It has been
shown (33, 34, 35) in model systems that N-substituted amines can
be formed by reaction between free amines and triglycerides.

Under appropriate conditions, they are readily N-nitrosated. The authors tentatively concluded that the major contribution of N-substituted amides, if they were present in foods, could be as precursors of N-nitroso compounds formed by in vivo N-nitrosation reactions.

Some work has been completed on reaction of proteins with nitrite followed by hydrolysis and analysis for amino acids. It has been shown that 3-nitrotyrosine and 3,4-dihydroxyphenylalanine are formed from bovine serum albumin when nitrosation occurs under conditions similar to those found in the human stomach (36). Direct demonstration that nitrite reacts with protein has been made by using $NaNO_2$ with bovine serum albumin (pH 5.5, 20°C and 200 ppm nitrite). A 60% loss of the originally added nitrite was observed in one week and nearly half of the nitrite (labelled ^{15}N) could be recovered from the protein. Similar work with myosin revealed that 10-20% of the incorporated label was present as 3-nitrotyrosine (37).

Direct proof has been given for the earlier comment (25) that nitrite reacts with the indolic nitrogen of tryptophan (38, 39, 40). When the conditions of pH, temperature and concentration found in cured meat are considered, then the reaction rate is extremely slow. Evidence has also been given (41) that the reaction can occur while the tryptophyl residue is in the protein structure.

Reaction with Lipid, Carbohydrate and Metal Ions

Some proof is now available to show that nitrite reacts with lipid components in meat. Evidence for this was first obtained (42) with the use of ^{15}N labelled nitrite and bacon in which 20-25% of the label was recovered from the adipose tissue. It must be considered indirect evidence, however, because adipose tissue contains not only lipid but also insoluble connective tissue proteins as well as other soluble proteins. Goutefongea et al. (43) gave direct proof that reaction of nitrite occurred with lipid and suggested that it reacted with unsaturated carbon - carbon bonds. This conclusion substantiated an earlier suggestion by Frouin (44), and more recently good evidence has been published (45) that nitric oxide reacts with palmitodiolein to form a pseudo-nitrosite across the double bonds of the triglyceride.

The carbohydrate content of meat is generally low and therefore reaction of this constituent with nitrite has received little attention. In some curing procedures, carbohydrates are added and this is worthy of further study. It has been reported that nitrite can react with sugars to produce nitrogen and carbon dioxide (46). It has also been reported (47) that nitrite reacts with amino sugars to produce 2,5-anhydroderivates which are unstable and convert easily to the corresponding aldehydes. This reaction may be of significance since glucosamine and galactosamine are associated with the connective tissue. There are also reports that nitrite reacts with the compounds formed from the browning reaction in foods (48, 49).

The fact that nitrite reacts with the iron of the heme compound was described earlier. Because such a large number of metal ions are present in meat, and because some occur in relatively high concentration, there has been considerable interest in them. For the most part, studies have dealt with how metal ions influence reactions of nitrite. The role of sodium chloride (which is used extensively in meat processing) must also be recognized both in terms of its functional role in making reactants in the meat more or less available, and in terms of reports that it directly influences nitrosation reactions (50). Ando (51) studied the effect of several metal ions on decomposition of nitrite, and in the absence of ascorbate, only Fe^{++} caused a loss of nitrite but in its presence, the effect of Fe^{++} was more pronounced and Fe^{+++}, Mg^{++}, Ca^{++} and Zn^{++} showed similar effects. Lee et al. (52) used a meat system containing 2% NaCl and 156 ppm $NaNO_2$ to study the effect of various metal ions on residual nitrite. They found that Fe^{++}, Fe^{+++}, Cu^{++} and Zn^{++} caused a depletion of nitrite while Ca^{++} and Mg^{++} did not have an effect.

Fate of Nitrite

It is obvious from the foregoing that to select and quantitate a reaction of nitrite with a given constituent of meat is a tremendous challenge. Another approach has been to conduct bookkeeping type experiments with a label in order to balance nitrite loss from analysis against detection of the label in another form. Sebranek et al. (53) used ^{15}N-labelled nitrite in a commercial type canned luncheon meat and analyzed for the label in various fractions of the meat at different time intervals following processing. Residual nitrite was lost rapidly, and as it declined, the label appeared in various fractions of the meat. A small amount of the label (about 5%) was given off as gas during processing. The pigment fraction contained 9-12% of the label and the amount changed little during storage of the meat.

A series of papers using ^{15}N, has been published by the Japanese workers (54, 55). They added ^{15}N-labelled nitrite to a model system containing myoglobin, nitrite and ascorbate and reported nearly complete recovery of the ^{15}N in the forms of residual nitrite, nitrate, nitrosyl group of denatured nitrosomyoglobin and gaseous nitrogen. In the Sebranek experiments (53), recovery was in the range of 72-86%. Subsequent work by the Japanese (55) in a meat instead of model system resulted in recovery of label in the range 66-90%.

The results from the ^{15}N studies, together with information from other experiments, can be brought together (56) to give the following accountability for nitrite added to a typical cured meat and expressed as percentage of original amount of nitrite added. Some of the nitrite reacts with myoglobin and accounts for 5-15% of the lost nitrite, 1-16% is converted to nitrate, 5-20%

remain as nitrite, 1-5% is evolved as gas, 5-15% reacts with sulfhydryls, 1-5% reacts with lipid, and 20-30% reacts with non-heme protein. The conclusions are that when nitrite is added to meat, it reacts with various of the constituents, with protein accounting for most of that reacted.

Acknowledgements

Muscle Biology Laboratory manuscript number 151 and Meat and Animal Science manuscript number 794. The work of the authors has been supported by the College of Agricultural and Life Sciences of the University of Wisconsin, by the Food Research Institute and by the American Meat Institute.

Literature Cited

1. Cassens, R. G.; Greaser, M. L.; Ito, T.; Lee, M. Food Tech. 1979, 33, (7), 46.
2. Suzuki, A.; Cassens, R. G. Histochem. J. 1980, 12, 687.
3. Bailey, M. E.; Swain, J. W. Proc. Meat Ind. Res. Conf. 1973, p. 29.
4. Greene, B. E.; Price, L. G. Ag. Food Chem. 1975, 23, 164.
5. MacDonald, B.; Gray, J. I.; Stanley, D. W.; Usborne, W. R. J. Food Sci. 1980, 48, 885.
6. Price, J.; Schweigert, B. S., Eds.; "The Science of Meat and Meat Products"; W. H. Freeman & Co., San Francisco, 1971; p. 452.
7. Ridd, J. H. Quart. Rev. 1961, 15, 418.
8. Challis, B. C.; Butler, A. R.; Substitution at an amino nitrogen In "The Chemistry of the Amino Group"; S. Patai, ed. Interscience Publishers: New York, 1968.
9. Bonnett, R.; Martin, R. A. I.A.R.C. 1976, 14, 487.
10. Giddings, G. G. J. Food Sci. 1977, 42, 288.
11. Dryden, F. D.; Birdsall, J. J. Food Tech. 1980, 34, (7), 29.
12. Stryer, L.; "Biochemistry"; W. H. Freeman and Co., San Francisco, 1975, p. 46.
13. Tarladgis, R. G. J. Sci. Food Agr. 1962, 13, 481.
14. Lee, S. H.; Cassens, R. G. J. Food Sci. 1976, 41, 969.
15. Fox, J. B.; Thomson, J. S. Biochemistry 1963, 2, 465.
16. Fox, J. B. J. Agri. Food Chem. 1966, 14, 207.
17. Fox, J. B.; Ackerman, S. A. J. Food Sci. 1968, 33, 364.
18. Ehrenberg, A.; Szczepkowski, T. W. Acta Chem. Scand. 1960, 14, 1684.
19. Orii, Y.; Shimada, H. J. Biochem. 1978, 84, 1543.
20. Izumi, K.; Cassens, R. G.; Greaser, M. L. Submitted, 1981.
21. Izumi, K.; Cassens, R. G.; Greaser, M. L. Submitted, 1981.
22. Challis, B. C.; Osborne, M. R. J. Chem. Soc. Perkin II. 1973, p. 1526.
23. Burglass, A. J.; Challis, B. C.; Osborne, M. R. I.A.R.C. 1975, 9, 94.

24. Ito, T.; Cassens, R. G.; Greaser, M. L. Submitted, 1981.
25. Kurosky, A.; Hofmann, T. Can. J. Biochem. 1972, 50, 1282.
26. Mirna, A.; Hofmann, K. Fleischwirtschaft 1969, 49, 1361.
27. Olsman, W. J.; Krol, B. 18th Eur. Meet. Meat Res. Work. 1972, Guelph, Canada.
28. Kubberød, G.; Cassens, R. G.; Greaser, M. L. J. Food Sci. 1974, 39, 1228.
29. Olsman, W. J.; In "Proc. 2nd Int. Symp. Nitrite Meat Prod." B. J. Tinbergen and B. Krol, eds. 1977, p. 101.
30. Dennis, M. J.; Massey, R. C.; McWeeny, D. J. J. Sci. Food Agric. 1980, 31, 1195.
31. Kanner, J.; Juven, B. J. J. Food Sci. 1980, 45, 1105.
32. Bonnett, R.; Holleyhead, R. J.C.S. Perkin I 1974, p. 962.
33. Kakuda, Y.; Gray, J. I. J. Agric. Food Chem. 1980, 28, 580.
34. Kakuda, Y.; Gray, J. I. J. Agric. Food Chem. 1980, 28, 584.
35. Kakuda, Y.; Gray, J. I.; Lee, M. J. Agric. Food Chem. 1980, 28, 588.
36. Knowles, M. W.; McWeeny, D. J.; Couchman, L.; Thorogood, M. Nature, 1974, 247, 288.
37. Woolford, G.; Cassens, R. G.; Greaser, M. L.; Sebranek, J. G. J. Food Sci. 1976, 41, 585.
38. Bonnett, R.; Holleyhead, R. I.A.R.C. #9 1975, p. 107.
39. Brown, T. R.; Stevens, M. F. G. J.C.S. Perkin I. 1975, p. 2357.
40. Nakai, H.; Cassens, R. G.; Greaser, M. L.; Woolford, G. J. Food Sci. 1978, 43, 1857.
41. Ito, T.; Cassens, R. G.; Greaser, M. L. J. Food Sci. 1979, 44, 1144.
42. Woolford, G.; Cassens, R. G. J. Food Sci. 1977, 42, 586.
43. Goutefongea, R.; Cassens, R. G.; Woolford, G. J. Food Sci. 1977, 42, 1637.
44. Frouin, A.; Jondeau, D.; Thenot, M. 21st Eur. Meet. Meat Res. Work. 1975, Berne, Switzerland.
45. Walters, C. L.; Hart, R. J.; Perse, S. 2. Lebensmittel Undersuch Forsch. 1979, 168, 177.
46. Ten Cate, L. Fleischwirtschaft 1963, 15, 99.
47. Mirna, A.; Coretti, K.; In "Proc. 2nd Int. Symp. Nitrite Meat Prod." B. J. Tinbergen and B. Krol, eds. 1977, p. 39.
48. Coughlin, J. R.; Wei, C. I.; Hsieh, D. P. H.; Russell, G. H. A.C.S. Chem. Cong. 1979.
49. Sakaguchi, M.; Aitoku, A.; Shibanlate, T. A.C.S. Chem. Cong. 1979.
50. Hildrum, K. I.; Williams, J. L.; Scanlan, R. A. J. Agri. Food Chem. 1975, 23, 439.
51. Ando, N.; In "Proc. Int. Symp. Nitrite Meat Prod."; B. Krol and B. J. Tinbergen eds. 1974, p. 149.
52. Lee, M.; Cassens, R. G.; Fennema, O. R. Submitted, 1981.
53. Sebranek, J. G.; Cassens, R. G.; Hoekstra, W. G.; Winder, W. C.; Podebradsky, E. V.; Kielsmeier, E. W. J. Food Sci. 1973, 38, 1220.

54. Fujimaki, M.; Emi, M.; Otitani, A. Agri. Biol. Chem. 1975, 39, 371.

55. Emi-Miwa, M.; Okitani, A.; Fujimaki, M. Agri. Biol. Chem. 1976, 40, 1387.

56. Cassens, R. G.; Woolford, G.; Lee, S. H.; Goutefongea, R.; In "Proc. 2nd Int. Symp. Nitrite Meat Prod."; B. J. Tinbergen and B. Krol eds. 1977, p. 95.

57. Cassens, R.G.; Ito, T.; Lee, S.H. J. Food Sci. 1974, 44, 306.

RECEIVED July 31, 1981.

ANALYSIS AND OCCURRENCE

N-Nitroso Compounds: Diet and Cancer Trends

An Approach to the Prevention of Gastric Cancer

JOHN H. WEISBURGER

American Health Foundation, Valhalla, NY 10595

Sensitive analytical techniques can detect nitrosamines in the environment. Thus far no human cases of cancer attributable unambiguously to short chain dialkylnitrosamines detected in foods have been observed in the Western world. Burnt and unburnt tobacco contains appreciable amounts of nitrosonornicotine and related compounds that have been incriminated in tobacco carcinogenesis. Certain nitrosamines were identified in foods consumed in China in areas with a high incidence of esophageal cancer and a cause-effect relationship has been suggested but not yet proven. Eastern Iran and Southern regions of the U.S.S.R. also have a high incidence of esophageal cancer but without such direct evidence of dietary nitrosamines. Nonetheless, all of the high risk regions have in common low intakes of foods with vitamin C. Gastric cancer can be induced in animals by alkylnitrosoureido compounds. These can form from nitrite and suitable substrates, a reaction that can be inhibited by vitamins C and E. The risk factors for human gastric cancer include foods pickled with salt and saltpeter or residence in areas with high geochemical nitrate. We have found mutagenic activity for Salmonella typhimurium TA 1535 with an extract of a pickled fish frequently eaten in high risk Japan, which also induced glandular stomach cancer in rats. The formation of the mutagen was inhibited by vitamin C. Thus, gastric cancer may stem from consumption of specific pickled foods or residence in areas with high soil nitrate without the simultaneous consumption of foods containing sufficient inhibitory vitamins C or E. This concept accounts for the pronounced decrease in stomach cancer in the United States.

0097-6156/81/0174-0305$05.00/0
© 1981 American Chemical Society

Are Nitrosamines Human Carcinogens?

The effect of the prototype dimethylnitrosamine or N-nitrosodimethylamine as a carcinogen was discovered about 1955 when
Barnes and Magee, then at the Toxicology Center of the Medical
Research Council of Great Britain, were asked to investigate in
animal models a finding of severe liver toxicity in three men in
a laboratory accident involving this chemical.   They found that
dimethylnitrosamine was indeed severely hepatotoxic (1,2). As an
extension of this research, it was discovered that this chemical
was also a powerful carcinogen for the liver and secondarily for
the kidney of rats.   In the next 25 years, nitrosamines as a
class were found to be potent carcinogens, the specific action of
which depended on their structure; frequency and route of administration; species, strain, and other variables.   These carcinogens permitted the induction of cancer at many target sites as a
function of structure and metabolism (1,3,4).
     Just as their chemistry would predict, it was also found
that nitrosamines could be formed from nitrite and the appropriate amines or amides at the pH found in the stomach (5).   Even
tertiary amines could interact with nitrite and form dialkylnitrosamines (6, 7).   In addition, the analytical chemistry used to
detect nitrosamines has improved importantly.   Whereas 25 years
ago techniques were suitable only to detect parts per thousand,
at this time, through instruments such as the Thermal Energy Analyzer designed specifically to detect nitrosamines, levels of
parts per billion are easily quantitated, and even parts per
trillion can be detected (8).   To be totally reliable, however,
such measurements need to be confirmed by other analytical systems including mass spectroscopy.   Occasionally artifacts seem to
activate the specific instrumentation used for nitrosamines.   In
any case, these very sensitive analytical systems have now demonstrated the occurrence of nitrosamines at low but definite levels
in the human environment.   It is probable, although not yet so
demonstrated, that 100 years ago, when saltpeter was used extensively in the preservation and pickling of many kinds of foods
(9), that the levels of nitrosamines in the human environment
were quite appreciable.
     Yet, except for specific cases of acute poisoning (2,10,
11), and thus mostly of liver damage, there appears to be no reliably documented evidence that there are any cases of human cancer which can be unambiguously associated with an antecedent exposure to nitrosamines in the Western world (12).

Naturally Occurring or Endogenously Produced Nitrosamines as
Human Carcinogens.

On the other hand, as discussed at this meeting by Hoffmann
et al. (13,14), there are certain tobacco-specific nitrosamines
such as nitrosonornicotine which form a substantial part of the

genotoxic carcinogens in tobacco smoke and in tobacco chews. Specific association of these kinds of carcinogens with cancer at sites related to the use of tobacco, such as cancer of the lung, pancreas, kidney, urinary bladder, and also, in the presence of heavy alcohol intake, cancers of the oral cavity and esophagus remains to be fully documented (15-22). It is reasonable to assume, however, that these carcinogens play a role for cancer in these target organs that accounts for a substantial part of the cancers in various countries.

In addition, in certain parts of the world, such as Eastern Iran, Southern Soviet Union, and Central China, cancer of the esophagus is seen in people who do not appear to smoke and drink heavily. Despite extensive cooperative efforts between investigators in Iran and the International Agency for Research on Cancer, the etiology of esophageal cancer in Iran remains obscure (23, 24, 25). Nitrosamines have been suspected as etiologic factors but have not been found as yet. Nonetheless, the nutritional intake is poor in green and yellow vegetables, fruits, and other vitamin C and E containing antidotes to nitrite (25), and thus, through this association, the endogenous formation of nitrosamines cannot be ruled out. This is especially true inasmuch as recent data from China (26-29), the easternmost extension of the belt of esophageal cancer incidence, does suggest that nitrosamines are present in that environment.

Evidence That Nitrosamides May Be Human Carcinogens.

While attempting to examine the question whether some powerful bacterial mutagens, in this instance N'-methyl-N'-nitro-N-nitrosoguanidine (MNNG), were carcinogenic, Sugimura and Kawachi (30) administered this mutagen, a water-soluble, rather polar substance, to rats in drinking water. The important discovery was made that this mutagen not only was a powerful carcinogen but reliably induced cancer in the glandular stomach, mimicking accurately the disease seen in man. A detailed study of the pathogenesis of this then new animal model revealed that this chemical taken orally not only induced the neoplasm but also induced all the antecedent and accompanying lesions such as gastritis and intestinalization of the gastric epithelium, as seen in man (30, 31, 32).

Salt exerts a promoting effect (33). The effect of salt as a promoter is linked to the epidemiologic information that populations with a high incidence of gastric cancer also tend to exhibit a high incidence of hypertension (34-40).

We have linked the gastric cancer model of Sugimura to the examination of the occurrence of human gastric cancer and to consideration of the underlying mechanisms.

Nitrosamides and Gastric Cancer.

Between 1900 and 1950, gastric cancer was one of the major
types of cancer in the United States. Its incidence and mortality have decreased sharply since 1950 (41), and the concepts presented in this paper account for this decline. In other parts of
the world, including Japan, the mountainous interior regions of
Central and Western Latin America, and in Northern and Eastern
Europe, as well as in Iceland, gastric cancer remains one of the
major cancers. There is a North-South gradient, or South-North
in the Southern hemisphere, with greater incidence in more frigid
zones.

From geographic pathology and migrant studies, stems the
view that nutrition and specific dietary factors were associated
with gastric cancer (34,42,43,44). Typically, high risk groups
eat diets rich in carbohydrates, without any direct role in the
etiology, but with the significant inclusion of dried and salted
fish or certain pickled or smoked foods. Also, there was a variable intake of fresh fruits and vegetables, namely, a seasonal
low consumption of such foods during winter and spring. The geo-
chemistry often involved elevated soil nitrate and, thus, foods
and water rich in nitrate (45,46,47).

We have suggested that an alkylnitrosoureido type of compound, such as the one, mentioned above, developed by Sugimura to
induce glandular gastric cancer in animal models, may be involved
in human gastric carcinogenesis (48). The required nitrite might
be derived from: (1) its reductive formation in foods rich in
nitrate and stored at room temperature after cooking, (2) the
consumption of pickled foods, and (3) consumption of smoked food,
in some parts of the world. It has also been observed that siz-
able amounts of nitrite are produced by the reduction of nitrate
secreted by the salivary glands through the oral microbiologic
flora. The nitrate stems from the oral intake of foods rich in
this compound either through soil chemistry or by addition in the
pickling process (49-52). Precursors such as amides can be converted by nitrite to carcinogenic nitrosamides in the acidic environment of the stomach or during pickling in the presence of
acids such as acetic acid (vinegar) or lactic acid (5).

Mirvish (53,54) discovered that vitamin C could inhibit nitrosation reactions. The purely chemical interaction of ascorbic
acid with nitrite has been studied for theoretical reasons and
because of its importance in the preservation of foods. This interaction has received increased attention for minimizing the
presence of nitrosamines and nitrosamides in the environment, and
especially in foods. We have studied the relationship in gastric
carcinogenesis between high levels of nitrite, including pickling, and of vitamin C as a protective and inhibiting element.

Vitamin E can also inhibit nitrosation reactions but the
mechanisms may be somewhat different than those for vitamin C
(55,56). Of course, vitamin C is water soluble while vitamin E

is lipid soluble. Certain nitrosated substrates are lipid soluble, and thus vitamin E may offer certain advantages as an inhibitor.

We used an experimental model, the fish Sanma, which is eat-
en in a region of Japan at high risk for gastric cancer. Treatment of homogenates of this fish with nitrite at pH 3 led to the
development of a direct-acting mutagen for Salmonella typhimurium
TA 100. The formation of mutagens could be completely blocked by
vitamin C (57). This property is similar to the inhibition of
the formation of nitrosamines and nitrosamides, discussed above,
and is compatible with the view that the mutagens from fish are
compounds with a nitrosamide type of functional group. Sanma
yielded more mutagenic activity than other types of fish, but it
is important that several types of meat failed to produce such
direct-acting mutagens (57).

We demonstrated that the mutagenic activity from the reaction of nitrite and Sanma was carcinogenic (48). It is quite
relevant that this product induced glandular stomach cancer in
rats. Since the formation of the mutagen can be blocked by vitamin C, it would seem that the formation of glandular stomach cancer can also be so inhibited.

Materials and Methods

Treatment of Fish Extracts and Assays. The preparation of
the fish extract has been described (57). Briefly, the fish were
homogenized and incubated with nitrite (5 g of homogenate plus
5,000 ppm NaCl and 5,000 ppm NaNO$_2$) at pH 3 and 25°C for 3
hours. These levels were used to mimic the historically employed
doses of saltpeter and salt in pickling (9). Moreover, nitrosation can occur at lower levels of nitrite (see Mirvish, p.271, in
ref. 8). The reaction was stopped by the addition of 5,000 ppm
ammonium sulfamate. A control sample was treated likewise but
without adding nitrite. The mixture was extracted twice with
hexane (removal of lipid fraction, which was mutagen-free) and
four times with ether. The combined ether extracts were taken to
dryness. The residue was redissolved in water (adjusted to pH
3). The mutagenic activity of this solution was determined by
conventional Ames assay and the volume was adjusted so that 10 ul
would yield 125 His+ revertants/plate (spot test). The extracts
were prepared a week before use and were deep-frozen. The mutagenic activity of the extracts was stable during storage for 3
weeks.

The mutagenic activity of the extracts was assayed using
Salmonella typhimurium strain TA 1535, as described by McCann et
al. (58). Ten ul of fish extract were spot-tested without addition of an S9-enzyme preparation (57). MNNG (20 ug/plate) was
used as a positive control.

Bioassays-- Male random-bred Wistar rats, 50 days old, on
lab chow and water ad libitum, were divided into two groups. The

control group (I) was given extracts from fish homogenates not treated with nitrite three times per week for six months by gavage. The experimental group (II) received the extract from fish treated with nitrite. The weekly amount of extract given was calculated to have mutagenic activity (Ames assay) equal to a weekly oral dose of MNNG active in inducing adenocarcinomas of the glandular stomach in rats (59). Thus, the rats received 0.5 ml per gavage of the extract during the first two months of the treatment period and 1.0 ml per gavage of extract during the remaining four months. A rat received per week the extract from about 100 g (initial two months of treatment period) or about 200 g (remaining four months of treatment period) of fish. The animals were held for an additional observation period of 18 months. The surviving rats were killed and all animals were carefully autopsied with emphasis on the gastrointestinal tract. The tissues were fixed and processed by conventional histologic techniques and studied microscopically.

Results

Two of the ten rats from Group I died from causes unrelated to treatment, mainly pneumonia or other infectious processes. Three rats died after living more than 18 months. Five rats were alive at the 24-month point. In Group II, three of the rats given the extract of nitrite-treated fish died early, due to causes unrelated to treatment. Five rats died after 18 months and their tissues were available for study. Finally, seven rats alive at 24 months were killed.

Thus, there was a total of 8 rats for evaluation in the control group and 12 in the experimental group. In the control group given an extract of fish alone, there were no neoplastic changes in the stomach or pancreas except an epithelial hyperplasia of the glandular stomach of one rat (Table 1). Instead, there were tumors in the testes, kidneys, and soft tissues, as are often seen in aged rats.

In the group given the extract of nitrite-treated fish, glandular hyperplasia of the stomach was seen in six rats (Table 1). In most, there was also intestinal metaplasia. In addition, five rats had squamous cell hyperplasia of the forestomach epithelium. Five tumors, including two adenocarcinomas and one adenosquamous cell carcinoma, occurred in the glandular stomach of the four animals of the experimental group (Table 2). One rat had a tubular adenoma and simultaneously a well differentiated tubular adenocarcinoma. Another histologic type of adenocarcinoma was a poorly differentiated carcinoma (scirrhous type). In the forestomach, squamous cell carcinoma developed in two of these rats. Of interest also is the finding of an adenocarcinoma in the small intestine of one rat, and pancreatic acinar cell tumors in three animals, all of which may be significantly related to the treatment. In addition, there were several other types of neoplasms in this group.

Table I.
Incidence of Stomach Epithelium Hyperplasia
and Intestinal Metaplasia (66)

Experimental Group and Treatment	Effective Number of Rats	Forestomach Squamous Cell Hyperplasia	Glandular Stomach	
			Glandular Hyperplasia	Intestinal Metaplasia
I Fish extract alone	8	0	1[a]	1
II Fish extract + NaNO$_2$	12	5[b]	6[c]	6[d]

[a] One rat had both lesions.
[b] One rat had only forestomach hyperplasia; 3 also had glandular hyperplasia; 1 had intestinal metaplasia.
[c] One rat had only glandular hyperplasia; 3 also had forestomach hyperplasia; 2 had intestinal metaplasia.
[d] One rat had only intestinal metaplasia; 1 also had only forestomach hyperplasia; 2 also had only glandular hyperplasia; 2 had all three lesions.

Table II.
Sites and Incidence of Tumors in Rats Given Extracts
of Fish Treated with Nitrite or Fish Alone (66)

Site and Type of Tumor	Number of Rats In Experimental Group I (Fish Extract Alone)	Number of Rats In Experimental Group II (Fish Extract + NaNO$_2$)
Effective Number of Rats–	8[a]	12[b]
Forestomach		
Papilloma	0	0
Squamous cell carcinoma	0	2[c]
Glandular Stomach		
Adenoma	0	2[d]
Adenocarcinoma	0	2[e]
Adenosquamous carcinoma	0	1[f]
Pancreas		
Adenoma	0	2
Adenocarcinoma	0	1
Small Intestine		
Adenocarcinoma	0	1[f]

[a] Three rats had interstitial cell ademona of testis, 1 renal nephroblastoma, 1 rhabdomyosarcoma.

[b] Eight rats had at least one of the tumors tabulated here. In addition, the following miscellaneous tumors were noted: 1 rat had cortical adenoma of adrenal gland, 1 had follicular adenoma of thyroid gland, and 1 had pulmonary adenoma.

[c] One rat also had a pancreatic adenoma, and the other had an adenocarcinoma.

[d] One rat had only an adenoma, but the other also had an adenocarcinoma.

[e] One rat also had an adenocortical adenoma, and the other had a gastric adenoma and a thyroid adenoma.

[f] The only tumor in 1 rat.

Discussion

Treatment of fish of a type eaten in Japan, a high risk region for gastric cancer, with nitrite at pH 3 yielded not only an extract with mutagenic activity, but one which induced cancer in the glandular stomach. The experimental design essentially mimicked the conditions of migrants who have lived in a high-risk region in the early part of their lives, and then moved to a low-risk region. The animals were treated with mutagenic substances for only 6 months and were then held on a control diet without further exposure to known carcinogens. Had the treatment been longer, as would be the case for anyone living continuously in a high-risk region, the cancer incidence might have been higher.

Glandular stomach cancer is rare in rats, and the control group had no neoplastic lesions in the stomach. Also, a high incidence of epithelial hyperplasia or intestinalization of glandular stomach cancer was observed only in the experimental group. Thus, the neoplastic lesions in the glandular stomach were no doubt due to the treatment administered. As is true for rats given MNNG (59), we also noted tumors in the small intestine and in the pancreas.

We had found earlier that similar pickling of beans, as eaten in high-risk Latin America, or of borscht, as consumed in Eastern Europe, led to mutagenic activity, and thus presumably carcinogenic activity (57)

The nature of the mutagens responsible for carcinogenesis is not yet known and is being investigated. It is clear from the inhibition of its formation by vitamin C (57), and from the fact that, like alkylnitrosoureido compounds, it produced cancer in the glandular stomach, that we are dealing with this kind of compound. However, in contrast to compounds such as methylnitrosourea, we noted that the mutagenic activity was rather stable at pH 3 and even at higher pH values. Nonetheless, Mirvish et al. (60, 61) have evidence that Bonito fish contains some nitrosatable alkylureas.

Conclusion

Studies in migrants who maintained their risk for gastric cancer when going from high-risk to low-risk regions demonstrated clearly that the nature of the carcinogens operating in man is probably quite similar to what we have identified through mutagen and carcinogen bioassays. With such compounds, exposure throughout life is not required to yield eventual stomach cancer. Thus, it seems logical to conclude that if gastric cancer were to be prevented, exposure to agents causing this disease must be minimized or avoided from the earliest age. This means, in turn, that foods providing the necessary vitamin C, such as fresh fruits, vegetables, and salad, or supplementary vitamin C, ought to be consumed with every meal. It is known that some vegetables and salads are also good sources of nitrate. They are consumed fresh and are usually refrigerated, so that their content in bacterially-produced nitrite (2) is minimal. The nitrite, possibly stemming from reduction in the oral cavity (51), is not important since those foods contain substantial amounts of the antidote, vitamin C, and also phenolic inhibitors (8, 62).

Salt appears to have a promoting effect, although in some regions of the world it may also lead to mucosal damage, gastritis, rise in stomach pH, bacterial overgrowth, and thence to another source of nitrite through bacterial reduction of nitrate in the stomach (33). However, within the MNNG model, the antecedent precancerous lesions occur with or without salt (63,64,65), and thus we believe that salt use augments rather than causes gastric cancer. Nonetheless, a lowered salt intake would not only reduce the risk of gastric cancer, but also lower the occurrence of hypertension and stroke, associated with salt in genetically sensitive individuals (35, 39).

If the concepts and facts presented in this paper are correct, a major kind of human cancer in many regions of the world, cancer of the stomach, is due to a type of nitroso compound, a nitrosoureido derivative, even though not a nitrosamine. It is quite certain that the formation of such compounds can be blocked by vitamin C and vitamin E, as well as by some other substances such as gallates. Thus, the primary prevention of cancer caused by nitroso compounds is readily accomplished through an adequate intake of such harmless inhibitors with every meal from infancy onwards.

Acknowledgements

The research described is supported in part by U.S. Public Health Service grants CA-12376 and CA-29602 from the National Cancer Institute, and by a gift from Hoffmann-LaRoche, Inc.
My thanks to Mrs. C. Horn for editorial assistance.

Literature Cited

1.   Magee, P.N.; Montesano, R.; Preussmann, R.   IN Searle, Charles E., Ed.; "Chemical Carcinogens"; American Chemical Society: Washington, DC, 1976; pp 491-625.
2.   Weisburger, J.H.; Rainieri, R.   Toxicol. Appl. Pharmacol. 1975, 31, 369-74.
3.   Druckrey, H. Gann Monograph 1975, 17, 107-32.
4.   Preussmann, R.   IN Altmann, H.W.; Buchner, F.; Cottier, H.; Grundman, E.; Seifert, G.; Siebert, G., Ed.; "Handbuch der Allgemeinen Pathologie"; Springer-Verlag: Berlin, 1975; pp 421-594.

5. Sander, J.; Schweinsberg, F.; LaBar, J.; Burkle, G.; Schweinsberg, E. Gann Monograph 1975, 17, 145-60.
6. Singer, G.M. IN Walker, E.A.; Griciute, L.; Castegnaro, M.; Borzsonyi, M.; Davis, W., Eds.; "N-Nitroso Compounds: Analysis, Formation and Occurrence", vol.31; Internatl. Agency for Research on Cancer: Lyon, France, 1980; pp 139-151.
7. Lijinksky, W.; Taylor, H.W. Food Cosmet. Toxicol. 1977, 15, 269-74.
8. Walker, E.A.; Griciute, L.; Castegnaro, M.; Borzsonyi, M.; Davis, W., Eds.; "N-Nitroso Compounds: Analysis, Formation and Occurrence", vol. 31; Internatl. Agency for Research on Cancer: Lyon, France, 1980
9. Binkerd, E.F.; Kolari, O.E.   Food Cosmet. Toxicol. 1975, 13, 655-60.
10. Cooper, S.W.; Kimbrough, R.D.   J. Forensic Sci. 1980, 25, 874-82.
11. Fussgaenger, R.D.; Ditschuneit, H.   Oncology 1980, 37, 273-77.
12. Higginson,J.   J. Environ. Pathol. Toxicol. 1980, 3, 113-126.
13. Hoffmann, D.; Adams, J.D.; Brunnemann, K.D.; Hecht, S.S.   IN This Volume.
14. Hoffmann, D.; Adams, J.D.   Cancer Research 1982
15. Lowenfels, A.B.; Tuyns, A.J.; Walker, E.A.; Roussel, A.   Gut 1978, 19, 199-201.
16. Deckers, C.   J. Belge Radiol. - Belgisch Tijdschr. Radiol. 1978, 60, 543-549.
17. Tuyns, A.J.; Pequignot, G.; Jensen, D.M.   Front. Gastrointest. Res. 1979, 4, 101-110.
18. Kolonel, L.   NCI Monograph 1979, 53, 81-87.
19. Kono, S.; Ikeda, M.   Br. J. Cancer 1979, 40, 449-455.
20. Graham, S.; Dayal, H.; Rohrer, T.; Swanson, M.; Sultz, H.; Shedd, D.; Fischman, S.   J. Natl. Cancer Inst. 1977, 59, 1611-1615.
21. McMichael, A.J.   1978, The Lancet 1978, ii, 1244-1246.
22. McCoy, G.D.; Hecht, S.S.; Wynder, E.L.   Prev. Med. 1980, 9, 622-629.
23. Hewer, T.; Rose, E.; Ghadirian, P.; Castegnaro, M.; Bartsch, H.; Malaveille, C.; Day, N.   The Lancet 1978, ii, 494-496.
24. Joint Iran-International Agency for Research on Cancer Study Group.   J. Natl. Cancer Inst. 1977, 59, 1127-1138.
25. Cook-Mozaffari, P.J.; Azordegan, F.; Day, N.E.; Ressicaud, A.; Sabai, C.; Aramesh, B.   Br. J. Cancer 1979, 39, 293-309.
26. Cheng, S.-J.; Sala, M.; Li, M.H.; Wang, M.-Y.; Pot-Deprun, J.; Chouroulinkov, I.   Carcinogenesis 1980, 1, 685-692.
27. Lu, S.H.; Camus, A.M.; Ji, C.; Wang, Y.L.; Wang, M.Y.; Bartsch, H.   Carcinogenesis 1980, 1, 867-870.
28. Huang, D.P.; Ho, J.H.; Webb, K.S.; Wood, B.J.; Gough, T.A.   Fd. Cosmet. Toxicol. 1981, 19, 167-171.

29. Lin, J.Y.; Wang, H.I., Yeh, Y.C.   Fd. Cosmet. Toxicol. 1979,
     17, 329-331.
30. Sugimura, T.; Kawachi, T.   IN Lipkin, M.; Good, R., Eds.;
     "Gastrointestinal Tract Cancer"; Plenum: New York, 1978; pp
     327-42.
31. Cuello, C.; Correa, P.; Zarama, G.; Lopez, J.; Murray, J.;
     Gordillo, G. Am. J. Surg. Pathol. 1979, 3, 491-500.
32. Piper, D.W.   "Stomach Cancer"; UICC Tech. Rept. 34; Union
     Internationale Contre le Cancer: Geneva, 1978, 1-138.
33. Tatematsu, M.; Takahashi, M.; Fukushima, S.; Hanaouchi, M.;
     Shirai, T. J. Natl. Cancer Inst. 1975, 55, 101-4.
34. Bjelke, E. Scandinavian J. Gastroenterol. 1974, 9, Suppl.
     31, 1-235.
35. Joossens, J.V.; Kesteloot, H.; Amery, A. New Eng. J. Med.
     1979, 300, 1396.
36. Altschul, A.M.; Grommet, J.K.   Nutrit. Rev. 1980, 38,
     393-402.
37. Ueshima, H.; Tanigaki, M.; Iida, M.; Shimimoto, T.; Konishi,
     M.; Komachi, Y. Lancet 1981, 1, 504.
38. Armstrong, B.; Clarke, H.; Martin, C.; Ward, W.; Norman, N.;
     Masarei, J. Am. J. Clin. Nutrit. 1979, 32, 2472-76.
39. Meneely, G.R.; Battarbee, H.D.   Am. J. Cardiol. 1976, 38,
     768-81.
40. Weisburger, J.H. Lancet 1980, ii, 424-5.
41. Devesa, S.S.; Silverman, D.T. J. Natl. Cancer Inst.   1978,
     60, 545-71.
42. Fraumeni, J.F., Jr.   "Persons at High Risk of Cancer";
     Academic: New York, 1975; pp 1-544.
43. Haenszel, W.; Kurihara, M.; Locke, F.B.; Shimuzu, K.; Segi,
     M. J. Natl. Cancer Inst. 1976, 56, 265-74.
44. Hiatt, H.H.; Watson, J.D.; Winsten, J.A, Eds.; "Origins of
     Human Cancer"; Cold Spring Harbor Lab.: Coldspring Harbor,
     N.Y., 1977.
45. Armijo, R. Natl. Cancer Inst. Monogr. 1979, 53, 115-118.
46. Tannenbaum, S.R., Moran, D.; Rand, W.; Cuello, C.; Correa,
     P. J. Natl. Cancer Inst. 1979, 62, 9-12.
47. Zaldivar, R.; Wetterstrand, W.H. Z. Krebsforsch 1978, 92,
     227-34.
48. Weisburger, J.H.; Marquardt, H.; Hirota, N.; Mori, H.;
     Williams, G. J. Natl. Cancer Inst. 1980, 64, 163-67.
49. Hayashi, N.; Watanabe, K.; Ishiwata, H.; Mizushiro, H.;
     Tanimura, A.; Kurata, H. J. Food Hyg. Soc. Japan 1978, 19,
     392-400.
50. Spiegelhalder, B.; Eisenbrand, G.; Preussman, R. Food
     Cosmet. Toxicol. 1976, 14, 545-48.
51. Tannenbaum, S.R.; Weisman, N.; Fett, D.   Food Cosmet.
     Toxicol. 1976, 14, 549-52.
52   Fraser, P.; Chilvers, C.; Beral, V.; Hill, M.J. Int. J.
     Epidemiol. 1980, 9, 3-11.
53. Mirvish, S.S. J. Toxicol. Environ. Health 1977, 2, 1267-77.

54. Mirvish, S.S.; Karlowski, K.; Sams, J.P.; Arnold, S.D. IN "Environmental Aspects of Nitroso Compounds"; Sci. Pub. 19, Internatl. Agency for Research on Cancer: Lyon, France, 1978; pp 161–74.
55. Mergens, W.J.; Kamm, J.J.; Newmark, H.L., Fiddler, W.; Pensabene, J. In "Environmental Aspects of N–Nitroso Compounds"; Sci. Pub. 19, Internatl. Agency for Research on Cancer: Lyon, France, 1978; pp 199–212.
56. Kamm, J.J.; Dashman, T.: Newmark, H.; Mergens, W.J. Toxicol. Appl. Pharmacol. 1977, 41, 575–83.
57. Marquardt, H.; Rufino, R.; Weisburger, J.H. Food Cosmet. Toxicol. 1977, 15, 97–100.
58. McCann, J.; Choi, E.; Yamasaki, E.; Ames, B. Proc. Nat. Acad. Sci. USA, 1975, 72, 5135–39.
59. Lipkin, M.; Good, R.A. "Gastrointestinal Tract Cancer"; Plenum: New York, 1978, 602 pp.
60. Mirvish, S.S.; Karlowski, K.; Cairnes, D.A.; Sams, J.P.; Abraham, R.; Nielsen, J. J. Agric. Food Chem. 1980, 28, 1175–1182.
61. Mirvish, S.S. This Volume.
62. Kurechi, T.; Kikugawa, K.; Fukuda, S. J. Agric. Food Chem. 1980, 28, 1265–1269.
63. Matsukura, N.; Kawachi, T.; Sugimura, T.; Nakadate, M.; Hirota, T. Gann 1979, 70, 181–185.
64. Matsukura, N.; Itabashi, M.; Kawachi, T.; Hirota, T.; Sugimura, T. J. Cancer Res. Clin. Oncol. 1980, 98, 153–163.
65. Quimby, G.F.; Eastwood, G.L. J. Natl. Cancer Inst. 1981, 66, 331–337.
66. Weisburger, J.H.; Marquardt, H.; Mower, H.F.; Hirota, N.; Mori, H.; Williams, G. Prev. Med. 1980, 9, 352-361.

RECEIVED August 10, 1981.

The Microecology of Gastric Cancer

PELAYO CORREA

Department of Pathology, Louisiana State University Medical Center, New Orleans, LA 70112

STEVEN R. TANNENBAUM

Department of Nutrition and Food Science, Massachusetts Institute of Technology, Cambridge, MA 02139

Evidence is reviewed supporting the proposition that stomach cancer etiologic factors operate by modifying the microenvironment of the gastric mucosa. Three main components of the gastric microecology probably related to gastric carcinogenesis are reviewed: the progressive histological lesions of the mucosa, the altered gastric secretions and the dietary items which apparently have an overriding influence in the process. A discussion and an update of the etiologic hypothesis based on intragastric synthesis of N-nitroso compounds are presented.

Introduction

The present state of our knowledge concerning the etiology of gastric cancer has focused on the microecology of the gastric mucosa. This point has been reached after decades of research by scientists working in different disciplines and different parts of the world. Epidemiologic observations on the distribution of cancer and precancerous conditions of the stomach throughout the world and observations of the pathology of gastric cancer and related lesions have been correlated with biochemical observations and led to the formulation of an etiologic hypothesis which is being tested in different ways. The acceptance or rejection of this hypothesis hinges mainly on the results of the work on the chemistry of the process. We will briefly review the basic data related to the evolution of these ideas.

Geographic Distribution

Marked contrasts in the frequency of gastric cancer have been recognized for a long time, as shown in Figure 1 which gives the rank order of the magnitude of the incidence rate for populations throughout the world (1, 2). It is immediately apparent that some populations in Japan, Chile, Colombia, Iceland and

0097-6156/81/0174-0319$05.00/0

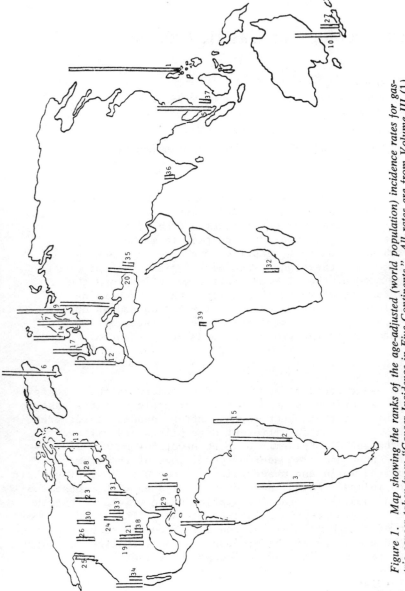

Figure 1. Map showing the ranks of the age-adjusted (world population) incidence rates for gastric cancer taken from "Cancer Incidence in Five Continents." All rates are from Volume III (1) except the rate for Chile, which is from Volume I (2). The following regions are represented by more than one racial group: Singapore (Chinese rank 5, Malay rank 37); New Zealand (Maori rank 10, nonMaori rank 27); Hawaii (Japanese rank 11, Caucasian rank 22); California Bay area (black rank 18, white rank 34); New Mexico (Indian rank 19, Spanish rank 21, other white rank 38).

Europe have rates that are several times greater than those found
in Africa or India. Geography by itself cannot explain these
differences. People of Chinese extraction living in Singapore
(rank 5) have several times the rates of Malays in the same city
(rank 37). Similar contrasts are found between whites and Maori
living in New Zealand (ranks 10 and 27), blacks and whites in
Detroit and California (ranks 18 and 34), or Indians and whites
in New Mexico (ranks 19 and 38).

Race by itself does not explain this geographic distribution.
Nonmigrating Japanese have several times the rate of people of
Japanese extraction living in Hawaii, especially the second
generation. Whites in eastern and northern Europe have consider-
ably higher rates than whites in New Zealand and the United
States. Chinese in Singapore have rates approximately five times
greater than those of Chinese in Hawaii. Blacks in Jamaica and
Louisiana have rates much higher than those of African blacks.

These observations clearly indicate that the disease is
determined mostly by the environment and not by the race of the
population. Since the geographic environment per se is ruled out,
it seems that the cultural environment is the one that matters.
The most obvious component of the cultural environment is the
diet, and it may explain the interpopulation distribution. A
carcinogen in the diet has been sought for decades. As is often
the case when carcinogens are looked for in the diet, they are
found and that was the case in several countries, especially in
Iceland where Dungall and others isolated 3,4-benzpyrene from
smoked salmon (3). Those findings, however, remain unconvincing
if not linked to epidemiologic or other biologic observations
which indicate their relevance to the human cancer situation.

A crucial observation concerning the distribution of gastric
cancer has been that the populations migrating from a high-risk
area (such as Russia, Norway, Japan, Latin America) to a low-risk
area (such as the United States or Australia) keep their original
high risk in the host country (4). They continue to have high
rates of incidence and mortality in spite of the fact that they
live practically all of their adult lives in a country whose
indigenous population displays low rates. This migration effect
cannot be adequately explained by dietary patterns, and it has
led to the idea that gastric cancer is the result of forces set
in motion many years before cancer becomes clinically evident.
Abundant support for this idea has come from the studies of the
pathology and cell kinetics of the neoplastic process. Fujita,
studying the rate of growth (doubling time) of gastric cell populations, estimates that the early (intramucosal) phase of gastric
cancer lasts from 16 to 24 years (5). The multistage nature of
the carcinogenic process has been well documented in experimental
gastric cancer, especially with the MNNG model which can be
accelerated with well-known promoters such as croton oil (6).
Whether the process of carcinogenesis can be inhibited or slowed
down by chemopreventive agents still needs to be demonstrated,

but there is suggestive epidemiologic evidence that this may occur in humans (7).

Microenvironment

Epidemiologic studies of the macroenvironment focused on the diet and the long incubation period have led us to postulate that long-lasting disturbances in the normal gastric mucosa may determine the final outcome of the neoplastic transformation (8). This explains the present interest in precursor lesions rather than in cancer itself. We, therefore, need to scrutinize the gastric microenvironment and attempt to point out the components that may be relevant to neoplasia. The microenvironment could be considered to be determined by three basic elements:
A. What is the structure of the gastric mucosa
B. What is secreted into the cavity
C. What is ingested into the stomach
We will analyze these components basing our discussions mostly on our observations in Nariño, Colombia, where gastric cancer is extremely frequent (9, 10, 11).

Mucosal Lesions

The gastric mucosa surrounding most gastric carcinomas (those of "intestinal" or "epidemic" type) is the site of profound changes which take many years to develop. The normal glands disappear and are replaced by new glands which resemble those found normally in the intestine. This process is called intestinal metaplasia and is considered a result of chronic inflammation leading to loss of glands and abnormal regeneration, all components of the so-called chronic atrophic gastritis. These lesions are found in the majority of adults in populations at high gastric cancer risk. It is characteristically a focal process starting around the union of the antrum and the corpus of the stomach. The independent small foci of metaplasia gradually expand and coalesce until extensive areas of the surface are covered. Only rarely the metaplasia covers all the surface so that most of the time there are atrophic and normal areas coexisting side by side. The metaplastic cells in most patients are "mature" in that they have all the characteristics of normal adult small intestinal cells. In some patients, groups of metaplastic cells begin to lose some of the "differentiated" characteristics, produce less mucus and increase their mitotic activity. These changes are usually called dysplasia, which includes distortions of the architecture of the glands (12). As this process becomes more advanced, the gastric mucosa more closely resembles cancer tissues. The final transformation consists of the loss of dependency of the dysplastic cells on other tissues, which enables them to invade the neighboring structures, the hallmark of cancer.

The process of cancerization in the stomach mucosa, therefore, seems to be a continuum of changes which turns a normal cell into a metaplastic cell and then to a progressively dysplastic cell and finally a cancerous cell. The changes usually take several decades, are gradual, and the change from one cell type to the other does not seem obligatory. They seem to reflect sequential mutations or cell transformations.

The morphologic changes just described are accompanied by profound histochemical alterations especially related to glycoproteins and cytoplasmic enzymes. The normal surface epithelial cell of the stomach accumulates in its cytoplasm multiple droplets of a neutral glycoprotein which constitutes the gastric mucus (13). This is changed in metaplastic mucosa which accumulates in their cytoplasm a large globus (goblet) of acid glycoproteins (mucin). It appears that in the early stages of metaplasia the mucin resembles small intestinal mucin in that it stains with Alcian Blue but not with HID (high-iron diamine) which classifies it as a sialomucin (14). This type of metaplasia has been called complete metaplasia. In later stages the mucin resembles that of the colon in that the mucin stains with HID and, therefore, corresponds to sulfomucin. This type of metaplasia has been called incomplete (15-18). Intestinal metaplasia brings to the gastric mucosa enzymes which are foreign to it. In complete metaplasia, sucrase, trehalase, leucine aminopeptidase (LAP) and alkaline phosphatase (ALP) are present; in incomplete metaplasia, sucrase but not trehalase is present in the surface: LAP is found in the tissue sections and ALP is present but only in very small amounts (16).

Cancer is mostly surrounded by "incomplete" metaplasia, probably another indication of the progressive loss of differentiation of the metaplastic cells (19).

Secretions

The gastric juice is a mixture of compounds mostly secreted by the different compartments of the gastric mucosa and it varies in composition according to the physiologic needs and other circumstances. Changes in the mucus and enzyme secretions which take place in cancer precursor states have already been described. Fasting gastric juice in normal persons has a pH of 1 to 2 as a result of the secretion of HCl by the parietal cells. In chronic atrophic gastritis the progressive loss of parietal cells gradually decreases HCl secretion but for a long time this is compensated by overstimulation due to an excess of gastrin secretion and G cell hyperplasia (20). When the loss of parietal cells is insufficient to respond to the excessive stimulation by gastrin, the secretion of HCl is decreased and the pH of the gastric juice becomes elevated. Most patients with chronic atrophic gastritis have a pH of 5 or above in their fasting gastric juice. Whatever HCl is secreted comes from islands of

well-preserved mucosa which lie side by side with the atrophic
mucosa. The gastric juice pH, therefore, represents well the
contents of the gastric cavity but not the microecology of the
gastric mucosa. The well-preserved areas of the mucosa, with
their normal or exaggerated HCl secretion, probably have a low
pH at their surface. In the immediate vicinity, the atrophic
metaplastic mucosa, deprived of HCl secretion, has a much higher
pH. This type of microenvironment, therefore, lends itself for
chemical reactions that may take place in a wide range of pHs,
from 1 to 7.

Alterations in other gastric secretions, such as pepsinogens
and blood group substances also take place in chronic atrophic
gastritis. The secretion of pepsinogen I has been used as an
indicator of intestinal metaplasia and gastric cancer (21, 22).

Ingested Materials

An important component of gastric microecology derives from
ingested materials: food items, bacteria, saliva and the large
series of chemical compounds ingested with our foods and our
drinking water. There have been many attempts to characterize
the diet of populations at high gastric cancer risk and of
patients with gastric cancer (23). No individual food item has
been found universally associated with gastric cancer. This
probably should have been anticipated when considering dietary
habits in high-risk areas as disimilar as those of England,
Finland, Japan, Iceland, Hawaii and Latin America. The similarities of these diets can only be described in general terms.
Most diets associated with high gastric cancer risk have the
following characteristics:
- They are low in animal fat and animal protein
- They are high in complex carbohydrates
- They obtain a substantial part of their proteins from
  vegetable sources, mostly grains
- They are low in salads and fresh, green, leafy vegetables
- They are low in fresh fruits, especially citrus
- They are high in salt content

Exactly how this type of diet predisposes to gastric cancer
has not been determined. Its high bulk obviously influences the
gastric microenvironment, and its low fat content probably
interferes with the absorption of lipid soluble substances such
as vitamin A and vitamin E. The scarcity of fresh, green leafy
vegetables and of citric fruits may also point to some vitamin
deficiencies. The high salt content may damage the mucous
barrier that protects the mucosa from food items, as well as from
the acid-pepsin secretions. There are many populations in the
world with high bulk diet and low gastric cancer rates, and in
some populations such as Finland, the gastric cancer rate has
been steadily declining without apparently drastic changes in the
bulk or the vegetable content of the diet. The epidemiologic

observations suggest that in some populations a protective factor
is being added to the diet.

Another ingested substance which has been linked to the diet
is nitrate. It has been found that in some, but not all populations at high gastric cancer risk, the ingestion of nitrates in
the food and drinking water is higher than in similar populations
with lower risk (9, 24). The correlation of nitrate intake and
gastric cancer rates is not observed in some other populations.
This is not unexpected since nitrate by itself has not been
implicated in carcinogenesis. Nitrate is reduced to nitrite in
the saliva and may then be ingested into the stomach. Ingested
nitrate is probably not reduced to nitrite to any significant
degree in the gastric cavity of normal individuals. In patients
with chronic gastritis, however, a significant amount of nitrite
is detected in the gastric juice, especially when the pH is above
5, as can be seen in Figure 2.

The elevated concentrations of nitrite in gastric juice may
be a result of the multitude of bacteria ingested with the diet
and the saliva. Instead of being destroyed by the normal acid
medium, these bacteria survive and proliferate at the higher pH.
Many species of bacteria have been obtained from fasting gastric
juice in patients with chronic atrophic gastritis and many of
them have nitrate reductases (25, 26). These surviving bacterial
colonies come in contact with the gastric mucosa and would find
a favorable environment for their survival in the atrophic areas
with their higher than normal pH. Since these areas are adjacent
to the HCl-secreting areas with low pH, the microecology of the
junction of the two areas provides the unusual circumstances of
having nitrate, nitrite, abundant bacteria and a wide range of
pHs.

Etiologic Hypothesis

Based on the above data and on other available experimental
work showing that nitroso compounds can induce gastric cancer
in animals (6) and that nitrosation reactions leading to synthesis
of carcinogens can take place in the gastric cavity environment
(27, 28, 29), we have formulated an etiologic hypothesis for
gastric cancer (8).

We have proposed that gastric cancer may be the end result
of a series of mutations (or similar cell transformations) begun
many years before cancer becomes clinically apparent. The
mutagens could be nitroso compounds synthesized in situ from
nitrite and naturally-occurring nitrogen-containing compounds.
The nitrite results from reduction of nitrate by bacteria
abnormally present in the gastric mucosa and the gastric cavity.
The bacteria grow in situ because the pH is elevated as a result
of loss of HCl secretion secondary to the loss of parietal cells
and their replacement by intestinal-type epithelium. Parietal
cells are lost as a result of chronic atrophic gastritis. What

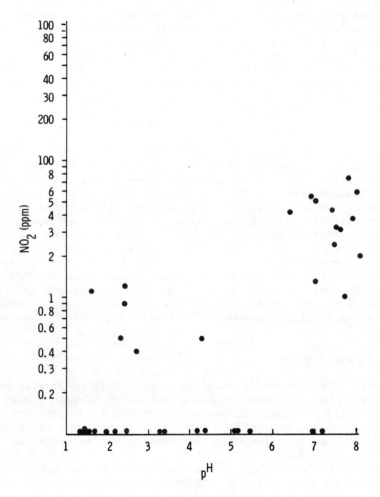

Figure 2. Correlation between pH and gastric juice of patients in Nariño, Colombia.

initiates chronic atrophic gastritis is not clear; it may be related to alterations in the mucous barrier of the stomach brought about by abrasives or irritants such as hard grains, high concentrations of salt in the diet, or surfactants.

The process, once initiated, is self-sustaining and may become more accelerated with time because the atrophy and intestinal metaplasia are progressive lesions and lead to further loss of parietal cells and incrased bacterial colonization of the mucosa. The initial mutations transform gastric cells into mature intestinal-type cells. Further superimposed mutations transform metaplastic cells into progressively dysplastic cells and eventually into neoplastic cells. This is a process of loss of differentiation which implies a multihit phenomenon which could be explained on the basis of continued formation of minute amounts of nitroso compounds over many years.

Recent Work

After our formulation of the hypothesis in 1975, many pieces of evidence have come to support the basic theme. A positive association of dysplasia, representing advanced stages of the process, with nitrite levels in the gastric juice has been reported in England and Colombia (12, 30). Higher than expected rates of cancer have been reported in patients with gastrectomy performed 5 or more years previously (31). Of special interest is the type of gastrectomy called Bilroth II in which intestinal reflux to the gastric cavity is facilitated. In such patients, high levels of nitrite are found in the gastric cavity (26). The incomplete (less differentiated) foci of metaplasia surround foci of cancer, suggesting progressive loss of differentiation (16, 19). High nitrite levels in the gastric cavity have been associated with acceleration of the gastric cancer induction with MNNG in experimental animals (26). Gastric cancer has been induced in experimental animals with products derived from nitrosated fish (32).

So far, we have been considering the nitrosation to occur in the lumen of the stomach. Recently, Stemmermann et al. (33) have reported that nitrosation may occur within the gastric mucosa itself.

Literature Cited

1. Waterhouse, J.; Muir, C.; Correa, P.; Powell, J., Eds.;
    "Incidence in Five Continents. Vol. III"; International
    Agency for Research on Cancer: Lyon, France, 1976.
2. Doll, R.; Payne, P.; Waterhouse, J., Eds.; "Cancer
    Incidence in Five Continents"; International Union Against
    Cancer: Springer Verlag, Berlin, 1966.
3. Dungal, N. J.A.M.A. 1961, 178, 789-98.
4. Haenszel, W. J. Natl. Cancer Inst. 1961, 26, 37-132.
5. Fujita, S.; Takanori, H. In: Pathophysiology of Carcino-
    genesis in Digestive Organs. University of Tokyo Press,
    1977.
6. Sugimura, T.; Fujimura, S.; Baha, T. Cancer Res. 1970,
    30, 455-65.
7. Bjelke, E. Scand. J. Gastroenterol. 1974, 9, 1-235.
    (Suppl.).
8. Correa, P.; Haenszel, W.; Cuello, C.; Tannenbaum, S.;
    Archer, M. Lancet 1975, 2, 58.
9. Cuello, C.; Correa, P.; Haenszel, W.; Gordillo, G.; Brown,
    C.; Archer, M.; Tannenbaum, S. J. Natl. Cancer Inst. 1976,
    57, 1015-20.
10. Haenszel, W.; Correa, P.; Cuello, C.; Guzmán, N.; Burbano,
    L. C.; Lores, H.; Muñoz, J. J. Natl. Cancer Inst. 1976,
    57, 1021-6.
11. Correa, P.; Cuello, C.; Duque, E.; Burbano, L. C.; García,
    T.; Bolaño, O.; Brown, C.; Haenszel, W. J. Natl. Cancer
    Inst. 1976, 57, 1027-35.
12. Cuello, C.; Correa, P.; Zarama, G.; López, J.; Murray, J.;
    Gordillo, G. Am. J. Surg. Pathol. 1979, 3, 491-500.
13. Lev, R. Lab. Invest. 1965, 14, 2080-2100.
14. Spicer, S. S. J. Histochem. Cytochem. 1965, 13, 211-34.
15. Abe, M.; Ohuchi, N.; Sokano, H. Acta Histochem. Cytochem.
    1974, 7, 282-9.
16. Matsukura, N.; Suzuki, K.; Kawachi, T.; Soyagi, M.;
    Sugimura, T.; Kitaoka, H.; Numajiri, H.; Slivota, A.;
    Itahashi, M.; Hirota, T. J. Natl. Cancer Inst. 1980, 65,
    231-40.
17. Jass, J. R.; Filipe, M. I. Histopathology 198, 4, 271-9.
18. Sipponen, P.; Seppala, K.; Varisk, L.; Ihamaki, T.;
    Kekki, M.; Siurala, M. Acta Pathol. Microbiol. Scand.
    1980, Sect. A 88, 217-24.
19. Heilmann, K.; Hopker, W. Pathol. Res. Pract. 1979, 164,
    249-58.
20. Correa, P. Front. Gastrointest. Res. 1980, 6, 98-108.
21. Nomura, A.; Stemmermann, G. N.; Samloff, M. Ann. Intern.
    Med. 1980, 93, 537-40.
22. Stemmermann, G. N.; Ishidate, T.; Samloff, M.; Masuda, H.;
    Walsh, J.; Nomura, A.; Yamakawa, H.; Glober, G. Am. J.
    Dig. Dis. 1978, 23, 815-20.

23. Haenszel, W.; Correa, P. Cancer Res. 1975, 35, 3452-9.
24. Fraser, P.; Chilvers, C.; Beral, U.; Hill, M. Int. J.
    Epidemiol. 1980, 9, 3-11.
25. Hawksworth, G.; Hill, M.; Gordillo, G.; Cuello, C.
    "N-nitroso Compounds in the Environment"; International
    Agency for Research on Cancer, Scientific Publication No.
    9: Lyon, France, 1975.
26. Herfarth, C.; Schlag, P., Ed.; "Gastric Cancer"; Springer
    Verlag, Berlin, 1979; p 44, 120.
27. Druckery, H.; Steinhoff, D.; Beuthner, H.; Schneider, H.;
    Klarner, P. Arzneim. Forsch. 1963, 13, 320-5.
28. Sander, J.; Burkle, G.; Schweinsberg, F. "Topics in
    Chemical Carcinogenesis"; University Park Press: Baltimore,
    USA, 1973 ; p 297.
29. Endo, H.; Takahashi, K. Nature 1973, 245, 325.
30. Jones, S. M.; Davis, P. W.; Savage, A. Lancet 1978, 1,
    1355.
31. Nichols, J. C. World J. Surg. 1979, 3, 731-6.
32. Weisburger, J. H.; Marquardt, H.; Hirota, N.; Mori, H.;
    Williams, G. M. J. Natl. Cancer Inst. 1980, 64, 163-7.
33. Stemmermann, G. N.; Mower, H.; Rice, S.; Ichinotosubo, D.;
    Tomiyasu, L.; Hayeshi, T.; Nomura, A.; Mandel, M.
    "Gastrointestinal Cancer - Endogenous Factors"; Banbury
    Report No. 7: Cold Spring Harbor Laboratory, N. Y., 1981;
    p 175.

RECEIVED August 10, 1981.

Analytical Methods for Nitrosamines

PHILLIP ISSENBERG

Eppley Institute for Research in Cancer and Allied Diseases,
University of Nebraska Medical Center, Omaha, NE 68105

Reliable analytical methods are available for
determination of many volatile nitrosamines at
concentrations of 0.1 to 10 ppb in a variety of
environmental and biological samples. Most
methods employ distillation, extraction, an
optional cleanup step, concentration, and final
separation by gas chromatography (GC). Use of the
highly specific Thermal Energy Analyzer (TEA) as
a GC detector affords simplification of sample
handling and cleanup without sacrifice of
selectivity or sensitivity. Mass spectrometry
(MS) is usually employed to confirm the identity
of nitrosamines. Utilization of the mass spectrometer's capability to provide quantitative data
affords additional confirmatory evidence and
quantitative confirmation should be a required
criterion of environmental sample analysis.
Artifactual formation of nitrosamines continues
to be a problem, especially at low levels (0.1 to
1 ppb), and precautions must be taken, such as
addition of sulfamic acid or other nitrosation
inhibitors. The efficacy of measures for prevention of artifactual nitrosamine formation
should be evaluated in each type of sample
examined.

Reliable methods are available for determination of
nitrosamines, especially volatile nitrosamines, in a variety of
foods, environmental samples, commercial products, blood and
animal tissues. Reviews of these methods are available (1, 2)
and descriptions of some state-of-the-art procedures are
included in papers on nitrosamine occurrence in this volume.
This paper is not intended to be a comprehensive review of
historical developments or of the many variations of procedures

0097-6156/81/0174-0331$05.00/0
© 1981 American Chemical Society

currently in use, but rather will focus upon some of the major
remaining problems. Examples of these problems and some
solutions to them are taken from analytical studies in progress
in our laboratory and from the recent literature. The continued
existence of real problems in nitrosamine analysis is illustrated by the recent report (3) of up to 23 ppb of N-nitrosodimethylamine (NDMA) and 28 ppb of N-nitrosodiethylamine (NDEA)
in commercial vegetable oils, while levels less than 1 ppb were
found when special precautions were taken to avoid artifacts (4).

Major emphasis in studies of N-nitroso compounds in foods
has been placed upon volatile nitrosamines, in part because
these compounds are relatively easy to isolate from complex
matrices by virtue of their volatility. Procedures utilizing
atmospheric pressure or vacuum distillation have been used by
most investigators, with variations of the method of Fine et al.
(5) being among the most popular. This procedure employs vacuum
distillation of a mineral oil suspension of the sample with
optional addition of water to improve nitrosamine recovery from
low moisture content samples (6). The usual approach to prevention of nitrosamine formation during analysis involves adding
sulfamic acid or ascorbate to destroy residual nitrite at an
early stage of sample preparation.

Aqueous distillates are extracted, usually with dichloromethane (DCM), concentrated to small volumes, generally in a
Kuderna-Danish evaporator, and examined by gas chromatography
(GC) using a specific detection system. Additional chromatographic cleanup may be required, depending on the complexity
of the sample and specificity of the chromatographic detector.
The subnanogram sensitivity and specificity of the Thermal
Energy Analyzer (TEA) (7) make additional purification of
concentrates unnecessary for most samples. The TEA has become
the detector of choice in laboratories performing large numbers
of nitrosamine analyses because of the minimal sample preparation required prior to GC-TEA determination and the relative
freedom from false positive results experienced in its use in
many laboratories. The TEA responds to some compounds other than
those containing the N-nitroso group, but methods for tentatively confirming the identity of TEA-positive chromatographic
peaks have been described. Elimination of TEA response by
photolytic decomposition of nitrosamines by ultraviolet
radiation (366 nm) is effective in providing tentative confirmation and in identifying response from compounds other than
nitrosamines (8). Methods for classifying TEA response based on
ultraviolet irradiation and chemical procedures appear to
eliminate most ambiguities other than possible interference from
C-nitroso compounds (9).

Confirmation of identity of nitrosamines via an independent
detection system is desirable since a higher level of confidence
is achieved if a different physical property or structural
characteristic is measured. Mass spectrometry (MS) has been used

for this purpose by many investigators. Sensitivity of low
resolution MS was sufficient to confirm the identity of NDMA at
the 4 ppb level in 400 g samples of smoked marine fish (10). At
least 50 ng NDMA was required for injection into the GCMS system
to produce a mass spectrum of sufficient intensity for
unambiguous confirmation. Use of high resolution MS
(M/ΔM ≈10,000) for determination of volatile nitrosamines was
described by Telling et al. (11). Limits of 1 ppb were achieved
by measuring the NO peaks at m/z 30 from nitrosamines utilizing
high resolving power to achieve specificity rather than chromatographic cleanup of sample concentrates. Developments based on
this approach provided results comparable to those obtained by
GC–TEA by measuring the molecular ions of NDMA and N–nitrosopyrrolidine (NPYR) at m/z 74.0480 and m/z 100.0637, respectively
(12). These authors reported detection limits for NDMA in foods
of 1 µg/kg by high and low resolution MS and 0.02 µg/kg by
GC–TEA: 250 g samples were used. When a high performance mass
spectrometer was employed, a detection limit of 0.3 pg,
corresponding to 0.001 ppb, was achieved (13).

  Liquid–liquid partition is used to extract volatile
nitrosamines from aqueous distillates and as the initial
isolation step in determination of non–volatile nitrosamines.
Since vacuum distillation is often the most time–consuming and
cumbersome step in the analytical procedure, direct extraction
is an attractive alternative when sample physical characteristics and composition are appropriate. Column extraction (14) is
attractive for this application, since it permits the use of
larger volumes of extraction solvent than is feasible in
conventional batch extraction apparatus. Columns containing
kieselghur (15) were used for DCM extraction of aqueous distillates from beer, and cellulose cartridges for direct extraction
of rumen fluid with DCM (16). Column–extraction permits use of
sequential solvents for sample cleanup, avoids subjecting labile
samples to high temperatures required in distillation, and is
readily adaptable to simultaneous processing of multiple
samples. Extraction and chromatographic separation may be
combined by appropriate choice of column packing and solvents
(17). It is important to remember that most published determinations of nitrosamines in foods and beverages were obtained
using distillation and the substitution of any new initial isolation method requires careful and thorough evaluation.

  Determination of nitrosamines in air samples has revealed
significant exposure of some groups of people in occupational
settings. A variety of approaches for isolating nitrosamines
from air have been used, but sorption on a solid has emerged as
a convenient and reliable method. The sorbents employed should
be chemically inert while possessing adequate capacity, and
should permit easy elution. Thermal (18, 19) and solvent elution
(20, 21) have been used. Artifactual formation of nitrosamines
during sampling or desorption can be a problem in atmospheres

containing amines and oxides of nitrogen. A nitrosamine sampling system which eliminates in situ nitrosation has been described (20).

An internal standard is desirable in any quantitative trace environmental analysis. The ideal internal standard should behave in a manner identical to that of the analyte in all the procedures followed for isolation, purification, and determination without producing interference. This is a difficult requirement to meet for nitrosamines, especially for NDMA.

Because of the special regulatory position occupied by foods and beverages, a great deal of attention has been given to development and application of analytical procedures for them. Improved procedures have resulted in quantitation and confirmation levels in the range of 1 to 10 ppb with sample amounts of 10 to 250 g. Detection limits for foods are in the 0.1 to 1 ppb range. Detection limits of 0.1 to 1 ppm appear adequate for pesticide formulations (17, 22), while sensitivity of 0.01, 0.005, and 0.2 ppb were reported for N-nitrosodi-n-propylamine in water, soil, and crops, respectively (23). Detection limits of 0.2 to 1 ppm for N-nitrosodiethanolamine (NDELA) in cosmetic products and ingredients (24) and higher limits in metal-working fluids (25) are probably too high, in view of recent evidence that NDELA is a more potent carcinogen than previously believed (26, 27). In general, detection limits have been set by the capabilities of available analytical methods rather than by consideration of toxicological potency.

Detection limits for most analytical procedures are determined by response of the detection system, potential interferences, and by the maximum sample quantity which can be conveniently processed in the analytical laboratory. In the special case of nitrosamines, precursor amines and nitrosating agents are usually present in the sample and artifactual formation of nitrosamines during sample handling is a problem which must be recognized and controlled by the analyst. Minimizing the number of steps and amount of sample handling (28) and removing at least one of the precursors prior to analysis are the most common approaches to minimizing artifact formation. Even with these precautions, practical detection limits may be determined by levels of nitrosamines produced artifactually or present as contaminants in solvents, reagents, and apparatus. Nitrosamine contamination has been reported in deionized water (29, 30), amines (31), blood collection tubes with rubber stoppers (32), and laboratory glassware washing detergents (33). Examination of procedural control (blank) samples will usually detect contamination, but artifactual formation of nitrosamines is a more difficult problem to detect and prevent.

Experimental

Materials. All solvents were "Distilled in Glass" from
Burdick and Jackson (Muskegon, MI). All other reagents were ACS
grade, except where otherwise noted. Celite 560 (Johns Manville,
Lompoc, CA) was screened to remove particles smaller than 60
mesh and used without further treatment.
Nitrosamine standards were obtained from Thermo Electron
Corp. (Waltham, MA) and diluted to appropriate concentrations
with DCM. N-Nitrosobis(2-hydroxypropyl)amine (BHP) and NDELA
were prepared by the Eppley Institute Chemical Services Unit.

Distribution Ratios. Air–water, ether–water and DCM–water
distribution ratios were measured as described previously (34)
using GC–TEA.

Column Extraction. Aqueous samples and distillates were
added to glass chromatographic tubes or plastic syringe barrels
containing 0.5 g Celite 560 per g of sample. After 20 to 30 min
of equilibration, the columns were eluted with 100 ml of DCM
[for NDMA, NPYR and N–nitrosomorpholine (NMOR) or ethyl acetate
(for NDELA and BHP)]. Residual solvent was removed from the
columns by applying nitrogen pressure. Extracts were dried with
Na_2SO_4 and concentrated to 1 ml in a Kuderna–Danish apparatus
(NDMA, NPYR, and NMOR) in a $50°C$ water bath or in a rotary
evaporator for NDELA and BHP, using a $30°C$ water bath.
Triisopropanolamine (Eastman lot B8X) was prepared as a 1 M
solution with 0.01 M sulfamic acid added. Concentrated sulfuric
acid was added to adjust to pH 1 to 2. One portion was exposed
to ultraviolet radiation from a "Blak–Ray" B–100A lamp (Ultraviolet Products, Inc., San Gabriel, CA) for 4 hr. at a distance
of approximately 30 cm. The solution was stirred during
irradiation. Irradiated and untreated solutions were saturated
with ammonium sulfate and 10–ml samples were transferred to
columns containing 15 g Celite 560 and equilibrated for 30 min.
Columns were eluted with ethyl acetate, and the extracts were
dried and concentrated. The concentrate was transferred to a
5–ml Reacti–vial (Pierce Chemical Co., Rockford, IL) and volume
was reduced to approximately 0.1 ml in a stream of dry nitrogen.
After addition of 0.5 ml t–butyldimethylchlorosilane/imidazole
reagent (Applied Science, State College, PA), the vial was
capped and held at $60°C$ for 30 min in a heating block and then
cooled to room temperature. Hexane (0.5 ml) and either 0.5 ml
water or 0.5 ml 10% NaOH solution were added. After shaking and
separation of phases, the hexane layer was removed and transferred to a second Reacti–vial. The reaction mixture was
extracted twice more with hexane and the extracts were combined
and volume reduced to 1 ml in a stream of dry nitrogen. Portions
were taken for GC–TEA or GCMS analysis. The same derivatization
procedure was used to determine NDELA in metal working fluids

and cosmetic preparations. In some cases, it was necessary to
add a portion of Celite to these samples prior to transferring
them to the extraction columns.

Beer Samples. The beer samples were examined as part of
the American Society of Brewing Chemists (ASBC) and Association
of Official Analytical Chemists (AOAC) collaborative studies of
NDMA in beer. Duplicate samples were analyzed by the column
extraction procedure and the ASBC distillation procedure (35).
The AOAC procedure (36) was similar, except that a larger sample
(50 vs. 25 g) was examined and sulfamic acid was added to
minimize artifactual formation of nitrosamines. Both methods
utilize N-nitrosodipropylamine (NDPA) as an internal standard.
In the AOAC procedure, the beer was treated with dilute HCl and
sulfamic acid, and the added acid was then neutralized by
addition of dilute alkali. Volatile nitrosamines were collected
by atmospheric pressure distillation. The distillate was made
alkaline and extracted with DCM. The extract was dried and
concentrated to 1.0 ml and an aliquot was analyzed by GC-TEA. We
used these concentrates, without further cleanup, for evaluating
the GCMS high resolution selected ion monitoring procedure.

Direct GC Injection of Amines. Morpholine, 2,6-dimethyl-
morpholine, or pyrrolidine in methanol (2.6 to 2.8 g/ml)
solutions were injected directly into the GC inlet at 210°C.
Portions of these solutions were exposed to ultraviolet
radiation (366 nm) for 16 hours prior to injection.

GC-TEA Analysis. A Bendix model 2200 GC and Thermo
Electron model 502 TEA were used. The GC injector temperature
was 210°C. The TEA pyrolysis furnace was operated at 450°C and
the cold trap was held at -150°C in isopentane slush. Oxygen
flow to the ozonator was 20 cc/min and indicated pressure was
1.5 torr at a helium flow rate of 20 cc/min. TEA output was
processed by a digital integrator (Spectra Physics System I).
    For the AOAC beer samples, a 2 m x 2 mm glass column packed
with 8.5% Carbowax 20 M + 0.85% NaOH on 100/120 mesh Chromosorb
G was used at 130°C and a helium flow rate of 20 cc/min.
Retention times of NDMA and NDPA were 4.5 and 12.2 min,
respectively. For the ASBC collaborative study, a 1 m x 2 mm
glass column containing 6% Carbowax 20 M-TPA on 100/120 mesh
Chromosorb G was operated at 90°C with 20 cc/min helium flow
rate. Retention times were 3.6 and 11.3 min for NDMA and NDPA,
respectively. For determination of nitrosamines in amines, a 2 m
x 2 mm, 10% Carbowax 20 M-TPA on 100/120 mesh Chromosorb G
column was operated at 190°C with a carrier gas flow rate of 20
cc/min. Retention times were: NPYR, 6.6 min; NMOR, 7.4 min.
    For analysis of BHP and NDELA TBDMS derivatives, a 1 m x
2 mm nickel alloy column packed with 1% Carbowax 20 M-TPA on
100/120 mesh Chromosorb G was operated at 150°C and a carrier

gas flow rate of 20 ml/min. Retention times of BHP-TBDMS and NDELA-TBDMS were 8.1 and 10.0 min, respectively.

GCMS Analysis. The GCMS system utilized a Bendix model 2200 chromatograph connected to an AEI MS902 mass spectrometer, equipped with a high speed pumping system (9-Systems, Morristown, NJ) for the ion source. No carrier gas separator was used. For determination of nitrosamines and TBDMS derivatives of hydroxy-nitrosamines, columns and operating conditions were identical to those for GC-TEA analyses: For most work, the He flow rate was 15 cc/min and the column effluent was split 1:1 between a flame ionization detector and the mass spectrometer. The stainless steel splitter, solvent vent valve (Carle Instruments, Inc., Fullerton, CA), and associated plumbing were maintained at $190^{\circ}-200^{\circ}C$ in a Bendix valve oven which also contained a high vacuum isolation valve for the mass spectrometer and an adjustable heated glass entry tube (R.H. Allen Co., Boulder, CO). The mass spectrometer was bypassed for the time required to elute the solvent.

The mass spectrometer resolving power was set at 10,000 (10% valley). Accelerating voltage was 8 kv and electron current was 450 μa. Electron energy was adjusted to minimize helium ionization. Indicated potential was 30v. Electron multiplier voltage was 3 kv and the preamplifier resistor was 10^9 ohms. A narrow mass range, approximately 500 ppm, was scanned, using either the peak matching system of the MS902 or a micro-processor-controlled system built in our laboratory. The latter applied a linear voltage sweep to the electrostatic analyzer plates and, via the reference voltage control, to the accelerating voltage. Similar performance was obtained with both systems, but the microprocessor provided more flexibility. Perfluoroalkane-225 (PCR, Gainesville, FL) was admitted through a glass inlet system to provide reference peaks. Analytical and reference peaks for the nitrosamines studied are shown in Table I. Sample and reference peaks were scanned alternately at a repetition rate of approximately 1 sec and were monitored on an oscilloscope. When the nitrosamine peak appeared, the oscillo-graphic recorder chart drive was engaged and remained on until the peak disappeared. Nitrosamine quantities were estimated by comparing the sum of sample peak heights measured from the chart (usually 10 to 20 values) with values derived from injection of standard solutions.

Results and Discussion

Internal Standards. A compound selected as an internal standard ideally should behave in a manner identical to that of the analyte in all separation steps in the analytical process and should be measured by the same final determination method. Distillation from aqueous systems and solvent partition are the

Table I. Mass Spectral Peaks Employed for Identification
                    of Nitrosamines

Nitrosamine	Composition	m/z	Relative Intensity	Reference Peak
NDMA	$C_2H_6N_2O$	74.0480^1	100	69.9986
NMEA	$C_3H_8N_2O$	88.0637^1	70	80.9952
NPYR	$C_4H_8N_2O$	100.0637^1	90	99.9936
NDEA	$C_4H_{10}N_2O$	102.0793^1	100	99.9936
NDPA	$C_6H_{14}N_2O$	130.1106^1	20	123.9936
NDELA–TBDMS	$C_{12}H_{29}N_2O_3Si_2$	305.1716^2	50	304.9825
BHP–TBDMS	$C_{14}H_{33}N_2O_3Si_2$	333.2029^2	50	331.9871

[1] $M^+_\cdot$

[2] $M^+_\cdot - tBu$

most widely used separation methods in determination of volatile nitrosamines. Table II presents air-water and solvent-water distribution ratios measured for NDMA and two nitrosamines which have been employed as internal standards for NDMA determination.

Table II. Distribution Ratios of Some Volatile Nitrosamines

Nitrosamine	Distribution Ratio Air-Water (x 10^5)	DCM-Water	Ether-Water
NDMA	7.1	3.8	0.28
NMEA	12	15	0.99
NDPA	25	140	22

It is clear that neither NMEA nor NDPA is appropriate for an internal standard in NDMA determination if criteria are interpreted strictly, but both compounds have been used for this purpose. Addition of a nitrosamine, not normally present in the sample, is helpful in detecting any gross errors in the procedure, but the addition should not be considered to be internal standardization. Utilization of NMEA or NDPA to indicate recovery of NDMA can lead to significant errors. In most reports of the application of these "internal standards", recovery of all nitrosamines was close to 100%. Under these conditions, any added compound would appear to be a good internal standard, but none is necessary. NDMA is a particularly difficult compound for use of internal standardization because of its anomalous distribution behavior. If mass spectrometry is employed for quantitative determination, ^{2}H- or ^{15}N-labeled NDMA could be added as internal standard. Because the labeled material would coelute from GC columns with the unlabeled NDMA, this approach is unworkable when GC-TEA is employed or when high resolution MS selected ion monitoring is used with the equipment described above.

Radioisotope-labeled nitrosamines have proven valuable in development of analytical methods and for demonstrating efficiency of recovery of nitrosamines from tobacco products and smoke (37-39). The very high specific activity required for low part-per-billion determinations has discouraged most analysts from using this approach. Unless a radiochromatographic detector with adequate sensitivity is available, samples must be counted independently of the final chromatographic determination, and one of the advantages of internal standardization, correction for variation in volume injected, is lost.

Column Extraction. Results obtained by the standard ASBC distillation procedure and by Celite column extraction are compared in Table III. The ASBC method included

Table III. NDMA Concentrations Beer, Using ASBC Method and Celite Column Extraction

	NDMA Found (μg/l)	
NDMA[1] Added	ASBC Method	Celite Column Extraction
2.08	1.3	2.0
2.55	1.4	2.0
5.67	4.2	4.0
6.02	4.9	5.1
10.42	7.8	8.0
0	ND (<0.5)[2]	ND (<0.5)[2]

[1] ASBC Collaborative Study.

[2] Not Detectable.

addition of NDPA to the samples, and NDMA concentrations were computed directly using a standard curve for the GC-TEA results and also by correcting for NDPA recovery. Only uncorrected values are included in Table III. Results obtained by the Celite column extraction procedure are comparable to those for the atmospheric distillation method. We have not observed artifactual formation of NDMA on Celite 560 columns, but addition of sulfamic acid to samples prior to extraction is now a routine precaution in our laboratory. Extreme care is required in development of new methods for nitrosamines to ensure that artifacts are not produced, especially at the part per billion, or lower, levels.

Determination of Hydroxy-Nitrosamines. The column extraction procedure has proven flexible and convenient for isolating NDELA and BHP from a variety of matrices. No artifactual formation of these nitrosamines has been observed when sulfamic acid was incorporated with the sample. Addition of excess acid prevents elution of amines from the Celite column, minimizing nitrosation reactions at later stages. The triisopropanolamine sample examined contained approximately 250 mg/kg BHP (Table IV).

Table IV. Determination of BHP in Triisopropanolamine

BHP Added (μg/kg)	BHP Found (μg/kg)
0	250
0	240
500	740
500	680
UV. irradiated	ND (<70)

Recovery data indicate that little nitrosamine was lost in the procedure and none appears to have been formed. Exposure to ultraviolet radiation destroyed the BHP originally present and provided a blank sample which confirmed that no BHP was formed during the column extraction procedure.

Both NDELA and BHP may be determined by GC-TEA without derivatization, but we have observed, in agreement with the data of Ohshima et al. (40), that detection limits were 100 to 1,000 times lower when derivatives were used. Neither acetyl (40) nor trimethylsilyl (TMS) derivatives (37) provide high intensity molecular ions in the mass spectra of these compounds. We observed molecular ions of relative intensity less than 5% in the mass spectra of the TMS ethers of NDELA and BHP, and few structurally significant peaks at lower m/z values. The TBDMS ethers are as easy to prepare as TMS derivatives, but their mass spectra exhibit intense peaks (50% relative intensity) arising from loss of the t-butyl group from the molecular ion (Table I). These peaks occur in a mass range (m/z 305 for NDELA-TBDMS and m/z 333 for BHP-TBDMS) which is usually free of interferences, even in complex samples and provide for convenient confirmation of identity by high resolution selected ion monitoring.

Direct Injection of Amines. In the course of developing methods for investigation of in vivo formation of NMOR and NPYR in rats treated with precursor amines and nitrite, it was necessary to determine the contamination levels of the amines by the nitrosamines. Spiegelhalder et al. (31) reported the presence of nitrosamines in all secondary and tertiary amine samples which they examined, using vacuum steam distillation followed by extraction and GC-TEA determination.

Direct injection of 0.3 g/ml solutions of amines indicated that the morpholine contained 240 μg/kg of NMOR and the pyrrolidine contained 550 μg/kg of NPYR (Table V).

Table V. Determination of NMOR and NPYR in Amines

Treatment	NMOR in Morpholine (µg/kg)	NPYR in Pyrrolidine (µg/kg)
None	240	550
Ultraviolet[1]	ND (<70)	ND (<70)
2 ppm NaNO$_2$	190	490

[1] 366 nm for 16 hr.

The concentrations of nitrosamines were reduced to undetectable levels by ultraviolet treatment of the amine solutions and were not increased by addition of 2 ppm NaNO$_2$, indicating that the nitrosamines were present originally in the amines and were not formed in the GC injection port. Similar concentrations were found when the amine samples were analyzed using the column extraction method. Direct injection is appropriate for analysis of relatively simple mixtures, if adequate precautions are taken (41), but can result in significant artifact formation in more complex systems (42).

GCMS Analysis. High resolution MS may be used as the primary means for quantitative determination of nitrosamines (1, 11–13). Usually it is more convenient and efficient to use the TEA for this purpose and employ GCMS for confirmation of identity of nitrosamines tentatively identified by GC–TEA. Figure 1 shows narrow range high resolution scans of the molecular ion region of NDMA, recorded near the maximum of the GC peaks, present in one of the beer samples prepared in the AOAC collaborative study. The peak at m/z 74.0480 represents approximately 0.15 ng of NDMA injected on the column, corresponding to a concentration of 0.6 µg/kg of beer. Use of high resolution MS permitted confirmation of the identity and amount of nitrosamine without additional cleanup of the concentrate prepared by the AOAC method. Sample quantity requirements were comparable to those of the TEA.

At a resolving power of 10,000 or greater, sample purification is not usually required, and the combination of chromatographic and mass spectrometric resolving powers is sufficient to provide unambiguous identification. At the relatively low masses of the molecular ions of nitrosamines, interference from coeluting compounds is rare. In the special case of NDMA, (m/z 74.0480), a peak at m/z 74.0469, arising from (CH$_3$)$_3$Si^{2+}, may cause problems (43), since a resolving power of 70,000 is needed for complete separation from the NDMA molecular

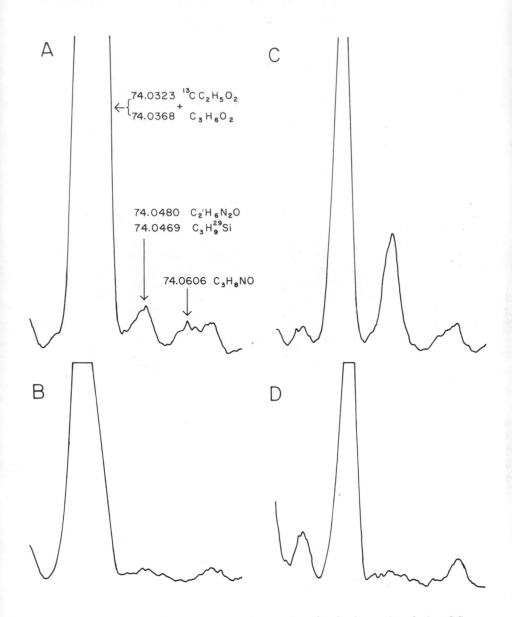

A

74.0323 $^{13}CC_2H_5O_2$
$+$
74.0368 $C_3H_6O_2$

74.0480 $C_2H_6N_2O$
74.0469 $C_3H_9^{29}Si$

74.0606 C_3H_8NO

B

C

D

Figure 1.   Narrow range mass spectral scan of molecular ion region during GC elution of NDMA.   A. Recorded at retention time of NDMA; beer sample containing 0.6 μg/kg NDMA (0.15 ng).   B. Background 1 min before elution of NDMA.   C. Standard solution containing 0.4 ng NDMA.   D. Background.

ion. The trimethylsilyl ion may arise from silicone antifoam
compounds, GC septum bleed, silanizing reagents, or other
sources, and these materials should be used cautiously or
avoided. NDMA may be separated from interfering silicone
compounds by modification of chromatographic conditions (44).
The problem is not serious when the TEA is used for initial
tentative identification and GCMS is used for confirmation,
since the TEA will not respond to the silicone compounds.

The major problem in this mode of operation is ion beam
suppression caused by extremely large quantities of coeluting
compounds. In some cases, total suppression of the ion beam has
been observed during monitoring of a GC effluent. Fortunately,
significant suppression did not occur at the retention times of
the nitrosamines of interest. Correction for moderate
suppression is achieved by measuring the decrease in intensity
of a reference peak and correcting the sample peak intensity
accordingly. In the cases of NDMA or NPYR, the sweep width may
be increased to include the PFA reference peaks at m/z 73.9968
or 99.9936 for use in normalizing observed intensities. This
approach will also improve precision by correcting for
fluctuations in ion beam intensity and electron multiplier gain.

Chromatographic resolution and mass spectrometer
performance are interrelated. Use of high resolution GC columns
minimizes the probability of ion beam suppression by coeluting
sample components. High chromatographic resolution increases
maximum concentrations of components entering the ion source and
can reduce minimum detection limits. The newer generation of
mass spectrometers provide higher sensitivity at high resolving
powers and with modern computer methods can provide highly
specific quantitative results for nitrosamines present in
femtogram quantities (13).

Low resolution MS yields specificity comparable to that of
high resolution MS, if a relatively pure sample is delivered to
the ion source. Either high resolution GC or additional sample
purification is required. To obtain sufficient specificity, it
is necessary to demonstrate that the intensities of the major
peaks in the mass spectrum are in the correct proportions.
Usually 10 to 50 ng of sample is required to establish identity
unambiguously. Use of preparative GC for purification of
nitrosamines detected by the TEA (45) is readily adaptable to
any nitrosamine present in a complex mixture and requires a
minimum of analytical method development when new types of
samples are examined.

The effort required to establish identity of a nitrosamine
in an environmental sample depends on the nature of the problem
and the specificity of the primary detection system. TEA
response is much stronger evidence of identity than response
from a flame ionization or nitrogen-specific detector. If TEA
response is supported by chemical (9) or ultraviolet photolysis
(8) supporting data, identification is adequate for many

purposes. Mass spectrometric evidence should be a requirement
for establishing identity of nitrosamines if important
commercial or regulatory decisions are based on the analytical
results.

Available detection and confirmation methods are adequate
for establishing identity and amounts of nitrosamines in most
environmental and biological samples. The validity of analytical
results, especially at levels in the part-per-billion and lower
range, depends upon the experience and skill of the analyst in
preventing or detecting contamination or artifactual formation
of nitrosamines.

Acknowledgements

The author is grateful to Mr. Doug Klein, Mr. James
Nielsen, and Ms. Evelyn Conrad for expert technical assistance.
This work was supported by Grant No. PO1-CA25100 from the
U.S. Public Health Service. We thank the National Cancer
Institute for the loan of the thermal energy analyzer under
Contract No. NO1-CP-33278.

Literature Cited

1. Egan, H.; Preussmann, R.; Walker, E.A.; Castegnaro, M.;
    Wasserman, A.E. Eds. "Environmental Carcinogens Selected
    Methods of Analysis Vol. 1 - Analysis of Volatile
    Nitrosamines in Food."; IARC Scientific Publications No. 18.
    International Agency for Research on Cancer: Lyon, 1978.
2. Walker, E.A.; Griciute, L.; Castegnaro, M.; Borzsonyi, M.,
    Eds. "N-Nitroso Compounds: Analysis Formation and
    Occurrence"; IARC Scientific Publications No. 31.
    International Agency for Research on Cancer: Lyon, 1980.
3. Hedler, L.; Schurr, C.; Marquadt, P. J. Am. Oil Chem. Soc.
    1979, 56, 681-684.
4. Fiddler, W.; Pensabene, J.W.; Kimoto, W.I. J. Food Sci.
    1981, 46, 603-605.
5. Fine, D.H.; Rounbehler, D.P.; Oettinger, P.E. Anal Chim.
    Acta 1975, 78, 383-389.
6. Hotchkiss, J.H.; Barbour, J.F.; Scanlan, R.A. J. Agr. Food
    Chem. 1980, 28, 678-680.
7. Fine, D.H.; Rufeh, F.; Lieb, D.; and Rounbehler, D.P. Anal.
    Chem. 1975, 47, 1188-1191.
8. Doerr, R.C.; Fiddler, W. J. Chromatogr. 1977, 140, 284-287.
9. Krull, I.S.; Goff, E.U.; Hoffman, G.G.; Fine, D.H. Anal.
    Chem. 1979, 51, 1706-1709.
10. Fazio, T.; Damico, J.N.; Howard, J.W.; White, R.H.; Watts,
    J.O. J. Agr. Food Chem. 1971, 19, 250-253.
11. Telling, G.M.; Bryce, T.A.; Althorpe, J. Agr. Food Chem.
    1971, 19, 937-939.

12. Gough, T.A.; Webb, K.S.; Pringuer, M.A.; Wood, B.J. J. Agr. Food Chem. 1977, 25, 663-667.
13. Webb, K.S.; Gough, T.A.; Carrick, A.; Hazelby, D. Anal. Chem. 1979, 51, 989-992.
14. Breiter, J.; Helger, R.; Lang, H. Forensic Science 1976, 7, 131-140.
15. Spiegelhalder, B.; Eisenbrand, G. Fd. Cosmet. Toxicol. 1979, 17, 29-31.
16. van Broekhoven, L.W.; Davies, J.A. J. Agric. Food Chem. 1980, 28, 957-959.
17. Day, E.W., Jr.; West, S.D.; Koenig, D.K.; Powers, F.L. J. Agric. Food Chem. 1979, 27, 1081-1085.
18. Pellizzari, E.D.; Bunch, J.E.; Bursey, J.T.; Berkley, R.E.; Sawicki, E.; Krost, K. Anal. Lett. 1976, 9, 579-594.
19. Issenberg, P.; Sornson, H. in "Environmental N-Nitroso Compounds Analysis and Formation"; Walker, E.A.; Bogovski, P.; Griciute, L. Eds. IARC Scientific Publications No. 14. International Agency for Research on Cancer, Lyon, 1976; pp. 97-108.
20. Rounbehler, D.P.; Reisch, J.W.; Coombs, J.W.; Fine, D.H. Anal. Chem. 1980, 52, 273-276.
21. Issenberg, P.; Swanson, S.E. "N-Nitroso Compounds: Analysis, Formation and Occurrence"; Walker, E.A.; Griciute, L.; Castegnaro, M.; Borzsonyi, M.; Eds. IARC Scientific Publications No. 31, International Agency for Research on Cancer: Lyon, 1980; pp. 531-540.
22. Bontoyan, W.R.; Law, M.W.; Wright, D.P., Jr. J. Agric. Food Chem. 1979, 27, 631-635.
23. West, S.D.; Day, E.W., Jr. J. Agr. Food Chem. 1979, 27, 1075-1080.
24. Rosenberg, I.E.; Gross, J.; Spears, T. J. Soc. Cosmet. Chem. 1980, 31, 237-252.
25. Williams, D.T.; Benoit, F.; Muzika, K. Bull. Environ. Contam. Toxicol. 1978, 20, 206-211.
26. Lijinsky, W.; Reuber, M.D.; Manning, W.B. Nature, 1980, 288, 589-590.
27. Habs, M.; Preussmann, R.; Schmahl, D. J. Cancer Res. Clin. Oncol. 1981, 99, A27.
28. Krull, I.S.; Fan, T.Y.; Fine, D.H. Anal. Chem. 1978, 50, 698-701.
29. Gough, T.A.; Webb, K.S.; McPhail, M.F. Fd. Cosmet. Toxicol. 1977, 15, 437-440.
30. Fiddler, W.; Pensabene, J.W.; Doerr, R.C.; Dooley, C.J. Fd. Cosmet. Toxicol. 1977, 15, 441-443.
31. Spiegelhalder, B.; Eisenbrand, G.; Preussmann, R. Angew. Chem. Int. Ed. 1978, 17, 367.
32. Lakritz, L.; Kimoto, W. Fd. Cosmet. Toxicol. 1980, 18, 31-34.
33. Pensabene, J.W.; Wasserman, A.E. Fd. Cosmet. Toxicol. 1980, 18, 329.

34. Mirvish, S.S.; Issenberg, P.; Sornson, H.C. J. Natl. Cancer Inst. 1976, 56, 1125-1129.
35. Marinelli, L. J. Am. Soc. Brew. Chem. 1980, 38, 111-113.
36. Sen, N.P., 1981. J. Assoc. Off. Anal. Chem. in press.
37. Schmeltz, I.; Abidi, S.; Hoffmann, D. Cancer Letters 1979, 2, 125-132.
38. Hoffmann, D.; Adams, J.D.; Brunnemann, K.D.; Hecht, S.S. Cancer Res. 1979, 39, 2505-2509.
39. Brunnemann, K.D.; Fink, W.; Moser, F. Oncology, 1980, 37, 217-222.
40. Ohshima, H.; Matsui, M.; Kawabata, T. J. Chromatogr. 1979, 169, 279-286.
41. Parees, D.M. Anal. Chem. 1979, 51, 1675-1679.
42. Fan, T.Y.; Fine, D.H. J. Agric. Food Chem. 1978, 26, 1471-1472.
43. Dooley, C.J.; Wasserman, A.E.; Osman, S. J. Food Sci. 1973, 38, 1096.
44. Gough, T.A.; Webb, K.S. J. Chromatog. 1973, 79, 57-63.
45. Hotchkiss, J.H.; Libbey, L.M.; Scanlan, R.A.; J. Assoc. Off. Anal. Chem. 1980, 63, 74-79.

RECEIVED August 11, 1981.

Pesticide-Derived Nitrosamines

Occurrence and Environmental Fate

JAMES E. OLIVER

Pesticide Degradation Laboratory, AEQI, USDA, SEA-AR, Beltsville, MD 20705

Possible exposure to pesticide-derived N-nitroso compounds depends on environmental processes that influence formation, movement, and degradation of the compounds. Although laboratory studies have shown the feasibility of environmental nitrosamine formation, there has been little evidence that it is an important process. Nitrosamines vary greatly in their environmental stabilities, but all seem to be susceptible to one or more modes of decomposition including photolysis, microbiological degradation, and plant metabolism.

This far into a nitrosamine symposium it should hardly be necessary to point out that nitrosamines are technically just one of a group of N-nitroso compounds that also includes nitrosamides, nitrosocarbamates, nitrosoureas, etc. Or that nitrosatable pesticides encompass all the categories just mentioned and more. Or that many diverse pesticides, including herbicides, insecticides, and fungicides have been converted to N-nitroso derivatives in the laboratory (a recent review[1] contained a 3-page, probably incomplete, compilation), or that some of the N-nitroso compounds thus synthesized were determined to be carcinogenic in test animals or mutagenic in various assays.

Although the word nitrosamine may at times be used loosely to include other classes of N-nitroso compounds, most of the nitroso compounds, with the exception of occasional reference to nitrosocarbamates, discussed in this paper are in fact nitrosamines. Beyond that they are structurally quite variable, and their environmental behaviors can be as variable as their structures.

There seem to be three general areas of possible concern over pesticide-derived nitrosamines: 1., that nitrogen-containing pesticide residues will be consumed with food products or water and subsequently nitrosated in vivo; 2., that pesticide products, as applied to plants or soils, may contain N-nitroso compounds which will in some way subsequently present a health problem; and 3., that nitrogen-containing pesticides (or impurities, metabolites, formulating agents, etc.) will somewhere in the environment be nitrosated to nitroso derivatives which will then enter the food chain or groundwater.

In Vivo Formation

Hard evidence for the first category seems to be nonexistent. An educated guess on the potential hazards would combine residue technology [how much of a secondary amine (or amide, urea, carbamate, etc.) might a person consume or otherwise be exposed to?], nitrosation chemistry (what would be the yield of in vivo nitrosation of the pesticide thus consumed?), and toxicology (what would be the toxicological effect and potency of the nitroso compound thus formed?). Frequently, these questions, which simplify to, "What dose--eg., in mg/kg--of a pesticide-derived nitroso compound might a person be exposed to and what would be the result if he were?" are not carefully considered. It has been popular for several years to dose rats or other animals with nitrosatable amines and sodium nitrite and demonstrate that the treated animals develop tumors not seen in the controls. Pesticides, in particular the carbamate insecticides, have not been exempt from these experiments; the difficulty comes in translating the results into environmentally meaningful information. In a recent Ph.D. thesis[2], R. W. Rickard calculated that since the tolerence for carbaryl residues on many fruits and vegetables was 10 ppm, consuming 100 g of produce containing permissible levels of carbaryl would expose the consumer to 5 μmol of the insecticide. He acknowledged that market-basket surveys showed that actual exposure to carbaryl from food would be much less than the tolerences allow, but he suggested that exposure during application, formulation, etc. could in certain instances be greater than the intake via food. Measured in vivo yields from nitrosation of carbaryl were 0.09% to 1.54% in guinea pigs' stomachs containing 145 μmol to 1160 μmol, respectively, of $NaNO_2$. A 1% nitrosation yield (most of the yields in guinea pigs were lower) in a 70 kg man having consumed the 100 g of produce containing 5 μmol of insecticide would present the subject with a 0.16 μg/kg dose of nitrosocarbaryl. Any discussion of the effect of 0.16 μg/kg doses of nitrosocarbaryl is far beyond the scope of this paper; however, for the purpose of comparison, single oral doses of nitrosocarbaryl in rats as high as 1500 mg/kg failed to produce any tumors within 21 months[3] although twice weekly doses of 130 mg/kg (average total doses 5000

mg/kg) over 200 days killed most of the treated rats, many of
which had developed carcinomas, papillomas or hyperkeratoses.[4]

Coapplication of Nitrosamines and Pesticides

Several aspects of the problem of herbicides being contaminated with nitrosamines, and the resulting inadvertent introduction of nitrosamines into the environment, will be discussed in
other papers in this symposium. Unrecognized until less than
five years ago, the situation has inspired intense debate and
prompted several of the environmental chemistry studies mentioned
in this paper. Like the presumed threat from the in vivo nitrosation of pesticide residues, discussions sometimes lack the type
of anticipated dose and effect calculations just mentioned. Unlike the active ingredients, whose benefits can justify residue
tolerances and acceptable daily intakes, nitrosamine contaminents
afford no known benefits, and the desirability of minimizing
their levels is undisputed.

Environmental Formation

The bulk of this paper will be concerned with the prospects
of N-nitroso compound formation in the environment, and with environmental behaviors of selected nitroso compounds, as best we
can describe or predict them from experimental work completed
thus far. Obviously, once in the environment, a compound will to
a large extent be subject to the same conditions whether it was
formed there or introduced as a pesticide contaminant, and in
this manuscript no attempt to differentiate between the two modes
of introduction has been made. Such distinctions could, however,
influence the location of a compound in the environment--say on
a plant or soil surface if sprayed with a pesticide, admixed with
soil if transported by leaching, etc., and some of the experiments cited will have been conceived with one or the other of the
introduction modes in mind.

Nitrite (NO_2^-), an intermediate in the soil nitrogen cycle,
would seem to be the most probable nitrosating agent if nitrogen-
containing pesticides are to be converted to nitroso derivatives
in the environment. Under normal conditions, however, nitrite is
short-lived in soils and its concentration is low--frequently undetectable. It has been suggested that relatively high NO_2^-
levels might result from applications of nitrate or ammonium
fertilizers[5], and that nitrite accumulates in alkaline soils
during nitrification.[6] Indeed, Cochran, et al[7] were able to
measure transient nitrite accumulation within 5 cm of sites of
application of gaseous ammonia to soils (but at greater distan-
ces NO_2^- was not detectable). Chapman and Liebig[8] found that
heavy applications of urea, which provided high levels of ammonia
and alkaline conditions, permitted nitrite to accumulate and persist for several months, especially in cold weather. In contrast,

Sander and Schweinsberg[9] reported that nitrite levels in soil
were low even in the presence of large amounts of nitrogen fertilizers. The general rule seems to be that NO_2^- levels in
normal, aerobic soils are undetectably low, but they may temporarily increase in special circumstances.

The chemical transformations of nitrite in soils has been
rather extensively studied by J. M. Bremner and coworkers[10]
who determined that, at least in sterile soils, pH and organic
matter largely determined NO_2^--fixation and decomposition.
Substantial amounts of nitrogen (N_2) and nitrogen dioxide (NO_2),
and small amounts of nitrous oxide (N_2O) were found in neutral
and acidic soils treated with nitrite. The amount of NO_2 found
was inversely related to pH and was not related to organic matter. In contrast, N_2-formation was dependent on both pH and
organic matter (lower pH's and higher levels of organic matter
favored N_2-production). In soils with pH's above 6, nitrite
was found to be reasonably stable. It has been noted[11] that
in aqueous solutions at high pH, nitrite is easily oxidized to
nitrate, even by molecular oxygen, but there doesn't seem to be
much reason to believe that reaction is important in soils.

The Van Slyke Reaction

Nitrosation of amines by soil nitrite is by no means a recent concept. The Van Slyke[12] reaction, summarized as

$$R-NH_2 + HNO_2 \longrightarrow ROH + H_2O + N_2,$$

was developed in 1911 as a method for quantitative determination
of primary amino groups, especially in amino acids, and has long
been proposed as a pathway by which soil nitrite would be consumed (urea, R = $CONH_2$, and ammonia, R = H would be just two
components of soil that could be nitrosated). Bremner and Nelson[10] concluded, however, that in fact the Van Slyke reaction
was not significant in soils. Several observations supported
their conclusion. For example, nitrite recovery was not affected by either urea or alanine in the soils, and no interaction
between nitrite and ammonia in acidic, neutral, or alkaline
soils was observed under incubation conditions. Only when neutral or alkaline soils pretreated with NH_4^+ and NO_2^- were air
dried was the reaction observed at all. Although self-decompos-
ition of nitrite in acidic soils could occur, the authors concluded that organic matter in the soil was very largely responsible for the fixation of nitrite-nitrogen and the formation of
nitrous oxide (N_2O) and nitrogen (N_2). Specifically, phenolic
components of the soil organic matter were responsible, with nitrosophenols being proposed as intermediates (phenolic compounds
have been recognized as inhibitors of amine nitrosation in various media--for a brief reveiw see Douglass, et al[11]--and also
as catalysts for certain nitrosations[13]). As I have pointed out

before[14], for a pesticide to be nitrosated in soil, not only would it have to coexist with the limited nitrite, it would have to compete with all of the other nitrogenous, sulfur-containing, and phenolic components also present. An example would be the circumstances just mentioned wherein NO_2^- temporarily accumulated following urea or NH_3-application. The secondary amine, presumably present at low levels, would have to compete with all the urea and ammonia for NO_2^- if a nitrosamine were to be formed.

Other Nitrosating Agents

Nitrosating agents other than nitrite are worthy of consideration. Nitric oxide (NO) may be capable of nitrosating amines under selected anaerobic conditions[15] but is rapidly oxidized to nitrogen dioxide (NO_2) in air. The latter, presumably via dimerization and/or subsequent transformations, can act as a nitrosating agent; in fact, a recent patent[16] describes the reaction of NO + NO_2 with aniline to produce triazenes and related products resulting from diazatized (i.e., nitrosated) aniline. The same combination (NO + NO_2) has been proposed to achieve nitrosation of amines in tobacco smoke[17] and of the triazine herbicides atrazine and simazine as well as the carbamate insecticide carbaryl.[18] In the latter experiments, dry powdered pesticides in petri dishes were exposed to atmospheres of NO + NO_2 (NO_x). Rather low yields of nitrosoatrazine and nitrososimazine, and lower yields of nitrosocarbaryl, were formed. The yields were inversely proportional to moisture levels in the air. Additional studies were conducted with NO_x and aqueous suspensions of the pesticides. At pH 1, 200, 500, and 1000 ppm of the nitroso derivatives of simazine, atrazine, and carbaryl, respectively were found; as the pH was increased, however, the yields diminished rapidly. The authors concluded that the nitrosations would not be expected to occur in soils under field conditions, and they related their experiments to possible situations where solid pesticides might be stored. The same authors mentioned that typical levels of NO_x in air in rural areas is less than 1 ppm, although ten-fold higher levels might be detected in polluted urban airs. Since NO_x is a byproduct of fossil fuel combustion, concern has been expressed over its reaction in the atmosphere with amines, but air samples from areas where concentrations of both amines and NO_2 were relatively high were largely negative when examined for nitrosamines.[19] Other nitrogen oxides--N_2O_3 and N_2O_4--readily nitrosate amines under laboratory conditions, but are not known to be widely distributed in the environment. Both can dissociate to the monomeric nitrogen oxides, and in the presence of water, would presumably end up as nitrite and/or nitrate. All of the nitrosating agents would be both oxidants and electrophiles, and their nitrosation reactions would be subject to all of the competitive reactions and reductions described for nitrite. Transnitrosation, from either other N-nitroso compounds or other nitroso compounds such as nitrosophenols [13], is another possibility, but, of course there still would have to be a source of nitrosating agent to produce the nitrosophenol.

Nitrosation in Waters

It was mentioned that nitrite does not seem to accumulate in soils under most conditions, and the same generalization seems to be true for natural waters. Dressel[20] reported NO_2^- levels of up to 1 ppm in rivers (although considerably higher levels were detected in effluents of an experimental sewage plant and of a galvanizing plant). Keeney, et al.[21] found that NO_2^- did not accumulate in lake sediments under conditions where nitrate rapidly disappeared. The low levels of both nitrite and nitrosatable amines in natural waters, along with expected third order kinetics[22] make extensive nitrosamine formation unlikely. The prospect has been discussed by Dressel.[20]

Microbiological Intervention

Most of the considerations developed to this point have concerned only chemical reactions, and some, particularly those of Bremner and coworkers, were derived in artificial or sterilized media. Since the environment is by no means sterile, the role of microorganisms in environmental nitrosamine formation (and disappearance) must be investigated. Many microorganisms are known to reduce NO_3^- to NO_2^-; for example, Bollag, et al.[23] isolated about 60 microorganisms from soil with nitrate reductase activity. Nitrate was completely consumed in 48-72 hours by all cultures under anaerobic conditions, but only six strains completely reduced the NO_3^- in the presence of oxygen. In some cultures NO_2^- tended to accumulate, but in others it was reduced further. The authors concluded that temporary or remaining NO_2^- accumlation could in certain cases occur when NO_3^- is reduced in a nonoptimal environment. Just as in the chemical nitrogen cycle, nitrite does not usually persist in microbiological systems. However, several factors were defined by Bollag and coworkers that affect the transformations of NO_2^- and thus promote its accumulation. Certain pesticides, including the fungicides nabam and maneb, were cited.[24] Chlorinated aniline derivatives of certain pesticides were reported[25] to cause temporary nitrite accumulation under anaerobic conditions. Certain heavy metals also caused NO_2^- accumulation in anaerobic soils.[26]

If the only role of soil microorganisms was to provide nitrite from nitrate, they could perhaps be dismissed, at least if one accepts the premise of Bremner and Nelson that whatever its source, nitrite is consumed via interactions with soil organic matter and not via nitrosations of nitrogen compounds. Some controversy remains, however, over the precise role of microorganisms in nitrosamine formation. Nitrosations have been observed at near neutral pH's in the presence of bacteria[27], and it was reported that nitrosamine formation was more rapid in the presence of growing bacteria than in their absence[28]. Although nitrate reductase was a requirement for nitrosamine formation in bacteria, not all nitrate-reducing bacteria were capable of nitrosating amines; this observation was cited as support for an active role of the bacteria in nitrosamine synthesis[29]. Ralt and Tannenbaum suggested in this symposium, however, that production of nitrite is in fact the main role of bacteria; they also pointed out that the medium can influence whether nitrosamines are formed. Alexander and coworkers[30,31] have examined nitrosamine formation in soils and natural waters and have attempted to define the roles of microorganisms, but truly definitive results have remained elusive.

Whatever the determining factors, examples of "environmental" nitrosations of nitrogen compounds have been observed in various laboratories. Small amounts of N-nitrosodimethylamine (NDMA) were observed in soils and sewage incubated with dimethylamine and nitrate; in autoclaved soils or sewage no NDMA was detected.[30] In subsequent experiments[31], nitrosamine formation (NDMA from added nitrite and dimethylamine) was observed in both sterile and nonsterile soil and sewage; it was not observed in soil in the absence of organic matter, however, and the organic matter was proposed to be the essential ingredient. Pancholy[32] also observed NDMA formation in soils incubated with both nitrite and dimethylamine, with little NDMA forming in autoclaved soils. N-Nitrosoglyphosate was formed in soils treated with high levels of glyphosate and nitrite[33], and that subject is being discussed elsewhere in this symposium by Dr. Khan. In our own laboratory, nitroso derivatives of both atrazine[34] and butralin[35] were formed in soils amended with the herbicides plus nitrite but not in soils containing the herbicides plus nitrate. Nitrosocarbarmates were not detected in three soils treated with sodium nitrite and any of three carbamate insecticides, carbaryl, carbofuran, and propoxur.[36] As mentioned earlier, literally dozens of pesticides of all types have been converted to N-nitroso compounds in various laboratories, but almost all of the nitrosamine formations observed in natural media have required substantial (i.e. artifically high) levels of added nitrite. Thus many publications have treated the feasibility of environmental nitrosamine formation, but few have really documented it.

Nitrosamines and Plants

This relationship, studies less than that of nitrosamines and microorganisms, offers similar possibilities. Plants could presumably affect nitrosation of secondary amines under certain circumstances; conversely, they might also cause their degradation. The former is an area of concern, because, of course, plants could provide a direct entry of nitrosamines into the food chain. An interesting example has been published by Schmeltz et al.[37] who found N-nitrosodiethanolamine in tobacco that had been treated with the diethanolamine salt of maleic hydrazide. Evidently the amine was taken up by the tobacco plants and nitrosated there, perhaps during the tobacco-curing process. The question of nitrosamines in crop plants is another on which very limited data are available (see also Dr. Khan's paper in this symposium). In principle, nitroso compounds could be taken up into plants from soil. Alternatively, the plant could acquire the amine from the soil or, some other source, and perform the nitrosation itself. The feasibility of nitrosamine uptake has been demonstrated. Nitrosamines were taken up from water by cress[38], but rapidly disappeared from the plant if not continuously replenished. Uptake of NDMA by lettuce and spinach was demonstrated.[39] On other hand, no nitrosamines were detected in wheat grown in soils treated with high levels of nitrogen fertilizers and either of three amines, dimethylamine, N-methyl-N-benzylamine, and N-methylaniline.[38] Essentially no radioactivity from NDPA-^{14}C was observed in soybean plants grown from seed in soil containing that nitrosamine[40], but when 10-day old beans were transplanted to nutrient solution containing NDPA-^{14}C, radioactivity was translocated into the plant parts.[41] From a 0.17 ppm concentration of NDPA in the nutrient solution, roots and shoots aquired 0.34 and 0.44 µg equivalent of NDPA per gram of fresh weight; however, upon fractimation, 82% of the radioactivity stayed in a polar aqueous fraction, 9% remained with plant residue, and only 9% was extracted into organic solvents. Furthermore, only part of the extractable 9% cochromatographed with NDPA on TLC. Thus the soybean plant was evidently able to transform the nitrosamine into more polar materials. Radioactivity from N-nitrosopendimethalin-^{14}C was observed in soybean plants grown from seed in soil containing that nitrosamine.[40] The radioactivity did not seem to be associated with the nitrosated herbicide, however, for thin-layer chromatography and autoradiography indicated a highly polar material. Since N-nitrosopendimethalin is relatively stable in aerobic soil[42], it appears that here, too, some transformation(s) had taken place in the plant. There is currently too little information available to draw any general conclusions about either formation or metabolism of nitrosamines by plants.

Metabolism of Nitrosamines by Microorganisms

The metabolism of nitroso compounds by microorganisms in the environment has attracted less attention than formation of the same compounds by the organisms. Tate and Alexander[43] reported that NDMA, N-nitrosodiethylamine (NDEA), and NDPA were resistant to metabolism by microorganisms from soil or sewage. the same authors[44] reported that the same three nitrosamines were stable in lake water for more than three months, that they disappeared slowly from soil after a 30-day lag period (the actual data indicate this to have been true only for NDMA, the others beginning to disappear earlier), and that they disappeared slowly from sewage. Comparisons between natural and autoclaved soils and sewage implied the involvement of microorganisms in these processes. N-Nitrosodiethanolamine was slowly metabolized by these microorganisms in sewage and, under some conditions, in lake water.[45] In the latter case, the stability of the nitrosamine depended on the time of year the water was collected; it was stable in water collected in winter, less stable in fall-collected water, and least stable in summer- collected water. NDMA was found to be stable in aerobic soil over a two month period.[43] Saunders et al.[46] found NDPA to degrade in anaerobic soil, but more slowly then in aerobic soil. Oliver, et al.[42] found that ^{14}C-labeled NDMA, NDEA, and NDMA were all converted to $^{14}CO_2$ in aerobic soils with half lives of approximately three weeks. Microorganisms were evidently involved, since no $^{14}CO_2$ was evolved from soils sterilized by either heat or ethylene oxide. The rate of $^{14}CO_2$-evolution from NDPA was independent of the location--i.e., C_1, C_2, or C_3--of the ^{14}C in NDPA which suggested that no stable intermediates were formed between the time of initial attack and final CO_2 production, and no evidence of intermediates was obtained by extraction of soils and TLC-autoradiography. The Saunders, et al.[46] experiments also showed NDPA to be degraded in aerobic soils. The N-nitroso derivatives of the carbamate insecticides carbaryl, carbofuran, and propoxur were very unstable in soil in the dark (about 80% disappearance in 12 hours) and even more unstable in the presence of sunlight.[36] The nature of the decomposition was not described. The N-nitroso derivatives of the dinitroaniline herbicides butralin[35] and pendimethalin[42] were quite stable in aerobic soils, persisting over several months. An aerobic actinomycete of the Streptomyces genus was isolated from soil that metabolized N-nitrosopendimethalin in laboratory culture.[47] Both oxidative (ring $CH_3 \longrightarrow CH_2OH$) and reductive ($-NO_2 \longrightarrow -NH_2$) transformations occurred, but the N-nitroso group was not cleaved or altered in the identified metabolites. N-Nitrosopendimethalin was rapidly metabolized in anaerobic soil[48]; here again, reduction of a nitro group was observed and again the N-nitroso functionality was unchanged. N-nitrosoatrazine was relatively unstable in aerobic soil (only 12% could be recovered after a month) as well as in an aquatic ecosystem.[34] A major decomposition pathway of nitrosoatrazine was denitrosation back to the original herbicide.[33] This may be the only reported example of denitrosation under "environmental" conditions (nitrosomethomyl, a particularly unstable derivative of an insecticidal oxime carbamate,

decomposed in part to methomyl in distilled water and in 0.1 N
HCl[49]) and it was not reported or suggested that the denitrosation was a microbiological reaction.

The safest conclusion to draw at this point is that it is
unsafe to generalize about the environmental stabilities of
N-nitroso compounds.

Volatilization

Other environmental properties of interest are those that
govern movement of chemicals, for these properties can influence
not only the possibility of human exposure but also the lifetime
and fate of the chemical. Clearly, if a nitrosamine is formed
in, or introduced into, the soil and stays there, it presents
little threat to man, and its lifetime will depend on the chemical or microbiological properties of the soil. If it should
move to the surface and volatilize into the atmosphere, on the
other hand, there will exist the possibility of human exposure
via inhalation and also the possibility of vapor-phase photodecomposition. If a nitrosamine were to leach from soil into
water, it could perhaps be consumed in drinking water; alternatively, exposure of the aqueous solution to sunlight could
provide another opportunity for photodecomposition.

Volatilization of NDPA, NDEA, NDMA, and N-nitrosopendimethalin were examined in a model system.[50] The nitrosamines were
either mixed into predetermined depths of the soil or applied to
the soil surface (the conditions were chosen to represent those
that would be encountered by nitrosamines coapplied with dinitroaniline herbicides). Volatilization of nitrosopendimethalin
was extremely slow regardless of application. The volatile nitrosamines, NDPA, NDEA, and NDMA, in contrast, volatilized so
rapidly after application to the surface of moist soil that we
predicted that a substantial proportion of the nitrosamine thus
applied would enter the atmosphere within a few hours. Incorporation of the nitrosamine in the top 7.5 cm of soil (as might
be the case when the herbicide was applied and incorporated in a
single operation) decreased total volatilization by at least an
order of magnitude.

If one pound of trifluralin containing 1 ppm NDPA were applied to, and left on, the surface of a one-acre field, 400 µg
NDPA could conceivably enter the atmosphere over a several hour
period. To calculate how much of that could be inhaled by a
single person would require, among other things, a knowledge of
dispersion rates, but without them it is apparent that some very
small numbers would be encountered. We must resist the temptation, however, to casually equate small numbers with negligible
risk without solid justification.

Leaching

The possibility of several nitrosamines--NDMA[39], NDPA[46], and nitrosoatrazine[34]--leaching from soil has been proposed. Deriving predicted concentrations in water would be difficult even if we had something more than predicted concentrations in soil to begin with. Here again, there might be the temptation to find comfort in the anticipated low concentrations, but the reported stabilities of some nitrosamines in water along with the (undemonstrated) possibility of renewable supplies might be warnings to not become too complacent.

Photolysis

Vapor phase photolysis of NDPA proceeds rapidly in the presence of oxygen[51,52]. A half-life of 10-20 min, depending on sunlight intensity, was estimated in one study[52] conducted outdoors; the half-life measured by Crosby, et al.[51] was longer, but the apparatus used in that study was not designed for determining environmental lifetimes. Both investigations showed the primary product to be N-nitrodipropylamine; the nitramine was more stable than the nitrosamine, but did eventually degrade upon further irradiation. These results were consistent with those obtained with vapor-phase photolysis of NDMA[53,54]. Saunders and Mosier[55] studied the photolysis of NDPA by sunlight in lake water and by sunlamps in distilled water. They found fairly rapid NDPA decomposition in both cases, and in contrast to published work with other nitrosodialkylamines, did not observe any pH effects in the range pH 3-9. The primary product in water was 1-aminopropane; dipropylamine was also observed. In another study, exposure of water containing NDEA to sunlight caused a rapid decrease in the NDEA concentration (to near zero after 16 h)[20]. Wolfe, et al.[56] found N-nitrosoatrazine to be photolabile in natural water with both denitrosation and dealkylation being identified as decomposition pathways.

Conclusion

Perhaps the ultimate challenge to an author writing on this subject would be to construct a summary that did not depend very largely on qualifying statements, moderating adverbs, or the subjunctive mood. Just as it is dubious science to prematurely warn of an imminent hazard where one may or may not exist, so is it equally unsatisfactory to rely on evasive and judgmental expressions such as "insignificant amounts", "negligible exposures", or "undetectable levels." That it is desirable to minimize exposure to N-nitroso compounds is noncontroversial. That little exposure to N-nitroso compounds is likely to result from normal pesticide use is controversial, not because the environmental chemistry studies have yielded threatening or equivocal results, but because no one can define how little exposure to a carcinogen is little enough. Such a definition would probably constitute the greatest breakthrough yet experienced in this area, but it does not appear imminent. Meanwhile, decisions have to be made. Regulatory decisions, which cannot wait for the big breakthrough, are discussed in this program by Dr. Garner. We are not in a position to announce that this is not a problem area, and are not convinced that it is, so, by admitting that the definitive answer doesn't exist, we try to postpone a judgment on whether pesticide-derived nitrosamines could constitute a hazard. But we still have to face the situation from another angle. Do these problems constitute a higher research priority than unrelated problems we might otherwise address with our limited resources? Based on our analysis of the data and the thoughts I've presented, we've placed our nitrosamine program on hold, probably to be reactivated only in the event of new evidence that would dictate further reevaluation of our priorities.

Literature Cited

1. Nitrosamines and Pesticides: A Special Report on the Occurrence of Nitrosamines as Terminal Residues Resulting from Agricultural Uses of Certain Pesticides (10), P. C. Kearney, Project Coordinator, Pure and Appl. Chem., 52, 499 (1980).
2. R. W. Rickard, Chemical Properties and Potential Toxicological Significance of N-Nitrosocarbamates, Ph. D. Dissertation, University of Kentucky, 1979.
3. G. Eisenbrand, O. Ungerer, and R. Preussmann, Food Cosmet. Toxicol., 13, 365 (1975).
4. G. Eisenbrand, D. Schmahl, and R. Preussmann, Cancer Lett., 1, 281 (1976).
5. R. L. Tate and M. Alexander, J. Nat. Cancer Inst., 54, 327 (1975).
6. B. J. Stojanovic and M. Alexander, Soil Sci., 86, 208 (1958).
7. V. L. Cochran, F. E. Koehler, and R. I. Papendick, Agron. J., 67, 537 (1975).
8. H. D. Chapman and G. F. Liebig, Jr., Soil Sci. Soc. Proc., 276 (1952).
9. J. Sander and F. Schweinsberg, Zbl. Bakt, I. Abt. Orig. B, 156, 299 (1972).
10. J. M. Bremner and D. W. Nelson, 9th Int. Congr. Soil Sci. Trans. Vol. II. pp. 495-503 (1968).
11. M. L. Douglass, B. L. Kabacoff, G. A. Anderson, and M. C. Cheng, J. Soc. Cosmet. Chem., 29, 581 (1978).
12. D. D. Van Slyke, J. Biol. Chem., 9, 185 (1911).

13. R. Davies, M. J. Dennis, R. C. Massey, and D. J. McWeeny in Environmental Aspects of N-Nitroso Compounds, IARC Scientific Publications No. 19, E. A. Walker, L. Griciute, M. Castegnaro, and R. E. Lyle, Eds., Lyon, 1978, p. 183.
14. J. E. Oliver, CHEMTECH, 9, 366 (1979).
15. B. C. Challis, A. Edwards, R. R. Hunma, S. A. Kyrtopoulos, and J. R. Outram in Environmental Aspects of N-Nitroso Compounds, IARC Scientific Publications No. 19, E. A. Walker, L. Griciute, M. Castegnaro, and R. E. Lyle, Eds., Lyon, 1978, p. 127.
16. J. M. Detrick and E. F. Herbes, Brit. 1,569,301, 11 June 1980; cf Chem. Abstr., 93: 23901p (1980).
17. G. Neurath, B. Pirmann, W. Luttick, and H. Wichern, Beitr. Tabakforsch., 3, 251 (1965).
18. C. Janzowski, R. Klein, and R. Preussmann. 6th Meeting on Analysis and Formation of N-Nitroso Compounds. Budapest, Hungary, Oct. 16-21, 1979. (in press).
19. J. B. Cohen and J. D. Bachman in Environmental Aspects of N-Nitroso Compounds, IARC Scientific Publications No. 19, E. A. Walker, L. Griciute, M. Castegnaro, and R. E. Lyle, Eds., Lyon, 1978, p. 357.
20. J. Dressel, Landwirtsch. Forsch. Sonderh., 28, 273 (1973).
21. D. R. Keeney, R. L. Chen, and D. A. Graetz, Nature, 233 5314, 66 (1971).
22. S. S. Mirvish, Toxicol. Appl. Pharmacol., 31, 325 (1975).
23. J. M. Bollag, M. L. Orcutt, and B. Bollag, Soil Sci. Soc. Amer. Proc., 34, 875 (1970).
24. J. M. Bollag and N. M. Henninger, J. Environ. Qual., 5, 15 (1976).
25. J. M. Bollag and E. J. Kurek, Appl. Environ. Microbiol., 39, 845 (1980).
26. J. M. Bollag and W. Barabasz, J. Environ. Qual., 8, 196 (1979).
27. J. Sander, Z. Physiol Chem., 349, 429 (1968).
28. M. J. Hill and G. Hawksworth in N-Nitroso Compounds in the Environment, IARC Scientific Publications No. 9, P. Bogovski and E. A. Walker, Eds., Lyon, 1975, p. 220.
29. M. J. Hill and G. Hawksworth in N-Nitroso Compounds Analysis and Formation IARC Scientific Publications No. 3., P. Bogovski, R. Preussmann, and E. A. Walker, eds., Lyon, 1972, p. 116.
30. W. Verstraete and M. Alexander, J. Appl. Bacterial. 34, IV (1971).
31. A. L. Mills and M. Alexander, J. Environ. Qual. 5, 437 (1976).
32. S. K. Pancholy, Agronony Abstr., Ann. Meeting, Houston, TX (1976).

33. S. U. Khan and J. C. Young, J. Agric. Food Chem., 25, 1430 (1977).
34. P. C. Kearney, J. E. Oliver, C. S. Helling, A. R. Isensee, and A. Kontson, J. Agric. Food Chem., 25, 1177 (1977).
35. J. E. Oliver and A. Kontson, Bull. Environ. Contamin. Toxicol., 20, 170 (1978).
36. J. Miyamoto, unpublished results cited in reference 1.
37. I. Schmeltz, S. Abidi, and D. Hoffmann, Cancer Lett. 2, 125 (1977).
38. J. Sander, F. Schweinsberg, J. LaBar, G. Burkel, and E. Schweinsberg, Gann, 17, 145 (1975).
39. D. Dean-Raymond and M. Alexander, Nature, 262, 394 (1976).
40. P. C. Kearney, J. E. Oliver, A. Kontson, W. Fiddler, and J. W. Pensabene. J. Agric. Food Chem., 28, 633 (1980).
41. D. F. Berard and D. P. Rainey, Bull. Environ. Contamin. Toxicol., 23 136 (1979).
42. J. E. Oliver, P. C. Kearney, and A. Kontson, J. Agric. Food Chem., 27, 887 (1979).
43. R. L. Tate, III, and M. Alexander, J. Environ. Qual., 5, 131 (1976).
44. R. L. Tate and M. Alexander, J. Natl. Cancer Inst., 54, 327 (1975).
45. J. R. Yordy and M. Alexander, Appl. Environ. Microbiol., 39, 559 (1980).
46. D. G. Saunders, J. W. Moiser, J. E. Gray, and A. Loh, J. Agric. Food Chem., 27, 584 (1979).
47. W. R. Lusby, J. E. Oliver, R. H. Smith, P. C. Kearney, and H. Finegold, J. Agric. Food Chem., 29. 246 (1981).
48. R. H. Smith, J. E. Oliver, and W. R. Lusby, Chemosphere, 855 (1979).
49. J. C. Y. Han, J. Agric. Food Chem., 23, 892 (1975).
50. J. E. Oliver, J. Environ. Qual., 8, 596 (1979).
51. D. G. Crosby, J. R. Humphrey, and K. W. Moilanen, Chemosphere, 9, 51 (1980).
52. P. Mazzocchi, unpublished results cited in ref. 1.
53. P. L. Hanst, J. W. Spence, and M. Miller, Environ. Sci. Technol., 11, 403 (1977).
54. J. N. Pitts, Jr., D. Grosjean, K. VanCauwenberghe, J. P. Schmid, and D. R. Fitz, Environ. Sci. Technol., 12, 946 (1978).
55. D. G. Saunders and J. W. Moiser, J. Agric. Food Chem., 28, 315 (1980).
56. N. L. Wolfe, R. G. Zepp, J. A. Gordon, and R. C. Fincher, Bull. Environ. Toxicol., 15, 342 (1976).

RECEIVED July 30, 1981.

Reduction of Nitrosamine Impurities in Pesticide Formulations

G. W. PROBST

Lilly Research Laboratories, Elanco Products Company, Indianapolis, IN 46206

The discovery of nitrosamine contaminants in some pesticides has led to a major technological effort to prevent, reduce or eliminate formation of these inadvertent impurities in pesticide formulations. Although the nitrosamine contaminants occur at low levels (in parts per million), many have been found to cause cancer in laboratory animals and may present a hazard to pesticide users. In pesticides, nitrosamines are formed in several ways: by the action of nitrosating agents on secondary amines in manufacturing processes, the use of nitrite or nitrate as container corrosion inhibitors or as impurities in amine reagents used in synthesis. Reduction of nitrosamine formation can be accomplished by removal of nitrosating agents, for example, nitrogen oxides or their precursors, nitrite and nitrate. Aeration and scrubbing of the intermediate nitration products effectively removes potential nitrosating sources. Destruction of trace quantities of nitrosamines in technical materials, like dinitroanilines, is accomplished by treatment with hydrogen chloride gas, hydrochloric acid, and hydrobromic acid, or by the use of aliphatic ketones and aldehydes in the presence of a strong acid. Agents which are less efficient include halogens (e.g. molecular bromine and chlorine gas), inorganic acid halides (e.g. phosphorus oxychloride and thionyl chloride), N-bromo-succinimide and others. The mechanisms, as well as scope and limitations of removal, will be reviewed.

0097-6156/81/0174-0363$05.00/0
© 1981 American Chemical Society

Formation of N-nitroso compounds has been observed from all types of amines, amides, quaternary ammonium compounds and many other nitrogen-containing compounds (Mirvish, 1975). Their carcinogenic potential (Magee and Barnes, 1967) has evoked much concern not only from preformed N-nitroso compounds in food (Lijinsky and Epstein, 1970), but as amine or amide precursors in the form of drugs (Lijinsky, 1974; Lijinsky et al, 1975) and as pesticide residues consumed in food (Ceaborn, Radeleff and Bushland 1960; Maier-Bode 1968). The presence of volatile nitrosamines in pesticide formulations was first reported by Ross (1976) at the 172nd National Meeting of the American Chemical Society in San Francisco. The nitrosamines were detected with Thermal Energy Analyzer (TEA) coupled with an isothermal gas chromatograph. The TEA was developed by Fine et al (1973) and Fine and Rufeh (1974). The instrument is highly specific and sensitive for nitrosamines and is adaptable for the direct testing of pesticide formulations. Of the pesticides tested by Ross et al (1976), three formulations were dimethylamine salts of acidic herbicides and were found to contain 0.3-640 ppm of N-nitrosodimethylamine (NDMA). One formulation of the dinitroaniline herbicide, trifluralin, was found to contain 154 ppm N-nitrosodipropylamine (NDPA) In conjunction with the observation, Ross et al (1977) observed no detectable nitrosamines in the air, water or crops as the result of application of TREFLAN (a registered trademark of Elanco Products Company, Division of Eli Lilly and Company, for the herbicide trifluralin, α,α,α-trifluoro-2,6-dinitro-N,N-dipropyl-p-toluidine), containing NDPA. Nevertheless, confirmation of the observations, evaluation of the potential exposure to man and the environment and a reduction of the impurity were pursued immediately.

PESTICIDES CONTAINING NITROSAMINES

Since the discovery of N-nitroso contaminants in certain pesticides, a wide variety of products has been analyzed in response to a request from the United States Environmental Protection Agency (Environmental Protection Agency, 1977). The analysis and survey reports by Cohen et al (1977), Bontoyan et al (1979) and Zweig et al (1980) includes dinitroanilines, dimethylamine and ethanolamine salts of phenoxyalkanoic acids, quaternary salts, amides, carbamates, organophosphates, triazines, urea derivatives and some miscellaneous pesticides. Many pesticides have been nitrosated under laboratory conditions for various research purposes. A list of 51 nitrosated pesticides reviewed by Kearney (1980) indicates the potential magnitude of the N-nitroso problem. However, the EPA analysis survey reveals that for positive detection, N-nitroso compounds in pesticide formulations is limited to dinitroanilines, amines and ethanolamine salts of acidic herbicides and quaternary salts. As might be expected, there are exceptions; in the survey the pesticide thiram contained 1.2 ppm N-nitrosodimethylamine (DMNA). Eisenbrand et al (1975) showed a rapid formation of N-nitrosamine from similar compounds ziram and ferbam, by both in vitro and in vivo studies. Thus, almost any compound belonging to the class of nitrogen-containing compound is a candidate, either primarily or secondarily, for N-nitroso derivatation. The scope of this review is not to deal with every eventuality that leads to the formation of N-nitroso compounds, but to examine existing knowledge for the prevention or the removal of N-nitroso compounds from technically-produced pesticides and their accompanying use formulations.

The chemistry and toxicology of nitrosamines have been adequately reviewed by Magee et al (1976), Mirvish (1975), Douglas et al (1978) and Fridman et al (1971). The problem of eliminating nitrosamines occurring as trace contaminants in pesticide formulations is markedly different than that of dealing with neat reactions of a given nitrosamine.

SOURCE OF NITROSAMINE CONTAMINATION

Nitrosamines in pesticides can occur as impurities or as inadvertent contaminants. They are formed in several ways:

Formation of Nitrosamines in Pesticides

1. Side reactions in the manufacturing synthesis
2. Contaminated reagents used in manufacture
3. Direct additives used as preservatives
4. Nitrosating agents in the environment
5. Intra-molecular rearrangements

Formation is dependent on the interaction of nitrogen-containing organic compounds with nitrosating agents, such as oxides of nitrogen.

The levels of N-nitroso compounds in pesticides have been reported to range up to 640 ppm. Considering the possibilities from synthesis, higher values are easily conceivable.

Factors Considered for Nitrosamine Elimination

Reduction of Nitrosamine

1. Prevent the nitrosating reaction with amine precursors.
2. Nitrosamine destruction by selective chemical action.

The literature has numerous citations on both the prevention and destruction of nitrosamines. Techniques, such as the use of scavengers or selective reactions, may be applied to commercial pesticide products.

Some Factors to Consider in Reduction of N-Nitroso Compounds in Pesticides

1. Effectiveness of prevention vs destruction
2. Effect on the pesticide product
3. Introduction of other chemical by-products
    a. By reactions with the pesticide or other existing impurities
    b. Newly created by-products and impurities
    c. The effect on stability of the formulation
4. Economics of production
    a. Special facilities requirements
    b. Cost of stabilizers, scavengers, etc.
    c. Cost of agents and reactants
5. Regulatory requirements
    a. Allowable nitrosamine levels (policy)
    b. Supplemental data requirements for product definition, chemistry, toxicology, environmental and use considerations

PREVENTION OF NITROSAMINE FORMATION

A. Synthesis and Manufacturing

The discovery of nitrosamines in pesticide formulations, Ross et al (1976), led to remarkable public and regulatory reaction. The widely used herbicide, trifluralin, because of the inadvertent nitrosamine impurity, received the most attention. All dinitroaniline herbicides have similarities; therefore, trifluralin, with appropriate exceptions, will serve as a suitable example.

Most dinitroanilines are synthesized from a precursor which is nitrated and subsequently aminated to yield the desired product. In the case of trifluralin, 4-chlorobenzo-trifluoride (PCBT) is dinitrated to yield 4-chloro-3,5-dinitro-α,α,α-trifluorotoluene (Dinitro PCBT),

Preparative Route to Trifluralin

which in turn is aminated with n-dipropylamine to yield the product, Probst et al (1975).

The formation of the impurity in trifluralin occurs when the amine is added to the reactive mixture of 4-chloro-3,5-dinitro benzotrifluoride (Dinitro PCBT) containing residual nitrosating agents arising from the nitration reaction.

$$CH_3CH_2CH_2 \!\!\diagdown \!\!\!\!\underset{CH_3CH_2CH_2 \diagup}{N}H + [NO]_x \longrightarrow CH_3CH_2CH_2 \!\!\diagdown \!\!\!\!\underset{CH_3CH_2CH_2 \diagup}{N}-NO$$

NDPA

Nitrosating reagents would include residual nitrate – nitrite or oxides of nitrogen. Laboratory studies with spiked samples of the Dinitro PCBT with low levels of nitrate and nitrite, then aminating to yield trifluralin, did not contribute significantly to nitrosamine formation. Therefore, the hypothesis was investigated that the active nitrosating agent present in the Dinitro PCBT mixture was oxides of nitrogen, dissolved in the reaction mixture. The active nitrosating agents, in this case oxides of nitrogen, are indicated along with possible precursors, Ingold (1953).

Nitrosating Agents	Precursors
NO^+	N_2O_4
$H_2NO_2^+$	N_2O_3
NO_2^-	NOBr
	NOCl

Many references suggest that the presence of any nitrogen oxides will result, under equilibrum or under specific pH conditions, in a nitrosating agent capable of converting amines or quaternary ammonium bases into nitrosamines, the rate and amount being a function of the existing conditions. Recognizing the conditions which result in the formation of nitrosamines in the manufacturing process, attention is directed to the purification of Dinitro PCBT to remove the nitrogen oxides resulting from nitration prior to the amination step in the synthetic sequence (Cannon and Eizember 1978). Therefore, modification of the process was directed to the purification of Dinitro PCBT to remove nitrogen oxides. If dinitro PCBT is prepared without special purification procedures, nitrosamines in the amount of 150-500 ppm are common. In the case of the synthesis of compounds, such as pendimethalin and butralin,

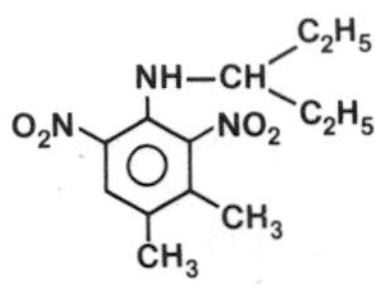

Pendimethalin
N-(1-ethylpropyl)-
3,4-dimethyl-2,6-dinitro
benzenamine

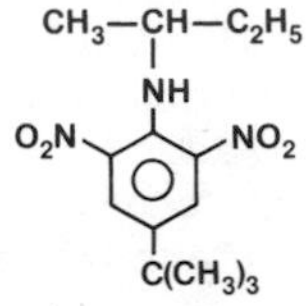

Butralin
4-(1,1-dimethylethyl)-N-
(1-methylpropyl)-2,6-
dinitrobenzenamine

which are secondary amines, formation of the nitroso analog can be a major reaction product. According to Diehl et al (1979) nitration of the aniline can yield 12-62% N-nitroso-N-(1-ethylpropyl)-3,4-dimethyl-2,6-dinitroaniline.

Synthesis of Pendimethalin

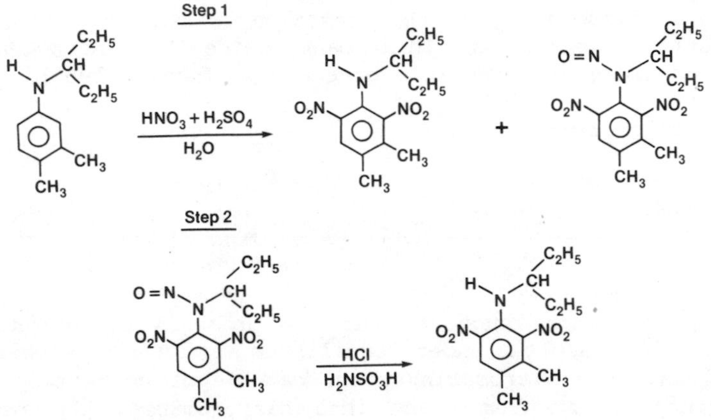

Step 1

$$HNO_3 + H_2SO_4 \xrightarrow{H_2O}$$

+

Step 2

$$\xrightarrow[H_2NSO_3H]{HCl}$$

Therefore, the N-nitroso derivative must be denitrosated to make the process economical as well as avoid high concentrations of the N-nitroso derivative in the product. The preferred method of nitration and denitrosation of N-(1-ethylpropyl)-3,4-dimethylaniline utilizes a nitration mixture of 35-53% water by weight, nitric acid and sulfuric acid in molar ratios of 3.25:1 and 2.25:1 of the aniline compound. The reaction period is about two hours with a temperature of 35-70°C. Thereafter, denitrosation is effected by adding hydrochloric and sulfamic acids to the mixture, maintaining a temperature of 70-100°C over a period of one to six hours, then recovering the N-(1-ethylpropyl)-2,6-dinitro-3,4-dimethylaniline product formed. Other denitrosating agents such as hydrochloric acid and ferrous chloride may also be employed.

Prevention of nitrosamine formation in other important commercial dinitroanilines such as trifluralin, benefin, isopropalin, profluralin, ethalfluralin and other tertiary amines is approached by first eliminating sources of nitrosating agents from the reaction mixture prior to amination.

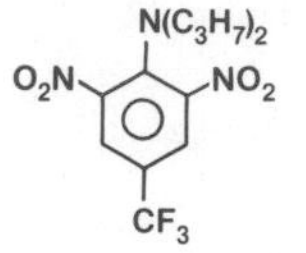

Trifluralin
α,α,α-trifluoro-
2,6-dinitro-N,N-dipropyl
p-toluidine

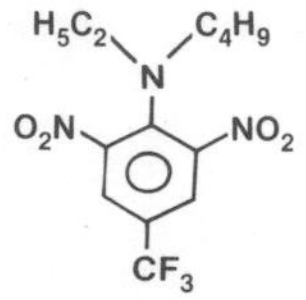

Benefin
N-butyl-N-ethyl-
α,α,α-trifluoro-2,6-dinitiro-
p-toluidine

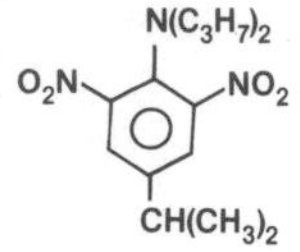

Isopropalin
2,6-dinitro-N-N-
dipropyl-cumidine

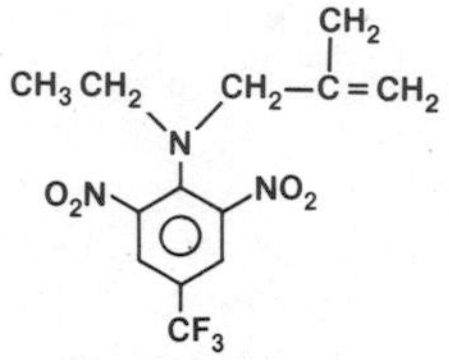

Ethalfluralin
N-ethyl-N-(2-methyl-2-
-propenyl-2,6-dinitro-4-
(trifluoromethyl) benzamine

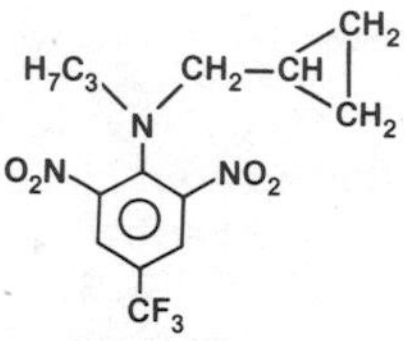

Profluralin
N-(cyclopropylmethyl)
α,α,α- trifluoro-2,6-dinitro-
N-propyl-p-toluidine

The method of Cannon and Eizember (1978) for the removal of nitrosating agents after the nitration of PCBT is simple and commercially acceptable. The process for reducing the concentration of nitrosating agents in Dinitro PCBT comprises bubbling a non-reactive gas, such as air, nitrogen or carbon dioxide through the reaction mixture containing an aqueous solution of base such as sodium carbonate, at a temperature of 50-100°C. The process can be carried out in several ways, that is, working the reaction mixture with base prior to sparging with gas, which reduces sources of nitrosating agents such as nitrate and nitrite ions, or simultaneous base treatment and gas sparging. However, gas sparging must follow base treatment to eliminate the oxides of nitrogen. The time of gas treatment is dependent on the conditions, equipment and facilities. Experience and care in the manufacturing technology for reduction of the nitrosating agents can result in more than a 95% reduction of nitrosamine formation.

B.   Prevention of N-nitroso Compounds In Formulations

As previously noted, certain commercial pesticides, such as the phenoxy or benzoic acids,

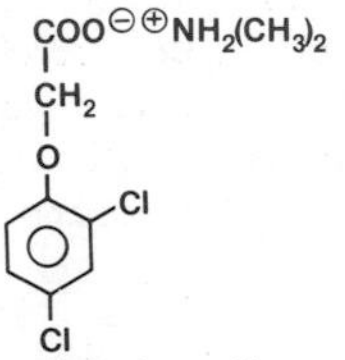

$COO^{\ominus}{}^{\oplus}NH_2(CH_3)_2$
|
CH_2
|
O

2,4-D (dimethylamine salt)
2,4-dichlorophenoxyacetic
acid

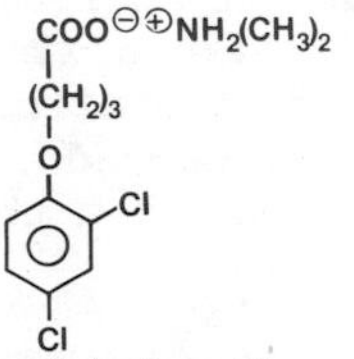

$COO^{\ominus}{}^{\oplus}NH_2(CH_3)_2$
|
$(CH_2)_3$
|
O

2,4-DB(dimethylamine salt)
2,4-dichlorophenoxybutyric
acid

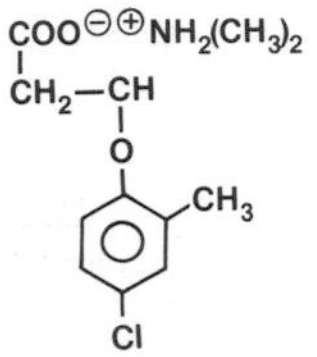

$COO^{\ominus}{}^{\oplus}NH_2(CH_3)_2$
|
CH_2-CH
|
O

MCPP
2-(4-chloro-2-methylphenoxy)
propionic acid

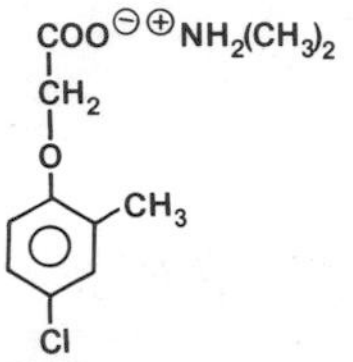

$COO^{\ominus}{}^{\oplus}NH_2(CH_3)_2$
|
CH_2
|
O

MCP
([4-chloro-o-tolyl]oxy)
acetic acid

have been formulated as secondary amine or ethanol amine salts. Ross et al detected 640 ppm of N-nitrosodimethylamine (NDMA) in formulated 2,3,6-trichlorobenzoic acid (Benzac) stored in metal containers. Sodium nitrite had been used as a corrosion inhibitor, which reacted with components of the mixture to yield a nitrosating species, hence nitrosamine formation. To further support this thesis, Bontoyan et al (1979) demonstrated that the addition of 0.1-0.5% sodium nitrite to dimethylamine salts of 2,4-D resulted in the formation of NDMA in 50-60 days in glass or metal containers. A simple solution seems to dictate: Find other corrosion inhibitors that do not produce nitrosating agents. However, the use of tin-plated containers for commercial pesticides susceptible to nitrosamine formation is unsatisfactory. Shown in the table below are the analysis results for NDPA from samples of trifluralin originally containing less than one ppm nitrosamine (Grove, 1979).

Nitrosamine Content of Technical Trifluralin*
Sampled in Glass and Tin Containers

Lots Sample #	Total Volatile Nitrosamine ppm	
	Sampled in Glass	Sampled in Tin Can
6683	.17	7.68
6684	.16	3.30
6685	.20	6.86
6697	.23	3.20
6719	.25	4.11

*Storage: 1 month at ambient temperature

The presence of increased amounts of NDPA in trifluralin was observed in a matter of days in the tin containers stored at ambient temperatures. No corrosion inhibitors were added to the tin containers. Container manufacturers use a flux in the tinning process, many contain either nitrite or nitrate salts. Thus, the potential for the nitrosating agents exists as part of the metallic film in a seemingly clean tin container. In addition, Archer and Wishnok (1976) demonstrated the formation of nitrosamines from constituents of polymeric liners of metal cans. Container specifications are vital for pesticide formulations prone to be nitrosated.

Another source of N-nitroso contamination of pesticides either in synthesis or formulation are commercially synthesized amines. Cohen et al (1978) examined different batches of dimethylamine and observed as much as 25 ppm NDMA, which could be carryover into the formulation.

Commercially produced amines contain impurities from synthesis, thus rigid specifications are necessary to avoid unwanted impurities in final products. Modern-day analytical capability permits detection of minute quantities of impurities in almost any compound. Detection in parts per million is routine, parts per billion is commonplace, and parts per trillion is attainable. The significance of impurities in products demands careful and realistic interpretation. Nitrosating species, as well as natural amines, are ubiquitous in the environment. For example, Bassow (1976) cites that about 50 ppb of nitrous oxide and nitrogen dioxide are present in the atmosphere of the cities. Microorganisms in soil and natural water convert ammonia to nitrite. With the potential for nitrosamine formation almost ever-present in the environment, other approaches to prevention should include the use of appropriate scavengers as additives in raw materials and finished products.

Scavengers inhibit nitrosamine formation by competing with the amine for the nitrosating agent. Ascorbic acid is a typical example of the competition by rapid reduction of the nitrosating agent as shown.

Ascorbic Acid   Dehydroascorbic   Excess
                          Acid   Ascorbic Acid

To prevent oxidation of nitrous oxide to nitrogen tetraoxide, excess ascorbic acid must be used. Douglas et al (1978) has reviewed the numerous literature reports on ascorbic acid inhibition of nitrosamine formation in amine-nitrite systems.

Some phenols can inhibit nitrosamine formation in a similar manner, (Challis and Bartlett, 1975),

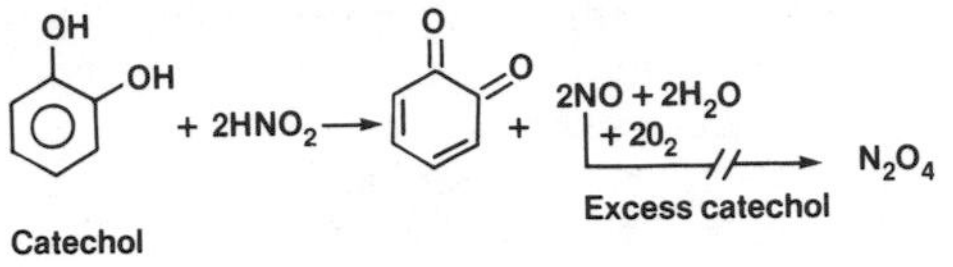

Catechol

Excess catechol

but under other conditions, serve to catalyze nitrosamine formation. For example, in the presence of excess nitrite, 4-methylcatechol catalyses the nitrosation of dimethylamine and piperidine. Sulfur compounds can inhibit nitrosamine formation. Bisulfite, sulfur dioxide or sulfamic acid reduce nitrite according to the following equations (Douglas et al, 1978).

$$SO_2 + 2HNO_2 \longrightarrow 2NO + H_2SO_4$$
$$SO_2 + 2NO + H_2O \longrightarrow N_2O + H_2SO_4$$
$$NaNO_2 + H_2NSO_3H \longrightarrow NaHSO_4 + N_2 + H_2O$$

Sulfamate reduces nitrite to molecular nitrogen and serves as a denitrosating agent in the synthesis of pendimethalin (Diehl et al, 1979).

Nitrosamine Formation by Intra-Molecular Rearrangement

Dinitroaniline herbicides have an unusual property of serving both as an amine contributor and nitrosating agent. The phenomenon (Grove, 1979) was observed when solid or liquid dinitroanilines are subjected to heat. For example, when trifluralin is heated to 70°C, N-nitrosodipropylamine (NDPA) is formed. Figure 1 shows the accumulation of NDPA with time. Increased nitrosamine formation is a function of temperature and time.

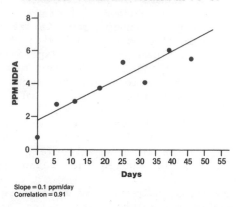

**Nitrosamine Generation from
Technical Trifluralin Heated at 70°C.**

Slope = 0.1 ppm/day
Correlation = 0.91

Figure 1.

The mechanism of the reaction is not understood at the present time, but is presumed to be an internal rearrangement effecting a nitro group. This results in the formation and liberation of NDPA. The presence of a proton would initiate an action on one nitro group yielding the nitrosating agent, which more than likely reacts with traces of residual dipropylamine in the mixture. The resulting mononitro compound has not been identified.

Equation for Hypothetical Intra-molecular Rearrangement

The reaction occurs with technical material only, and, once formulated, the nitrosamine content is not altered as a function of time at ambient formulation temperatures. This intra-molecular rearrangement requires careful attention to conditions in the formulation of dinitroaniline herbicides.

Destruction of Preformed Nitrosamines

Chemically, nitrosamines are considered to be quite stable compounds and are difficult to destroy once they are formed. Reducing or destroying preformed nitrosamines in pesticides offers special challenge, as they occur in trace amounts which require specific selective treatment without effecting the composition of the principal product. Experimental laboratory work revealed that reactions suitable for mass quantities of reactants, that is neat samples, are not necessarily analagous to micro reactions for the reduction of a given trace nitrosamine contaminant or impurity.

Eizember (1978) and Eizember et al (1979 and 1980) investigated other contingencies for the additional reduction of nitrosamine impurities in a wide variety of dinitroaniline analogs. Trifluralin is chosen as being typical of the dinitroanilines.

The action of acids, halogens and halogen-releasing agents, as well as phosphorus, sulphur and other inorganic halide reagents was carefully investigated. Typical laboratory parameters in the treatment of impure trifluralin with various acids follow.

Reaction Parameters in the Treatment of Impure Trifluralin with Various Acids

Phase:	Neat or in appropriate solvent
Temperature:	50° to 140°C
Nitrosamine Level:	10 to 300 ppm
Treatment Time:	5 min to 8 hrs
Amount of Acid:	Dependent upon

a)   nitrosamine level
b)   dinitroaniline used
c)   phase
d)   temperature
e)   treatment time

The amount of acid required for nitrosamine destruction is dependent on the level of the nitrosamine impurity, the dinitroaniline being treated, the organic solvent used, temperature, and time. Each reactive mixture was appropriately worked up to a final isolate of the product. Some typical results are shown in Table 1.

Table 1

Removal of NDPA from Trifluralin by
Different Acids[a]

Acid	NDPA, ppm
50% Sulfuric acid	22
70% Sulfuric acid	<1
85% Sulfuric acid	<1
10% Hydrochloric acid	81
33% Hydrochloric acid	<1
37% Hydrochloric acid	<1
Hydrogen chloride gas[b]	<1
50% Formic acid	74
98% Formic acid	58
70% Acetic acid	58
Oxalic acid[c]	9
40% Phosphoric acid	90
48% Hydrobromic acid	<1
Ascorbic acid[d]	85

[a]Conditions for the treatments were: time, 20 min;
temperature, 70°C; and amount of acid used was 20% w/w
of acid to trifluralin. Untreated trifluralin
contained 69 ppm of NDPA.
[b]Hydrogen chloride gas flow was 35 ml/min.
[c]Time, 2 h.
[d]Time, 3 h.

The acidic reagents vary widely in their ability to lower
NDPA levels in trifluralin. The concentration of the acid is
critical to produce the desired effect. In some instances, the
acid promoted additional nitrosamine formation, e.g. 10%
hydrochloric acid, 40% phosphoric acid, ascorbic acid, etc.
Hydrochloric acid and hydrogen chloride gas were the most
efficient at destroying NDPA impurity.

Figure 2 is a typical concentration curve that shows the
loss of NDPA upon treatment of trifluralin in a solvent with
hydrogen chloride.

**Removal of NDPA from Impure Trifluralin
with Gaseous Hydrogen Chloride[1]**

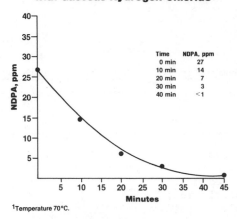

Time	NDPA, ppm
0 min	27
10 min	14
20 min	7
30 min	3
40 min	<1

[1]Temperature 70°C.

Figure 2.

Analysis of the organic layers, the acid layers, the neutralized organic layer and neutralized aqueous fractions revealed a significant lowering of NDPA levels. The mechanism of the reaction with either hydrochloric or sulfuric acid can be rationalized as follows.

The acid initially protonates the nitrosamine to give a charged intermediate having significant water solubility. When the agent is gaseous hydrogen chloride or concentrated hydrochloric acid, the protonated intermediate is rapidly attacked by the strong chloride anion to generate nitrosyl chloride and secondary amine. At the temperature of the reaction, the nitrosyl chloride is either removed by the hydrogen chloride gas flow or reacts with water present in the reaction mixture.

Studies have been extended to determine effect of halogens and halogen-releasing agents. Under similar laboratory conditions as previously described, the effect of these agents on NDPA destruction is shown in the following table.

Removal of NDPA from Trifluralin Containing 68 ppm NDPA[a]

	Grams Used	Temp. ^{o}C	Time Min.	NDPA ppm
Bromine	0.2	70	20	<1
Bromine	0.1	90	15	<1
Chlorine	35 ml/min	70	30	16
Chlorine	15 ml/min	110	120	~2
N-bromosuccinimide	0.5	70	30	1.7
Iodine	0.1	70	60	78

[a]Each reaction used 30g of trifluralin

Bromine is more efficient at removing NDPA from trifluralin than chlorine which requires higher reaction temperatures and longer reaction time. N-bromosuccinimide is also effective in removing nitrosamines from dinitroanilines.

Eizember and Vogler (1980) have examined a variety of inorganic halides to determine the efficiency of nitrosamine removal or destruction. Such inorganic halides as phosphorus trichloride, phosphorus tribromide, thionylchloride and sulfonyl chloride can be used to destroy nitrosamines in dinitroaniline pesticides. These agents should be effective in destroying nitrosamines in a wide variety of inert or relatively unreactive solvents or diluents.

If the nitrosamine content in the product is above acceptable levels after denitrosation, e.g. 5 to 10 ppm, the product can be further treated with these types of agents to reduce the nitrosamine content to less than 1 ppm. Several dinitroanilines can be successfully treated in this manner. For example, trifluralin technical product can be commercially produced with nitrosamine content of less than 1 ppm.

If the selected reaction for nitrosamine is rigidly controlled, the principal product is not modified or destroyed. However, the dinitroaniline, ethalfluralin, contains the N-ethylmethallyl group and is subject to the addition of HCl to the double bond as shown:

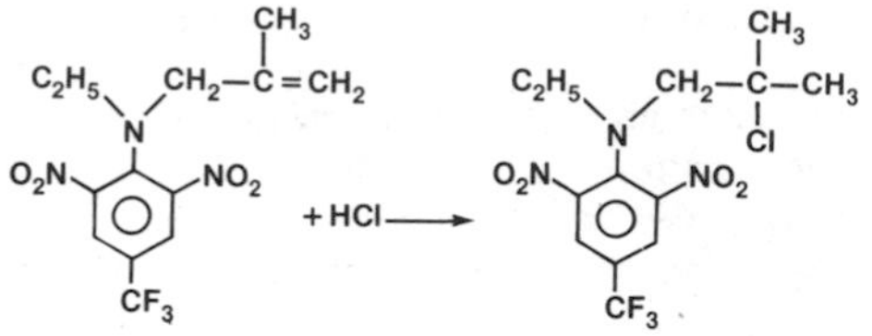

To avoid accumulation of this byproduct in ethalfluralin, time of exposure and temperature are important conditions of the reaction.

Ross and Chiarello (1979) have described a method for denitrosation of nitrosamines, such as dialkylnitrosamines e.g. N-nitrosodimethylamine, etc., and complex aryl-alkyl nitrosamines, e.g. N-nitroso-N-(1-ethylpropyl)-3,4-dimethyl-2,6-dinitroaniline. The products of nitration and amination containing nitrosamine as a substantial impurity are treated with an aldehyde or ketone in the presence of strong acids such as hydrochloric or hydrobromic acid. Under pressurized conditions at 105-110°C for one to two hours, the nitrosamine is destroyed. The desired product can be recovered after neutralization of the excess acid. The equation shows the denitrosation of pendimethalin:

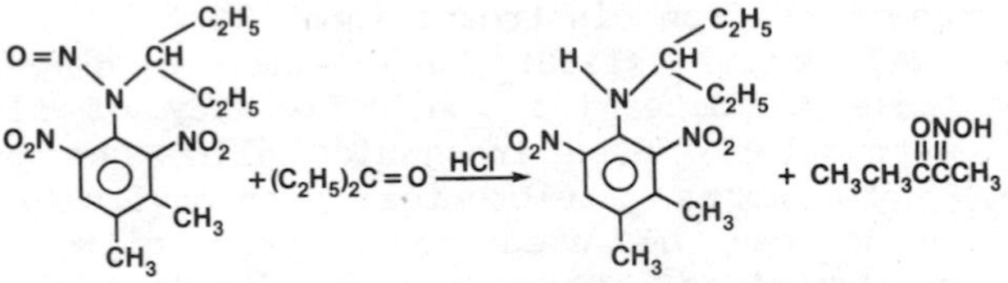

The reaction is applicable to a wide variety of dinitroanilines containing nitrosamine impurities but is particularly suitable to secondary amine type dinitroanilines, such as pendimethalin. Synthesis, as previously described, of pendimethalin yields the N-nitroso derivative in large amounts which must be converted to pendimethalin to make the synthesis commercially economical. A typical example of the denitrosation reaction is: Diethylketone is added to a mixture of N-(1-ethylpropyl)-3,4-dimethyl-2,6-dinitroaniline (approximately 40% by weight in solution) and N-(1-ethylpropyl)-N-nitroso-3,4-dimethyl-2,6-dinitroaniline (15% by weight in solution) in a solvent of ethylene dichloride. Concentrated hydrochloric acid is added and the containing vessel sealed. The reaction mixture is heated at

85-90°C. An analysis of samples withdrawn at intervals reveals the degree of denitrosation as shown:

Analytical Data, Showing the Degree of Denitrosation
Achieved in the Above Reaction

Sample	Sample Taken, Hours After Reaction Began	Found %A*	%B*
1	0	91.2	1.3
2	1	90.6	0.06
3	2	91.2	0.006
4	3	90.1	0.003
5	4	91.3	<0.003

*Wherein:
A = N-(1-ethylpropyl)-2,6-dinitro-3,4-xylidine
B = N-(1-ethylpropyl)-N-nitroso-2,6-dinitro-3,4-xylidine

Over a period of five hours, the nitroso derivative was reduced to less than 0.003% or 30 ppm.

Biggs and Williams (1976) have studied denitrosation over a wide range of acidities. They demonstrated that rate constants for denitrosation and rearrangements could be measured by limiting experimental conditions for N-alkyl-N-nitrosoanilines which is applicable to N-nitroso-dinitroaniline herbicides such as pendimethalin and butralin.

Nitrosamine Destruction in the Synthetic Process

As described by Eizember and Vogler (1980), many inorganic halides are effective in removal and destruction of nitrosamine impurities during in the synthesis of dinitroanilines. Nitrosamine destruction occurs during the synthesis of oryzalin and fluchloralin. Structures of oryzalin and fluchloralin are shown.

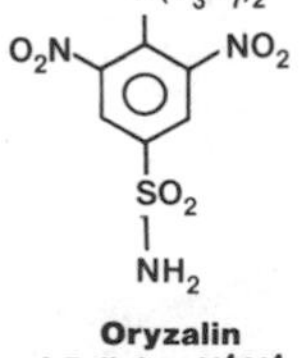

Oryzalin
3,5-dinitro-N⁴,N⁴,
dipropyl sulfanilamide

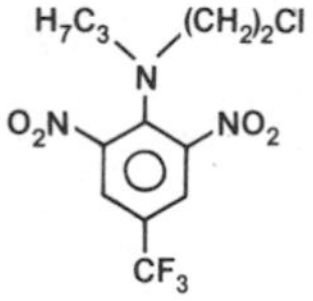

Fluchloralin
N-(Chloroethyl)-
α, α, α-trifluoro-2,6-dinitro
-N-propyl-p-toluidine

The synthesis makes use of inorganic halides to produce the desired intermediate or final product after the nitration and amination steps which are responsbile for nitrosamine formation. In the synthesis of oryzalin, 3,5-dinitro-N^4, N^4-dipropylsulfanilamide, the following reaction sequence is employed:

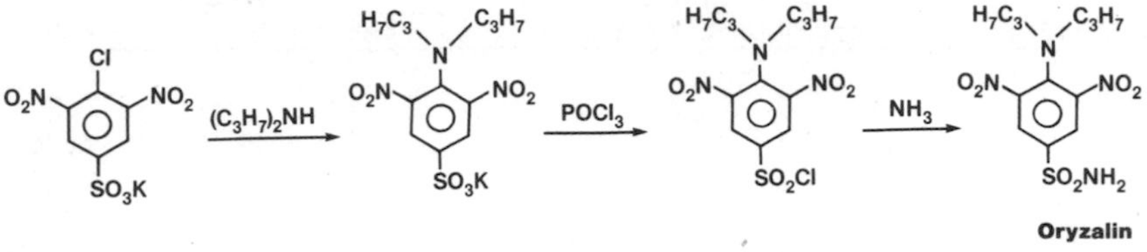

Oryzalin

$$(C_3H_7)_2NNO \xrightarrow{POCl_3} (C_3H_7)_2NH + NOCl$$

The use of phosphorus oxychloride not only produces 3,5-dinitro-N^4,N^4-dipropylsulfanilylchloride, but destroys contaminating N-nitrosodipropylamine existing in the reaction mixture after amination. Technical oryzalin contains less than 1 ppm NDPA due to the inorganic halide reaction.

Similarly, in the synthesis of fluchloralin, N-(2-chloro-ethyl)-α,α,α-trifluoro-2,6-dinitro-N-propyl-p-toluidine, the final synthetic step employes thionyl chloride, an inorganic halide, reacting with N-(2-hydroxyethyl)-α,α,α-trifluoro-2,6 dinitro-N-propyl-p-toluidine yielding fluchloralin. In the process of chloro substitution, the N-2-hydroxyethyl-N-propylnitrosamine formed during the amination reaction from nitrosating agents is effectively destroyed. Nitrosamine values reported for fluchloralin range from 0.5 to 1.7 ppm volatile nitrosamine (Zweig et al, 1980). The synthesis of fluchloralin is shown:

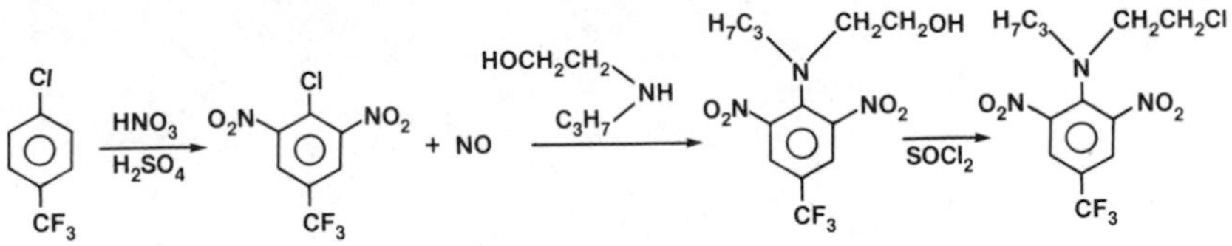

Conclusion:

The presence of volatile nitrosamine impurities in pesticide formulations was an unsuspected phenomenon until the advent of more sensitive and reliable analytical tools for their detection, such as the Thermo Energy Analyzer.

Reduction or removal of trace nitrosamine impurities in pesticide formulations can be accommplished by prevention of nitrosation or by direct chemical destruction. The ease of nitrosation reactions with appropriate amines demands careful attention to possible side reactions occuring in the synthetic process or from the various agents used in formulation. The ubiquitous presence of nitrosating agents in the environment is not readily amenable to solution. These natural agents will be a source of nitrosamine formation during manufacture and application as well as from those susceptable pesticides which remain as residuals in the environment. Reduction of nitrosamine contaminants in pesticides is an individual and unique situation which can be controlled by careful investigation of the sources. Total elimination in many instances can never be achieved.

LITERATURE CITED

Archer, M.C., Wishnok, J.S., J. Environ. Sci. Health, A11, 1976, 583, 10411.

Bassow, H., Air Pollution Chemistry. An Experimenter's Source Book, Hayden Book Company, Rochelle Park, N.J., 1976, 60.

Biggs, I.D., Williams, D.L.H., J. Chem. Soc. Perkin II, 1976, 601ff.

Bontoyan, W., Wright Jr., D., Law, M.W., J. Agric. Food Chem. 1979, 27, 631-635.

Cannon, W.N., Eizember, R.F., U.S. Patent 4,120,905, Oct. 17, 1978

Ceaborn, H.V., Radeleff, R.D., Bushland, R.C., USDA Agric. R. Serv., December 1960, 33.

Cohen, S.Z., Zweig, G., Law, M.W., Wright Jr., D., Bontoyan, W.R., IARC Scientific Publication, 1978, 19, 333.

Diehl, R.E., Levy, S.D., Gastrock, W.H., U.S. Patent 4,136,117, January 23, 1979.

Douglas, M.L., Kabacoff, B.L., Anderson, G.A., Cheng, M.C., J. Soc. Cosmet. Chem., 1978, 29, 581.

Eisenbrand, G., Ungerer, O., Preussman, R., Food Cosmet. Toxicol., 1975, 13, 365.

Eizember, R.F., U.S. Patent 4,127,610, November 28, 1978.

Eizember, R.F., U.S. Patent 4,226,789, October 7, 1980.

Eizember, R.F., Vogler, K.R., Souter, R.W., Cannon, W.N., Wege II, P.M., J. Org. Chem., 1979, 44, 784.

Eizember, R.F., Vogler, K.R., U. S. Patent 4,185,035,
    January 22, 1980
Fine, D.H., Rufeh, S., and Gunter, G., Anal. Lett., 1973, 6,
    731.
Fine, D.H., and Rufeh, S., IARC Scientific Publications, 1974,
    9, 40.
Fridman, A.L., Mukhametshin, F.M., Novikov, S.S., Russ. Chem.
    Rev., 1971, 40, 34.
Grove, D. R., Eli Lilly and Company, unpublished data, 1979.
Herbicide Handbook of the Weed Science Society of America, 4th
    Edition, 1979, 211.
Ingold, C.K., Structure and Mechanism in Organic Chemistry,
    Cornell University Press, 1953, 400.
Kearney, P.C., Pure and Appl. Chem., Pergamon Press Ltd.,
    IUPAC, 1980, 52, 499.
Lijinsky, W., Cancer Res., 1974, 34, 255.
Lijinsky, W., Taylor, H.W., Snyder, C., Nettsheim, P., Nature
    Lond., 1973, 244, 176.
Lijinsky, W., and Epstein, S.S., Nature Lond., 1970, 225, 21.
Magee, P.N., Barnes, J.M., Adv. Cancer Research, 1967, 10, 163.
Magee, P.N., Montesano, R., Preussman, R., ACS Monograph, 1976,
    173, 491.
Maier-Bode, H., Naturwissenschafter, 1968, 55, 470.
Mirvish, S.S., Toxicol. Appl. Pharmacol., 1975, 31, 325.
Probst, G.W., Golab, T., Wright, W.L., "Dinitroanilines",
    Herbicides, Chemistry, Degradation and Mode of Action, Vol
    I. Chap. 9, 1975, 453, Ed. P.C. Kearney and D.D. Kaufman,
    Marcel Dekker, Inc., N.Y., N.Y.
Ross, L.J., Chiarello, G.A., U.S. Patent 4,134,917, January 16,
    1979
Ross, R.D., Morrison, J., Rounbehler, D.P., Fan, S., Fine,
    D.H., Presented at the Division of Pesticide Chemistry,
    172nd National Meeting of the American Chemical Society,
    San Francisco, CA., Sept. 1976.
Ross, R.D., Morrison, J., Rounbehler, D.P., Fan, S., Fine,
    D.H., J. Agric. Food Chem.,1977, 25, 1416.
Zweig, G., Selim, S., Hummel, R., Mittelman, A., Wright Jr.,
    D., Law Jr., S.C., Regelman, E., "N-Nitroso Compounds:
    Analysis, Formation, and Occurrence", IARC, 1980, Pub. No.
    31, 555.

RECEIVED August 10, 1981.

Policy and Regulatory Aspects of *N*-Nitroso Contaminants in Pesticide Products

GUNTER ZWEIG[1] and WILLA GARNER

Office of Pesticide Programs, Hazard Evaluation Division,
U.S. Environmental Protection Agency, Washington, DC 20460

Since the finding five years ago of N-nitroso
contaminants in a number of pesticide products, the
EPA under the legal authority of FIFRA began to
regulate these pesticides on an ad hoc basis. The
initial concern that these contaminants affected
thousands of products was soon dispelled by the
finding that analyses revealed that only a few
classes of pesticides actually contained N-nitroso
impurities. EPA is now proposing an official policy
which describes in detail the philosophy for setting
priorities for detailed risk assessment through the
so-called RPAR process, based on determining the
risk to be above or below 10^{-6} for potential
carcinogenicity. This paper discusses the proposed
EPA policy and the procedures by which registrants
can obtain data to support their registration of
compounds containing detectable levels of N-
nitrosamine compounds.

Background

The basic authority for the regulation and use of pesticides
in the U.S. is spelled out in FIFRA (1972, amended in 1975 and
1978), the Federal Insecticide, Fungicide, and Rodenticide Act.
Section 3 of this law states that a pesticide shall be registered
if "it will perform its intended function without unreasonable
adverse effects on the environment..." The term "environment" in
this context has been interpreted by EPA to include fish, wild-
life, and man. More recently, a great emphasis has been placed
on the protection of pesticide applicators and farm workers. The
general public is protected through the Federal Food, Drug, and
Cosmetic Act, Sect. 408 which authorizes the Government (sic EPA)
to establish legal tolerances of pesticide residues in raw
agricultural crops.

[1] Current temporary address: School of Public Health, University of California,
Berkeley, CA 94720.

In addition, Sections 3 and 6 of FIFRA have given authority
to EPA to issue regulations establishing the Rebuttable Presumption Against Registration (RPAR) process to make the decisions
on whether to register, not to register, or to cancel the registration of a pesticide.

Nitroso Contaminants in Pesticide Products

When EPA became aware almost five years ago that N-nitroso
contaminants occurred in a number of pesticide products, the
Agency immediately acted on the authority of Section 3 of FIFRA
to place a moratorium on new registrations of pesticides suspected to contain N-nitroso contaminants at detectable levels (this
term is defined and explained later in the text). As has been
discussed in other papers of the Symposium, many N-nitroso
compounds are animal carcinogens and, consequently, suspected
human carcinogens.

N-Nitroso contamination in pesticide products was first reported by D. Fine and co-workers (1) who had developed a novel
and specific analytical method for N-nitroso compounds, called
thermal energy analysis.

During the next two years (1976 and 1977), registrants of
pesticides and EPA laboratories analyzed several hundred samples
of N- containing pesticides for N-nitroso contaminant occurrence
and concentration. Results of these analyses have been reported
elsewhere (2,3,4). What these analyses revealed was what was
first suspected to be a major problem concerning thousands of
pesticide products, seemed to affect only the following classes
of pesticides: substituted dinitroaniline derivatives;dimethyl-
amine salts of phenoxyalkanoic acid herbicides; di- and tri-
ethanolamine salts of several pesticides; some quaternary
ammonium compounds; and some morpholine derivatives. All other
nitrogen-containing compounds which were tested did not contain
detectable levels of nitrosamine contaminants.

There were now several options open to EPA to regulate the
safe use of these pesticides. One radical option came on the
recommendation of former Congressman Andrew Maguire to immediately suspend all uses of trifluralin (5). Products containing
this pesticide had been found to have levels of N-dipropylnitrosamine ranging from 2.1-115 ppm (4). The Agency carefully
considered the evidence and concluded that the request for suspension should not be granted for two reasons: The principal
manufacturer of trifluralin was able to reduce the nitrosamine
contamination considerably by changing the chemical process of
synthesis; and given the estimated exposure and risk to pesticide
applicators and the public-at-large, the Agency determined that
the benefits exceeded the risk.

Another case of high nitrosamine concentration in chlorinated
phenoxy- and benzoic acid herbicides was resolved by the
elimination by the manufacturer of nitrite salts in the formu-

lations. Nitrite was added as corrosion inhibitor in metal containers of the dimethylamine salt of these herbicides. Consequently, the high levels of dimethylnitrosamine first reported
by Ross (1), were eliminated by changing the metallic containers
to plastic-lined or non-metallic containers, thus obviating the
use of nitrite.

And yet, this left the six classes of pesticides which contained detectable levels of N-nitrosamines and had to be regulated.
The moratorium on these pesticides still remains in force until a
policy has been established. The Agency also could not ignore
that some regulatory action might have to be taken on those
pesticides that are already registered and contain detectable
levels of nitroso impurities.

Issuance of Proposed Nitroso Policy

For the above reasons, EPA's Office of Pesticide Programs has
proposed a policy on the procedures for regulating pesticides
contaminated with N-nitroso compound (6). Comments from the
public and involved parties were invited and will be considered
when EPA is preparing its final policy on N-nitroso compounds in
pesticides, later in 1981 or early 1982. It is the intent of
this paper to discuss (1) the proposed policy by EPA; (2) the
important issues raised during the public comment period; and (3)
suggest what the final policy might look like.

Proposed Policy on Limits of Detection

EPA will require that the registrant of pesticides belonging
to the classes of dinitroanilines, sec. and tert. alkylamines,
arylamines, and alkanolamines, and quaternary compounds containing
nitrite in their formulations, analyze his product qualitatively
and quantitatively for N-nitroso compounds by a scientifically
acceptable analytical method. The level of detection in technical,
manufacturing-grade or formulation product has been arbitrarily
set at 1 ppm. It is recognized that levels of probably 2 to 3
orders of magnitude in sensitivity is achievable, but the higher
level of 1 ppm has been chosen on the basis of praticality to
routinely analyze samples without clean-up or extraction.

Proposed Policy on Priority Setting

The basic tenets of the proposed policy is to encourage the
manufacturer to reduce or eliminate the N-nitroso contaminant
level in his product, or failing this to establish that the level
does not exceed an "acceptable risk" to the user of the pesticide
or the general public. This latter process of evaluation also
requires that the Agency consider the benefits of this pesticide
together with the risk in order to reach a final decision. Since
there might be hundreds of products falling under this policy, the

resources of the Government and the involved companies must be
harbored and a priority system devised.

The priority setting system is explained in the proposed
policy as follows (6):

"...the Agency must determine what level of risk from
exposure to N-nitroso compounds in pesticides warrants
the substantial resource expense of the RPAR (Rebuttal
Presumption Against Registration) process. ...as a
general rule the RPAR process will be initiated at
this time only for those products exceeding the
individual lifetime carcinogenic risk of $1x10^{-6}$ (that
is there is one chance in one million that an exposed
person will contract cancer as a result of the N-
nitroso exposure) for pesticide applicators and field
workers or other exposed populations." "It should be
clear that the Agency is not proposing that a risk
level of $1x10^{-6}$ for applicators is 'safe' or that it
is an acceptable risk level for all situations, or
that it is a permanent standard for all N-nitroso
decisions. Solely for the purpose of establishing
priorities, the $1x10^{-6}$ level appears from a policy
perspective to be a reasonable criteria to help
separate high risk (and high resource) situations
from low risk problems".

Proposed Policy on Reduction of Exposure

Before the EPA proceeds on a regulatory action, registrants
whose products exceed the 10^{-6} risk level from nitroso contamination, will be given the opportunity to lower potential
exposure to applicators and other users of their pesticides.
Reduction of exposure and, thus, reduction of risk can be accomplished by modification of the manufacturing process (see
below), improved packaging technology, modification of application
technique (e.g. closed systems), or deletion of high-exposure
uses.

Reduction of nitroso content of the product can be achieved
by several techniques:

1. Use of nitroso-free starting materials or intermediates
in the production of the final product.

2. Elimination or reduction of all potential nitrosating
agents formed or added during the manufacturing process (e.g.
nitrous acid resulting from nitration process during the process).

3. Elimination of nitrosating agents in formulations (e.g.
sodium nitrite as corrosion inhibitor for metal containers); use
of plastic-lined containers.

That all or some of these steps are practical has been demonstrated by the observation that many N-nitroso contaminants
have been removed from commercial products since they were first
discovered in 1977, or that the nitroso levels have been reduced
in these products below 1 ppm.

Examples of these improvements are the packaging of 2,4-D dimethylamine products in plastic-lined containers and the elimination of sodium nitrite in the final product. This change in packaging and formulating makes it highly probable that N-dimethylnitrosamine will not form during storage of the product. Another example of the improvement of products is that Treflan products today contain less than 1 ppm of N-dipropylnitrosamine (7), whereas some of the older products contained levels as high as 115-150 ppm several years ago (1,2).

If the manufacturing process, packaging and/or use pattern is changed by the manufacturer so that exposure to N-nitroso compounds is reduced so that the calculated risk is below 10^{-6}, the product probably will receive conditional registration, pending final analysis of all the facts and data. If, on the other hand, the risk is above 10^{-6} even with the instituted changes, the product will be a high priority item for the RPAR process, and registration cannot be granted outright. The next section explains how the manufacturer can obtain the data to make this type of risk assessment.

Proposed Registration Requirements

Sampling Instructions. Registrants who have not already submitted analytical results to the EPA, and whose products belong to one of the designated chemical classes, must submit analytical data on possible nitroso content of their product. Samples for analysis must come from fresh production batches, and in case a nitrosating agent is present or added in the final product, the same sample must be reanalyzed 3 and 6 months later, allowing the sample to be stored at room temperature. This requirement may be modified in the final policy statement to reduce the total number of samples. If the analyses show no contamination by N-nitroso compounds at the 1 ppm-sensitivity level, the product is cleared for registration.

Risk Analysis. For pesticides containing greater amounts than 1 ppm of N-nitroso compounds, the registrant must perform an exposure analysis for persons coming into contact with the pesticide product during normal pesticide application and farm practices. Actual field monitoring data on personnel is most desirable, although the Agency will consider experimental data from surrogate or similar compounds having the same use pattern as the pesticide under consideration. Detailed information on usage, application rate and technique, working conditions, size of acreage treated, duration of treatment -- all these data points are needed for a comprehensive exposure assessment. Non-agricultural use sites (e.g. hospital) will also be considered if applicable.

EPA has an active research program in progress on methodology and quantitation of pesticide exposure to pesticide handlers and

farm workers. It is fiscally unrealistic or legally mandated that
the Government (EPA) prove the safety of a pesticide. To the
contrary, it has been the Government's position that the registrant supply experimental data on the safety of a pesticide, e.g.
field monitoring data. Through laboratory audits and in-house
reviews by Government staff scientists, the experimental data
supplied by companies is carefully evaluated for scientific
validity. Any scientific questions are resolved at an early stage
before the risk analysis is undertaken.

   Toxicological Information. For each N-nitroso compound
occurring above 1 ppm, the registrant must provide data on the
oncogenic potential of the contaminant. When laboratory data is
available on the particular compound, this information will be
most useful and preferred. However, in the absence of such data,
the risk potential will be estimated by assuming that the oncogenic
potential of the contaminant is as great as that of N-nitrosodiethylamine, one of the most active known animal carcinogen (8).
If, on the other hand, it can be demonstrated that the particular
N-nitroso contaminant is not an oncogen, the product will be
cleared for registration without having to perform an exposure
analysis.

   Oncogenic Risk Calculations. On the basis of the exposure
analysis and potential oncogenic risk (oncogenic potency might be
more descriptive), a risk analysis will be performed according
to statistical methods like linear extrapolation (one-hit model)
or multistage estimation (9).

Conclusions

   Based on the experience gained during the past 4 to 5 years,
it was demonstrated that no pesticides were canceled on the basis
of unacceptably high levels of N-nitroso contaminants. It is
anticipated that the EPA Nitroso Policy will improve pesticide
products even further by a reduction or elimination of N-nitroso
contaminants from potential sources. In this way, the human risk
from these potentially hazardous chemicals will be minimized.

Literature Cited

1. Ross, R.; Morrison, J.; Rounbehler, D.P.; Fan, S.; Fine, D.H.
    J. Agric. Food Chem. 1977, 25, 1416.
2. Cohen, S.Z.; Law, M.; Wright, D.; Bontoyan, W.H.; Zweig, G.
    In "Environmental Aspects of N-Nitroso Compounds", IARC Sci.
    Publ. 19, International Agency for Research on Cancer, Lyon,
    France, 1978, 333-342.
3. Bontoyan, W.H.; Law, M.; Wright, D. J. Agric. Food Chem.
    1979, 27, 631.

4. Zweig, G.; Selim, S.; Hummel, R.; Mittelman, A.; Wright, D.; Law, S.; Regelman, E. In "N-Nitroso Compounds: Analysis, Formation and Occurrence", IARC Sci. Publ. 31, International Agency for Research on Cancer, Lyon, France, 1980, 555-564.
5. Maguire, A. "Request for Suspension of Treflan Products", Fed. Reg. Aug. 8, 1977, 40009.
6. EPA "Pesticides Contaminated with N-Nitroso Compounds, Proposed Policy", Fed. Reg., June 25, 1980, 42854.
7. E. Lilly and Co. "Present Levels of NDPA in Trifuluralin Products", Personal Communication, 1980.
8. Montesano, R.; Bartsch, H. Mutat. Res. 1976, 32, 179.
9. Fishbein, L. J. Toxicol. Environ. Hlth. 1980, 6, 1275.

RECEIVED August 30, 1981.

INDEX

INDEX

Jacket design by Carol Conway.
Production by Robin Giroux and V. J. DeVeaux.

Elements typeset by Service Composition Co., Baltimore, MD.
Printed and bound by Edwards Brothers, Inc., Ann Arbor, MI.